Vinyl Ester-Based Biocomposites

Vinyl Ester-Based Biocomposites provides a comprehensive review of the recent developments, characterization, and applications of natural fiber-reinforced vinyl ester biocomposites. It also addresses the importance of natural fiber reinforcement on the mechanical, thermal, and interfacial properties.

The book explores the widespread applications of natural fiber-reinforced vinyl ester composites ranging from the aerospace sector, automotive parts, construction and building materials, sports equipment, to household appliances. Investigating the moisture absorption and ageing on the physio-chemical, mechanical, and thermal properties of the vinyl ester-based composites, this book also considers the influence of hybridization, fiber architecture, and fiber-ply orientation.

The book serves as a useful reference for researchers, graduate students, and engineers in the field of polymer composites.

Vinyl Ester-Based Biocomposites

Edited by
Senthil Muthu Kumar Thiagamani,
Chandrasekar Muthukumar,
Senthilkumar Krishnasamy,
and Suchart Siengchin

CRC Press
Taylor & Francis Group
Boca Raton London New York

CRC Press is an imprint of the
Taylor & Francis Group, an **informa** business

Designed cover image: © Shutterstock

First edition published 2024
by CRC Press
4 Park Square, Milton Park, Abingdon, Oxon, OX14 4RN

and by CRC Press
6000 Broken Sound Parkway NW, Suite 300, Boca Raton, FL 33487-2742

CRC Press is an imprint of Taylor & Francis Group, LLC

British Library Cataloguing-in-Publication Data
A catalogue record for this book is available from the British Library

Library of Congress Cataloging-in-Publication Data
Names: Thiagamani, Senthil Muthu Kumar, editor. | Muthukumar, Chandrasekar, editor. |
Krishnasamy, Senthilkumar, editor. |
Siengchin, Suchart, editor.
Title: Vinyl ester-based biocomposites / Senthil Muthu Kumar Thiagamani,
Chandrasekar Muthukumar, Senthilkumar Krishnasamy, Suchart Siengchin.
Description: First edition. | Boca Raton : CRC Press, 2023. |
Includes bibliographical references and index. |
Identifiers: LCCN 2022061068 | ISBN 9781032220482 (hbk) |
ISBN 9781032220499 (pbk) | ISBN 9781003270997 (ebk)
Subjects: LCSH: Polymeric composites. | Fibrous composites—Materials. |
Vinyl ester resins. | Thermoplastic composites.
Classification: LCC TA455.P58 V57 2023 | DDC 620.1/18—dc23/eng/20230124
LC record available at https://lccn.loc.gov/2022061068

ISBN: 978-1-032-22048-2 (hbk)
ISBN: 978-1-032-22049-9 (pbk)
ISBN: 978-1-003-27099-7 (ebk)

DOI: 10.1201/9781003270997

Typeset in Times
by codeMantra

Dedicated to my family members

Elavarasi Thiagamani

Nandhini Nachiyar

Harsha Thiagamani

Vishwa Thiagamani

Dr. Senthil Muthu Kumar Thiagamani

Contents

Preface

Polymer composites are made up of several individual constituent materials, each with its own set of properties, that are combined to achieve better physical properties than the individual materials. They are typically made up of two distinct parts: a matrix and reinforcing materials. Reinforcing materials, such as fibers, give the composite strength, while the matrix functions as a binding agent to keep the composite together. Because of their lightweight, high stiffness, and high strength-to-weight ratios, fiber-reinforced polymer composites have become an appealing option. Polymer composites are now prevalently used in a broad array of applications, including aerospace, automotive, construction, and biomedical. Various matrix materials have been used in the formulation of polymer composites, including epoxy, polyester, and vinyl ester, to name a few commonly used polymers. Nonetheless, polymer composites are vulnerable to degradation by chemical, physical, and biological stressors. Given their widespread use in current and future human establishments, understanding and elucidating their long-term durability and survivability in natural environment is essential.

This book is organized in the following ways to highlight the performance characteristics of vinyl ester-based biocomposites reinforced with various natural fibers: Chapter 1 discusses the synthesis, curing behavior, and properties of vinyl ester resin. The influence of agro-wastes and hybridization on the thermal and mechanical properties of vinyl ester-based biocomposites is discussed in Chapters 5 and 6. The effects of infusing nanofillers on the mechanical and thermal properties of vinyl ester-based biocomposites are discussed in Chapters 7–9. The effects of accelerated weathering and soil burial on the physicochemical, thermal, and mechanical properties of vinyl ester-based biocomposites are discussed in Chapters 10 and 11. The various applications of vinyl ester-based biocomposites are discussed in Chapter 12. The thermomechanical, dynamic mechanical, and thermogravimetric analyses of vinyl ester-based hybrid composites are presented in Chapter 13.

Subject matter experts wrote each chapter of this book. It has been a pleasure to work with the authors who are well-known researchers in the field of biocomposites, and we would like to thank the publisher and the support staff for their assistance with this book. Undergraduate and postgraduate students, research scholars, academic researchers, professionals, and scientists seeking fundamental knowledge on the characterization of vinyl ester-based biocomposites, latest research trends, and the suitability of such composites in various applications may benefit from the content of this book.

Editors

Senthil Muthu Kumar Thiagamani, PhD, is an Associate Professor in the Department of Mechanical Engineering at Kalasalingam Academy of Research and Education (KARE), Tamil Nadu, India. In 2004, he earned a Diploma in Mechanical Engineering at the State Board of Technical Education and Training, Tamil Nadu. In 2007, he graduated from Anna University, Chennai with a Bachelor's degree in Mechanical Engineering. In 2009, he obtained his Master's degree in Automotive Engineering at Vellore Institute of Technology, Vellore. In 2018, he earned a PhD in Mechanical Engineering, specializing in Biocomposites, at KARE. He completed postdoctoral research in the Materials and Production Engineering Department at the Sirindhorn International Thai-German Graduate School of Engineering (TGGS), KMUTNB, Thailand in the year 2019. He started his academic career in 2010 as a Lecturer in Mechanical Engineering at KARE. He is a member of the International Association of Advanced Materials. His research interests include biodegradable polymer composites and their characterization. He has authored several articles in peer-reviewed international journals, book chapters, and in conference proceedings. He has edited books on the different themes of biocomposites published by CRC Press, John Wiley, and Springer. He is an Editorial Board Member of the *ARAI Journal of Mobility Technology* and a Reviewer Editor in the journal *Frontier in Materials*. He also serves as a reviewer for various journals, including the *Journal of Industrial Textiles*, *Journal of Cleaner Production*, *Materials Today Communications*, *Journal of Polymers and the Environment*, *SN Applied Sciences*, *Mechanics of Composite Materials*, and *International Journal of Polymer Science*.

Chandrasekar Muthukumar, PhD, is an Assistant Professor in the Department of Aeronautical Engineering, Hindustan Institute of Technology and Science, Chennai, India. He earned a Bachelor's degree in Aeronautical Engineering from Kumaraguru College of Technology, Coimbatore, India. He earned a Master's degree in Aerospace Engineering from the Nanyang Technological University – TUM Asia, Singapore, and a PhD in Aerospace Engineering from the Universiti Putra Malaysia (UPM), Malaysia. His PhD was funded through a research grant from the Ministry of Education, Malaysia. During his association with the UPM, he obtained internal research funding from the university worth 16,000 and 20,000 MYR. He has 5 years of teaching and academic research experience. His interests include fiber metal laminates (FMLs), natural fibers, biocomposites, additive manufacturing, and non-destructive testing. His publications are based on the fabrication and characterization techniques of biocomposites. He has authored and co-authored research publications in SCIE journals, book chapters, and articles in the conference proceedings. He is a peer reviewer for the *Journal of Industrial Textiles*, *Polymer Composites*, *Materials Research Express*, *Journal of Natural Fibres*, etc.

Senthilkumar Krishnasamy, PhD, is an Associate Professor in the Department of Mechanical Engineering at PSG Institute of Technology and Applied Research, Coimbatore, Tamil Nadu, India. He earned a Bachelor's degree in Mechanical Engineering at Anna University, Chennai, India, in 2005. He earned a Master's degree in CAD/CAM at Anna University, Tirunelveli, India, in 2009. He earned a PhD in Mechanical Engineering at Kalasalingam University, Krishnankoil, Tamil Nadu, India, in 2016. From 2010 to 2018, Dr. Senthilkumar Krishnasamy worked in the Department of Mechanical Engineering at Kalasalingam Academy of Research and Education (KARE), India. He completed a postdoctoral fellowship at Universiti Putra Malaysia, Serdang, Selangor, Malaysia and King Mongkut's University of Technology North Bangkok (KMUTNB) under the research topics of experimental investigations on mechanical, morphological, thermal, and structural properties of kenaf fiber/mat epoxy composites and sisal composites and fabrication of eco-friendly hybrid green composites on tribological properties in a medium-scale application. His area of research includes the modification and treatment of natural fibers, nanocomposites, 3D printing, and hybrid reinforced polymer composites. He has published research papers in international journals, book chapters, and at conferences in the field of natural fiber composites. He also edited books for several publishers.

Suchart Siengchin, PhD, is the President of King Mongkut's University of Technology North Bangkok (KMUTNB), Thailand. He earned a Dipl-Ing in Mechanical Engineering at the University of Applied Sciences Giessen/Friedberg, Hessen, Germany, in 1999; an MSc in Polymer Technology at the University of Applied Sciences Aalen, Baden-Wuerttemberg, Germany, in 2002; an MSc in Materials Science at the Erlangen-Nürnberg University, Bayern, Germany, in 2004; a PhD in Engineering (Dr-Ing) at the Institute for Composite Materials, University of Kaiserslautern, Rheinland-Pfalz, Germany, in 2008; and a postdoctoral research at Kaiserslautern University and the School of Materials Engineering, Purdue University, USA. In 2016, he received the habilitation at the Chemnitz University in Sachsen, Germany. He worked as a Lecturer for Production and Material Engineering Department at the Sirindhorn International Thai-German Graduate School of Engineering (TGGS), KMUTNB. He has been a full-time professor at KMUTNB and became the President of KMUTNB. He received the Outstanding Researcher Award in 2010, 2012 and 2013 at KMUTNB. His research interests include polymer processing and composite materials. He is the Editor-in-Chief of KMUTNB *International Journal of Applied Science and Technology* and the author of more than 150 peer-reviewed journal articles. He has participated with presentations at more than 39 international and national conferences on materials science and engineering topics.

Contributors

Tarkan Akderya
Faculty of Engineering and Architecture, Department of Biomedical Engineering
İzmir Bakırçay University
Izmir, Turkey

Tibor Alpár
Faculty of Wood Engineering and Creative Industry
University of Sopron
Sopron, Hungary

Anand G.
Department of Mechanical Engineering
Achariya College of Engineering Technology
Pondicherry, India

László Bejó
Faculty of Wood Engineering and Creative Industry
University of Sopron
Sopron, Hungary

Jie Chen
Institute of Chemical Industry of Forest Products
Chinese Academy of Forestry
Nanjing, P. R. China

W. S. Chow
School of Materials and Mineral Resources Engineering, Engineering Campus
Universiti Sains Malaysia
Penang, Malaysia

Csilla Csiha
Faculty of Wood Engineering and Creative Industry
University of Sopron
Sopron, Hungary

Czók Csilla
Faculty of Wood Engineering and Creative Industry
University of Sopron
Sopron, Hungary

Soubhik De
FRP Composites Laboratory, Department of Metallurgical and Materials Engineering
National Institute of Technology
Rourkela, India

Hossein Ebrahimnezhad-Khaljiri
Department of Materials Science and Engineering, Faculty of Engineering
University of Zanjan
Zanjan, Iran

K. M. Faridul Hasan
Faculty of Wood Engineering and Creative Industry
University of Sopron
Sopron, Hungary

B. N. V. S. Ganesh Gupta K
FRP Composites Laboratory, Department of Metallurgical and Materials Engineering
National Institute of Technology
Rourkela, India

R. Geetha Balakrishna
Centre for Nano and Material Sciences
Jain (Deemed-to-be University)
Bengaluru, India

Péter György Horváth
Faculty of Wood Engineering and Creative Industry
University of Sopron
Sopron, Hungary

Sangilimuthukumar Jeyaguru
Department of Automobile Engineering
Kalasalingam Academy of Research and Education
Krishnankoil, India

Athul Joseph
Faculty of Engineering Science, Department of Materials Engineering
Katholieke Universiteit Leuven
Leuven, Belgium

Jyothi M. S.
Department of Chemistry
AMC Engineering College
Bengaluru, India

Senthilkumar Krishnasamy
Department of Mechanical Engineering
PSG Institute of Technology and Applied Research
Coimbatore, India

Lavanya C.
Centre for Nano and Material Sciences
Jain (Deemed-to-be University)
Bengaluru, India

Le Duong Hung Anh
Faculty of Wood Engineering and Creative Industries
University of Sopron
Sopron, Hungary

M. Meena
Department of Physics
S.T. Hindu College
Nagercoil, India

Zsuzsanna Mária Mucsi
Faculty of Wood Engineering and Creative Industry
University of Sopron
Sopron, Hungary

Chandrasekar Muthukumar
Department of Aeronautical Engineering
Hindustan Institute of Technology and Science
Chennai, India

Xiaoan Nie
Institute of Chemical Industry of Forest Products
Chinese Academy of Forestry
Nanjing, P. R. China

Buket Okutan Baba
Faculty of Engineering and Architecture, Department of Mechanical Engineering
Izmir Katip Çelebi University
Izmir, Turkey

Pásztory Zoltán
Faculty of Wood Engineering and Creative Industries
University of Sopron
Sopron, Hungary

Rajesh Kumar Prusty
FRP Composites Laboratory, Department of Metallurgical and Materials Engineering
National Institute of Technology
Rourkela, India
and
Center for Nanomaterials
National Institute of Technology
Rourkela, India

Sabarish Radoor
Materials and Production Engineering, The Sirindhorn International Thai-German Graduate School of Engineering (TGGS)
King Mongkut's University of Technology North Bangkok
Bangkok, Thailand

Tanaya Sahoo
FRP Composites Laboratory, Department of Metallurgical and Materials Engineering
National Institute of Technology
Rourkela, India

Santhosh N.
Department of Mechanical Engineering
MVJ College of Engineering
Bengaluru, India

Shwetharani R.
Centre for Nano and Material Sciences
Jain (Deemed-to-be University)
Bengaluru, India

Suchart Siengchin
Materials and Production Engineering, The Sirindhorn International Thai-German Graduate School of Engineering (TGGS)
King Mongkut's University of Technology North Bangkok
Bangkok, Thailand
and
Institute of Plant and Wood Chemistry
Technische Universität Dresden
Tharandt, Germany

Senthil Muthu Kumar Thiagamani
Department of Mechanical Engineering
Kalasalingam Academy of Research and Education
Krishnankoil, India

Theivasanthi Thirugnanasambandan
International Research Centre
Kalasalingam Academy of Research and Education
Krishnankoil, India

Vishvanathperumal S.
Department of Mechanical Engineering
S.A. Engineering College
Thiruverkadu, India

Yatish K. V.
Centre for Nano and Material Sciences
Jain (Deemed-to-be University)
Bengaluru, India

1 Introduction to Vinyl Ester Resin

Synthesis, Curing Behaviour and Its Properties

Athul Joseph
Katholieke Universiteit Leuven

CONTENTS

1.1 INTRODUCTION

Organic polymers have played a crucial role in designing and developing modern-day composite materials through their versatile and reliable characteristics (Anderson & Messick, 1980; Cassis & Talbot, 1998; Li, 1998). Such composites have been beneficial in myriad ways owing to their superior weight-to-strength parameters, ease of design, affordability and applicability in a vast array of applications. Fundamentally, organic polymers are divided into thermoplastic polymers and thermosetting polymers based on their cross-linking capabilities. The former can be easily deformed to its initial liquid configuration by thermal means, while the latter is devoid of that due to the chemical reactions that occur during its synthesis, resulting in significant cross-linking. This cross-linking ability is also translated into superior mechanical strength

DOI: 10.1201/9781003270997-1

and thermal stability when compared to its thermoplastic counterparts. Some of the most common thermosetting polymers include unsaturated polyesters, vinyl ester resins (VERs), epoxy resins and phenolic resins. While polyesters and epoxy resins are widely used in numerous applications, they are often limited by several factors such as poor temperature resistance, poor chemical resistance, high rigidity, complicated processing and other factors. Such shortcomings are addressed by the usage of vinyl esters which offer the best of both worlds at an affordable and fairly easy synthesis process. Vinyl esters have the same ease of processing and swift curing of polyester resins while possessing the mechanical and thermal performance of epoxy resins. Furthermore, VERs offer additional advantages such as the use of non-toxic catalysts for curing and superior wettability to bond fibres to maximize composite performance.

The history of the development of VERs dates back to the early 1960s through the homopolymerization of acrylic polymers which were later copolymerized with a monomer (Fekete et al., 1965, 1966). The typical structure of a VER molecule is depicted in Figure 1.1. However, the curing process was fairly slow and required further optimization and modification to enable commercial production. Moreover, the resins thus manufactured were seen to be reactive and impeded their effective storage and transport. Bearden (1968) developed resins that cut in styrene monomers to produce more stable resin configurations. Furthermore, the study also emphasized the importance of heat ageing in the retention of the physical and electrical properties of the resins. Additionally, the processing route adopted by Bearden et al. was simple and rapid, which was synonymous with the processing of polyester resins. The use of methacrylic acid against convent acrylic acid also diversified the resin with higher resistance against hydrolysis while broadening its corrosion resistance. Since then, numerous studies have been published, which offer alternative possibilities and better processing routes for the production of VERs with major focus on the alteration of VER properties and their incorporation in modern-day composite and nanocomposite materials for superior performance for numerous applications (Atta, Abdel-Raouf, et al., 2006; Atta, El-Saeed, et al., 2006; Grishchuk & Karger-Kocsis, 2011; Jaswal & Gaur, 2014; Launikitis, 1982; Nodehi, 2022; Nouranian et al., 2011; Tu & Sodano, 2021; Vu et al., 2021; Yadav et al., 2018; Zhang et al., 2018). These studies also stress the possibility of more natural and less harmful alternatives to the chemicals generally used in the production of traditional VERs. Consequently, VERs are used in a diverse range of applications including adhesives (Ambrogi et al., 2002; Cui et al., 2014), coatings (Slama, 1996; Taillemite & Pauer, 2009), marine (Kandola et al., 2018; Shivakumar et al., 2006), boat-building (Marsh, 2007), structural composites (Burchill et al., 2001; Kuppusamy et al., 2020; Stanzione et al., 2013) and electrical applications (Cui et al., 2014; Pratap, 2002; Yurdakul et al., 2010), among many others.

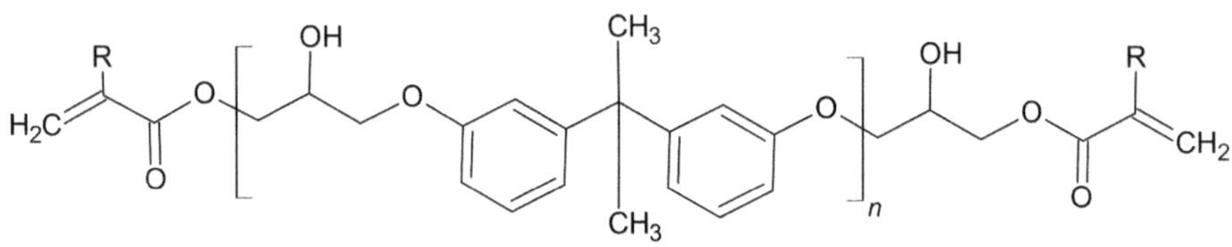

FIGURE 1.1 General structure of a vinyl ester resin.

1.2 COMMON TYPES OF VINYL ESTER STRUCTURES

Some of the most commonly found and traditional commercial VERs include bisphenol A-epoxy VER, sheet moulding compound VERs, epoxy-Novolac VERs, flame-retardant VERs, urethane-based VERs, Bisphenol A-Fumaric acid condensation polyester, radiation-curable resins, rubber-modified VERs and others.

1.2.1 Bisphenol A-Epoxy VERs

These VERs are characterized by rapid curing to provide green strength with a high degree of corrosion resistance due to the presence of terminal vinyl saturation present at the end of the molecules and partially due to the secondary hydroxyl group (Linow et al., 1966). In addition to this, these VERs have superior resistance to hydrolysis at a lower weight as a result of the addition of the methyl group and the presence of fewer ester groups within the molecule. Nevertheless, the ester groups within the molecule aid in the superior acid resistance capabilities of the VERs. Figure 1.2 represents a common bisphenol A-epoxy resin.

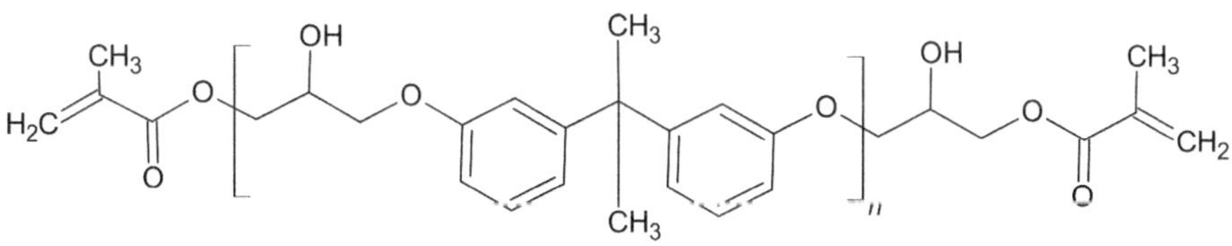

FIGURE 1.2 Structure of bisphenol A-epoxy resin.

1.2.2 Epoxy-Novolac VERs

The most notable properties of these VERs are their high-temperature capabilities while offering the desired mechanical integrity (Cravens, 1972). The reason behind such properties is the increased cross-link density within the resin imbibed by the curing of phenol-formaldehyde Novolac epoxy into the basic VER backbone. A common example of such a resin is illustrated in Figure 1.3.

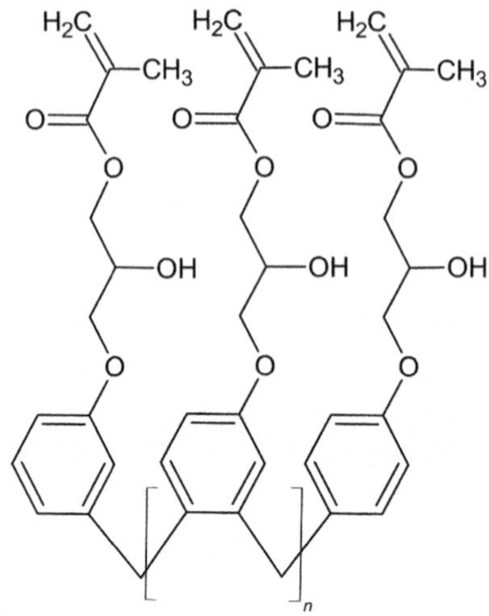

FIGURE 1.3 Structure of Novolac-epoxy resin.

1.2.3 Flame-Retardant VERs

Initially, flame-retardant VERs were required in ductwork and stack applications that required materials with decreased flammability (Anderson & Messick, 1980). This property was realized by the incorporation of bromine into the structure of VERs as witnessed in Figure 1.4. In addition to their flame-retardant properties, these VERs also preserve the corrosion resistance capabilities expected from VERs, which diversifies its spectrum of application.

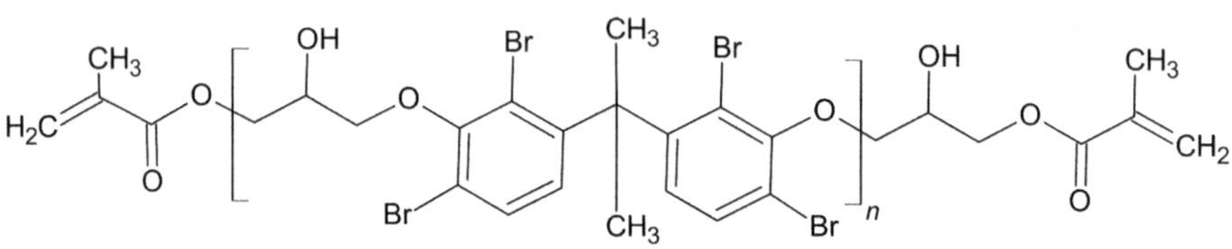

FIGURE 1.4 Structure of flame-retardant vinyl ester resin.

1.2.4 Urethane-Based VERs

In the context of better composite integrity, urethane-based VERs offer the most superior fibre–matrix bonding capabilities (Lewandowski et al., 1975). This is greatly evident in composites reinforced with glass fibres where the resin offers good wettability and adhesion due to the presence of internal and terminal unsaturation. However, a dispute arises in classifying these compounds as VERs due to a large amount of internal unsaturation. A typical urethane-based VER is represented in Figure 1.5.

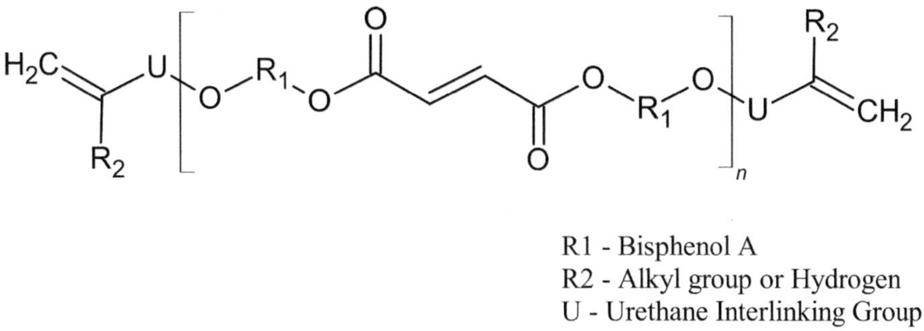

FIGURE 1.5 Structure of urethane-based vinyl ester resin.

1.2.5 Radiation-Curable VERs

These VERs are characterized by their ability to cure in the presence of ultraviolet (UV) or electron beam radiations. Such a possibility eliminates the need to use terminal methacrylate groups in VERs. Instead, they could be replaced by acrylate end groups as seen in Figure 1.6. Chemical compounds such as benzophenone and benzoin ethers are used for UV-based curing to enable vinyl polymerization. The cure times are quite fast as compared to conventional coatings. Such VERs primarily find applications in coating and printing inks.

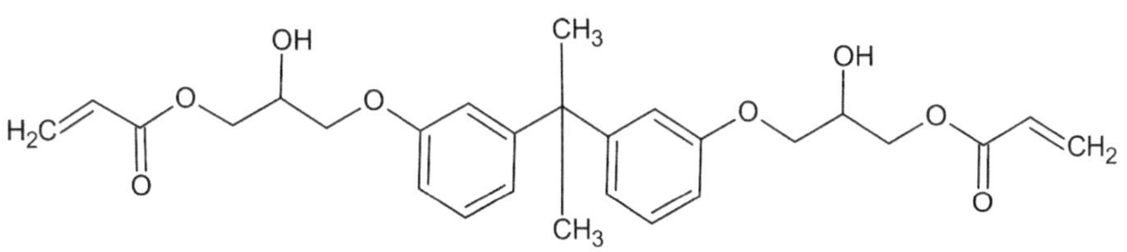

FIGURE 1.6 Structure of Dow XD-9002 experimental radiation-curable resin.

The physical and chemical configuration of VERs can be altered as per the requirement of the applications. The upcoming section expounds on the basic synthesis routes and the possible modifications imparted to introduce these alterations in the VERs.

1.3 SYNTHESIS OF VINYL ESTER RESINS

Vinyl esters are the reaction products of the exothermic esterification of epoxy resin and ethylenically unsaturated carboxylic acid with terminal unsaturation (Anderson & Messick, 1980; Launikitis, 1982; Yang et al., 2008). The reaction is often catalysed by compounds such as tertiary amines, phosphines, alkalis and onium salts that produce hydroxyl groups which are responsible for adhesion and further modification (Young, 1976). In most cases, VERs are often diluted by compounds such as styrene, vinyl toluene and others. The type of epoxy resin used varies with the mechanical, thermal and chemical requirements of the final product. The physical and rheological parameters of the resulting resin are a function of the diluent content, processing temperature and the type and fraction of the co-reactant. Numerous studies have utilized different materials for the synthesis and process control of VERs manufacturing, resulting in an array of modifications and novel techniques in synthesizing the same.

A study on a commercial VER, Derakane 441-400, was conducted to determine the exact reaction kinetics between the added styrene and epoxy resin base at low temperatures (Dua et al., 1999). It was observed that the rate of fractional double-bond conversion of the VER and styrene varied concerning time, where a high rate is observed initially for VERs. This is overtaken by the fractional conversion rate of styrene towards the end of the reaction cycle where it continues even after the conversion of the VER ceases. However, this is not an absolute reflectance of the conversion efficiency as it is dependent on the initial double-bond concentration in the styrene and the vinyl ester. Moreover, vinyl ester molecules used in the study possessed two double bonds which can mean that the fractional conversion of the double bonds does not reflect the actual fractional conversion of the vinyl ester molecule itself. In this regard, Padma et al. (1993) studied the effect of the addition of α-methyl styrene (α-MS) as a reactive diluent for the synthesis of Novolac-based VERs. It was observed that the addition of α-MS lowered the peak temperature required for the synthesis of the VER representing the reduced rate of polymerization. However, the addition of α-MS did not alter the flow behaviour of the produced VERs. On the contrary, the thermal and mechanical performance of the VERs altered with the addition of α-MS.

Improved reaction kinetics and gel times were observed when VERs were toughened with compounds such as rubber, polybutadienes and butadiene-acrylonitrile rubber modifiers (Pham & Burchill, 1995; Robinette et al., 2004; Ullett & Chartoff, 1995). Additionally, these substances also aided in reducing the temperature and time required during the curing and post-curing phases of the processing route. Similarly, the use of bimodal blends of vinyl ester monomers aids in the reduction of the viscosity of the solution with lower-molecular-weight compositions (La Scala et al., 2005). Nevertheless, the minimal viscosity could only be attained with at least 20% addition of styrene which is one of the limitations imposed by the use of bimodal blends for VER processing. Similar studies were conducted for VERs containing methacrylated fatty acid (MFA) comonomers where the viscosity decreased with the volume fraction of the MFA present (Dey, 2007). Additionally, the viscosity is observed to decrease with higher chain lengths of the MFA up to a threshold value after which there was an increase due to higher intermolecular interactions as a result of higher friction in the solution experienced by larger molecules. This is was similar to the characteristics exhibited by vinyl ester bio-copolymers derived from Dimer fatty acids (Li et al., 2013).

Experimental studies by Can et al. (2015) and Shah et al. (2015) put forward the possibility of synthesizing Novolac-based VERs from renewable resource materials. Cardanol-based epoxy VERs were synthesized using Novolac resin with a cardanol to formaldehyde ratio of 1:0.7 using PTSA as a catalyst in 2 ml methanol solution in warm conditions (Sultania et al., 2010). It was observed that the number of phenolic hydroxyl groups in the starting resin along with their molecular weight and extent of reaction affected the ultimate concentration of glycidal groups present in the resin. Furthermore, Garg et al. (2015) expounded on the effectiveness of adding toughening agents such as Carboxyl terminated butadiene acrylonitrile (CTBN) on the processing of cardanol-based VERs where vinyl-terminated CTBNs are produced when epoxy-terminated CTBN reacts with methacrylic acid. The addition of the toughening agent affected the morphology of the resin with dispersed rubber globules while only marginally affecting the curing process. In addition to the added components, the processing and properties of VERs can be affected by the impurities present in the reactive compounds as seen in Bassett et al. (2016).

Flame-retardant VERs for structural applications are typically synthesized from halogen-based precursor compounds. Tetrabromobisphenol-A is one such compound that is obtained by the bromination of bisphenol A. Tetrabromobisphenol-A along with glycidyl methacrylate and dimethyl benzamine was used to synthesize the flame-retardant VERs in an inert argon atmosphere under isothermal conditions as illustrated in Figure 1.7 (Dev et al., 2017). The VERs were later blended with Derakane 510-40A to assess their effect against traditional petroleum-based blends. Similar observations were made for acrylate made from phosphaphenanthrene and triazine-trione used in the synthesis of flame-retardant VERs (Tao et al., 2018).

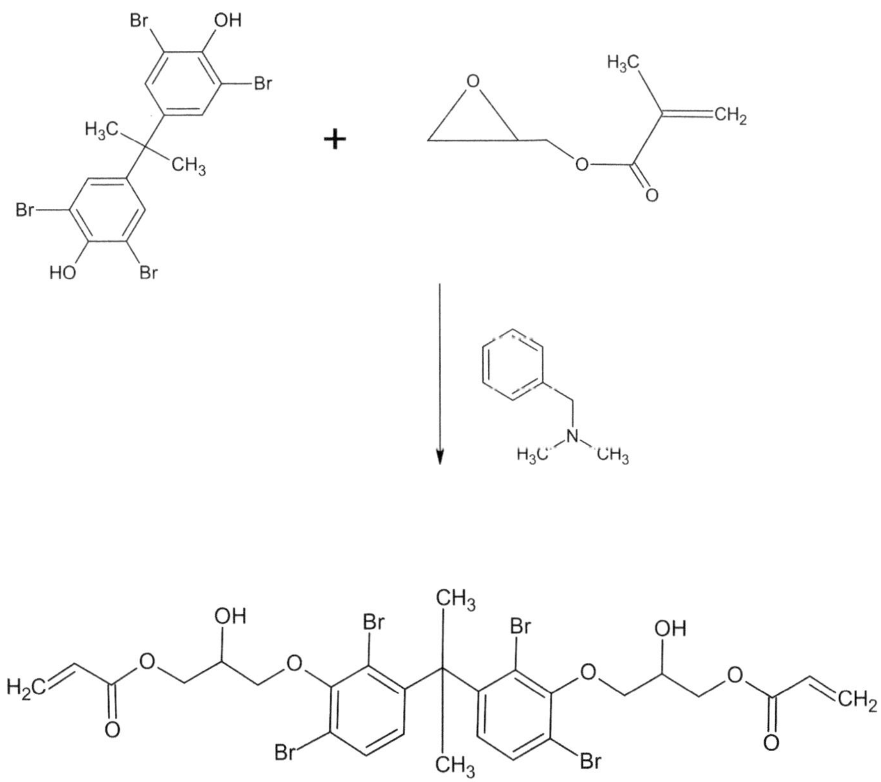

FIGURE 1.7 Synthesis of flame-retardant VERs.

1.4 CURING OF VINYL ESTER RESINS: MECHANISM AND KINETICS

Curing of a resin refers to the chemical process of turning a liquid resin into a solid-state by initiating and developing tridimensional polymeric networks through effective cross-linking of chains (Anderson & Messick, 1980). The catalytic agent for such a change can be physical or chemical based on the inherent nature and properties of the resin under consideration. The curing of VERs can take place when the double bonds present in the molecule react and cross-link in the presence of free radicals that are usually produced from a chemical source (Launikitis, 1982). The most common methyl ethyl ketone peroxide is used as a catalyst for the curing of VERs at room temperature (Thomas et al., 1977). At elevated temperatures, benzoyl peroxides are generally preferred (Cassoni et al., 1977). However, using certain "accelerators" like N,N-dimethyl aniline in combination with these catalysts can aid in quicker cure times at room temperatures (Brinkman et al., 1968). Nevertheless, the curing process is also dependent on numerous factors such as the residual monomer, its physical properties, amount of oxygen present in the working atmosphere, working time, required hardness, the exothermic temperature and others (Varco, 1975). All these parameters affect the physical integrity and the properties of the end product, sometimes resulting in cracked surface, poor chemical and corrosion resistance, premature mechanical failure and others.

Numerous studies have been conducted to explore different diluents, methods and cure conditions to provide for optimized cure processes for different VERs. In this regard, Kant et al. (1992) observed that VERs made bis(methacyloxy) derivatives of diglycidyl ether of bisphenol A had appreciable curing times at lower temperatures with different acrylates as reactive diluents. Additionally, the curing rate was seen to improve at lower temperatures with increased concentrations of the initiator material. The kinetics of the curing process of such materials are determined with the help of regression models in most cases. Such models require the understanding of the thermodynamic transitions of the VER across the temperature range of its curing process (Cook et al., 1997). A differential scanning calorimetry (DSC) thermogram is used to obtain the heat flow (dq/dt) which was later modified to obtain the fractional conversion rate ($d\alpha/dt$) using equation (1.1). However, Lee & Lee (1994) proposed the use of a phenomenological kinetic model to determine the cure kinetics of commercially available VERs. The kinetic parameters were estimated by drawing a relationship between the glass transition temperature and the degree of cure from a different point such as the zero initiation reaction rate, the conversion at vitrification point and the reaction ratio at other significant isothermal conditions. This model provided a more accurate understanding/prediction of the viscosity parameters of the VER under consideration.

$$d\alpha/dt = \left(dq/dt\right)/m \cdot \Delta H_p \tag{1.1}$$

where

m = sample mass in grams

ΔH_p = total heat of polymerization

In addition to the properties of the participating chemical compounds, the processing parameters also affect the cure characteristics of VERs (Abadie et al., 2002). Valea et al. (1998) stressed the importance of accurate cure schedules and appropriate solvents on the properties of the end resin product. The study explained that an incomplete or inadequate cure schedule alters the properties of the cured materials when they are in contact with solvents during the post-cure period. Additionally, care must be taken to ensure the usage of the right solvent and the most optimal thermal conditions to ensure proper product life (Rosu et al., 2006). In the event of an underestimation or overestimation, severe deterioration of the material is inevitable. The cure kinetics of VERs cured at low temperatures show that the decrease in the styrene content increased the rate of reaction (Li et al., 1999). However, at a higher concentration of styrene, a grafting reaction of styrene onto the vinyl ester chains was observed (Brill & Palmese, 2000). In some cases, the styrene underwent a homopolymerization reaction as a result of the high residual styrene content post the vinyl ester vinylene reactions.

Martin et al. (2000) and Scott et al. (2002) studied the cure kinetics of VERs using experimental techniques such as thermal scanning rheometry (TSR) and dynamic mechanical thermal analysis (DMTA) for isothermal processing conditions. The study evaluated the gel time which is seen to happen over a given period. The activation energy values determined by the gel time studies were seen to be dependent on the concentration of the initiator in the process (Li et al., 2008; Sultania et al., 2012). This may be associated with the fact that the treatment does not evaluate the whole gelation process. However, this contradicts the activation energy determined from the viscosity values obtained from TSR readings where the initiator and promoter concentrations do not affect the activation energy. In addition to the initiator content, the gel time of thermally cured resins are seen to be affected by the oxirane content in the resin (Malik et al., 2001). An increase in the oxirane content decreases the gel time while raises the heat of curing. Consequently, the activation energy of the curing reaction decreases when the oxirane concentration was increased by a significant amount while showing minimal fluctuations for small changes in the oxirane concentrations. Therefore, curing is a critical step in the preparation of resins that ensures that the desired properties are optimally imparted onto the material and demands careful control over the material selection, processing parameters and the concentration of the initiator, promoter and diluents.

1.5 PROPERTIES OF VINYL ESTER RESINS

As seen from the previous sections, the ultimate properties of VERs are highly dependent on the constituent materials, the curing process and the synthesis technique adopted. As a result, a variety of properties may be obtained for VERs based on the requirement of the application it is intended for. Primarily, VERs are known for their excellent mechanical strength, high thermal stability and good corrosion resistance properties. Therefore, the upcoming sections aim to describe the properties of VERs and how different parameters affect their performance.

1.5.1 Chemical and Rheological Properties

The rheological properties of resins are dependent on their internal molecular structure. DMTA studies by Han & Lem (1984) reveal that the VERs (Dow Chemical Co., XD-7608.05) exhibit two loss modulus peaks at higher angular frequencies as opposed to the single peak in lower frequencies. Additionally, the variation of the dynamical mechanical parameters is highly nonlinear when the temperatures exceed 40°C. However, for steady shear flow measurements, the differences and variations are not very profound, indicating that molecular structures influence the oscillatory measurements more than steady measurements of flow. Therefore, the rheological properties of VERs are dependent on the cure temperature and the molecular structure of the VER. In a similar study by Gaur & Rai (1993), the rheological properties of epoxy-Novolac resins containing methyl, ethyl and butyl acrylates as reactive diluents were investigated using a viscometer. It was observed that the viscosity increased with bulkier acrylate molecules at any given steady-state shear rate. Though a visible decrease in the viscosity is observed with increasing shear rates, independence on the shear rate is observed at very low values. Similar observations were made for the rheological properties of epoxy-vinyl ester interpenetrating polymer networks (Dean et al., 2001).

One of the major drawbacks of polymeric materials is their hydrophilic nature. Hence, this parameter needs to be accurately gauged to obtain a consensus on the applicability and performance of VERs particularly, in marine, boat-building, corrosion resistance and other similar applications. In this regard, Lee et al. (1992) investigated the interaction of VERs with water through a series of absorption–desorption–reabsorption experiments. Under a given condition of temperature and humidity, the moisture diffusion parameter depended on numerous factors including initiator type and chemistry, amount of initiator and the post-curing treatment process parameters. Moreover, higher temperatures and relative humidity values promoted accelerated weight loss rate which prevents the accurate calculation of the diffusion coefficient. However, this could be overcome by subsequent absorption–desorption cycles as the VERs were more stable then. Figure 1.8 compares the diffusion parameters of VERs in comparison with some of the most common resins during absorption, desorption and reabsorption under fixed temperature and humidity values.

Similar studies on glass fibre-reinforced composites of VERs revealed that when aged in water, the resin degrades as a result of hydrolysis (Boinard et al., 2000). However, VERs exhibit slower degradation in water than in air when compared to polyester resins due to the presence of a lesser number of pores. Seawater ageing of VER composites reinforced with glass and carbon fibres exhibited classical Fickian water diffusion which plateaued out after reaching the saturation point (Mouritz et al., 2004). The study pointed out the effect of curing on the water uptake behaviour of the composites. It was observed that under-cured composites gain weight more slowly when compared to fully cured composites. For such VERs aged in seawater,

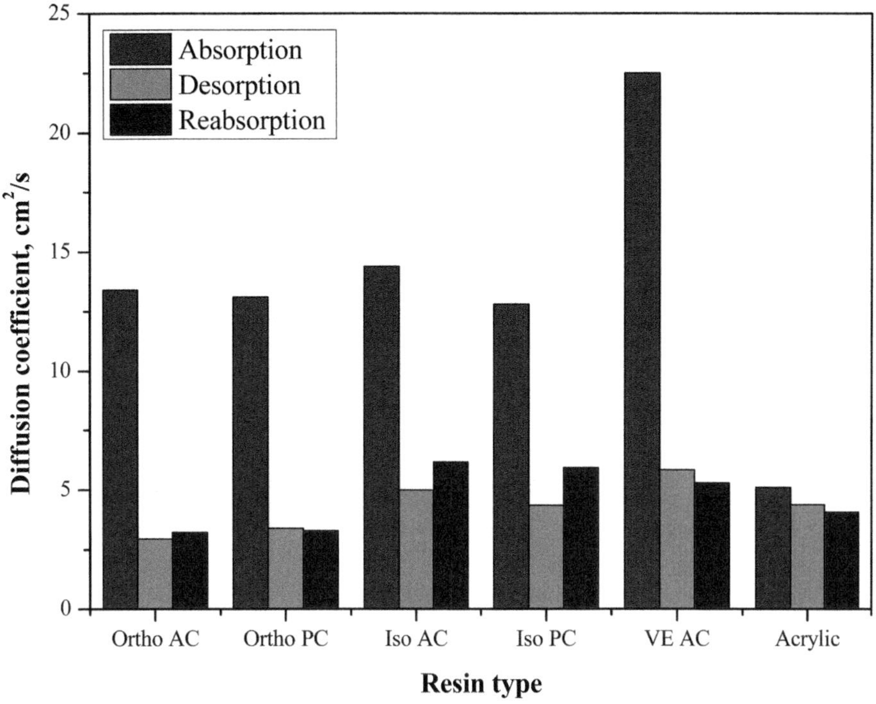

FIGURE 1.8 Diffusion parameters of VERs at 65° and 75% relative humidity during absorption–desorption–reabsorption cycles. Ortho = orthophthalic polyester; iso = isophthalic polyester; VE = vinyl ester; AC = as-cast; PC = post-cured.

it was observed that the mechanical properties such as the hardness and the impact resistance increased with ageing as seen in the works of Visco et al. (2012). A comparative understanding of this change is illustrated in Figure 1.9. One of the primary reasons for such a change may be attributed to the retention of its polymeric structure combined with the macromolecular structural configuration that denies water degradation within the bulk of the material. Likewise, for VERs aged in a climatic chamber, notable chemical and structural changes were observed (Alia et al., 2018). The degradation mechanics were dependent on the hydrolysis of the ester group in VERs where most structural modifications were seen in the first three days. This later translates to a dependence on diffusive processes where water seeps through the polymeric network initiating further hydrolysis. Furthermore, the chemistry of the resins is seen to alter when exposed to UV radiation where severe embrittlement and perforations were seen on the surface of the resins accompanied by reduced mechanical properties throughout the bulk of the material (Rosu et al., 2009; Signor et al., 2002).

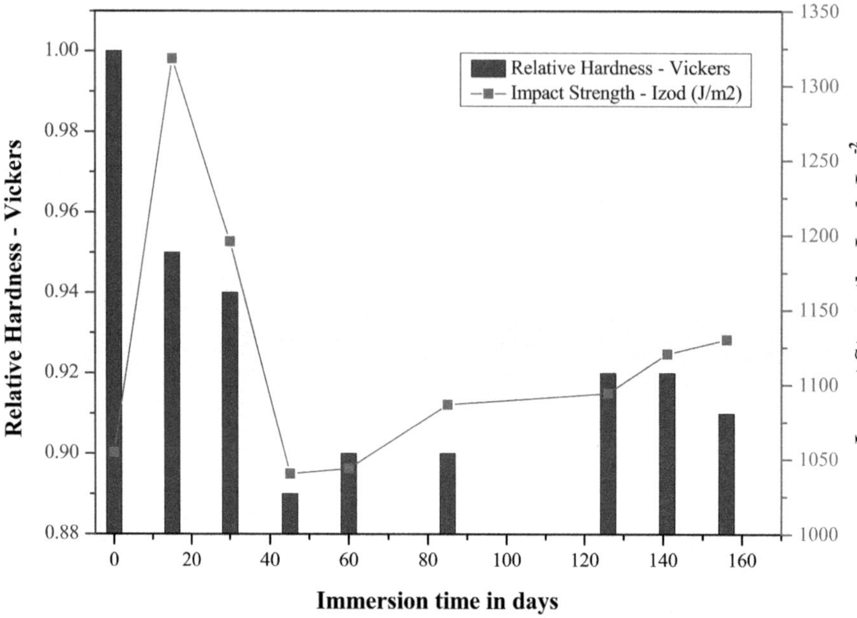

FIGURE 1.9 Variation of the mechanical properties of VERs with time when aged in seawater.

1.5.2 Mechanical Properties

Vinyl esters offer significant mechanical competence which is comparable to their epoxy counterparts while offering commendable thermal and corrosion resistance properties. However, the mechanical performance of the end resin is dependent on numerous factors as seen in the case of the chemical properties of VERs. Several studies have been conducted to assess the mechanical properties of VERs in terms of their tensile, shear, flexural, hardness and impact capabilities (Ardhyananta et al., 2019; Benmokrane et al., 2017; Jang et al., 2012; Qin et al., 2006; Rodriguez et al., 2006; Scott et al., 2008). In this regard, a study conducted by Sullivan et al. (1984) evaluated the shear properties of VER Iosipescu specimens. It was observed that the specimens failed in tensile mode despite the application of a pure shear field across the gauge section. The combined stress observed from these experiments was found to closely lie with the values obtained from pure tensile tests. This essentially coincides with the fact that brittle materials fail in tensile mode despite the type of stress applied. Similarly, Varma et al. (1985) found out that the mechanical properties of VERs such as the tensile and flexural strength along with the modulus of elasticity varied with the styrene content in the processing of the VER. While the tensile and flexural strengths of the resin increased, the moduli of elasticity and tensile modulus decreased with the increase in the styrene content as depicted in Figure 1.10. Such behaviour may be accounted for by the reduction in the cross-link density of the resin.

In addition to this, the mechanical properties of VERs also vary with the structure and the constituent materials as seen in the case of vinyl ester blends of VER

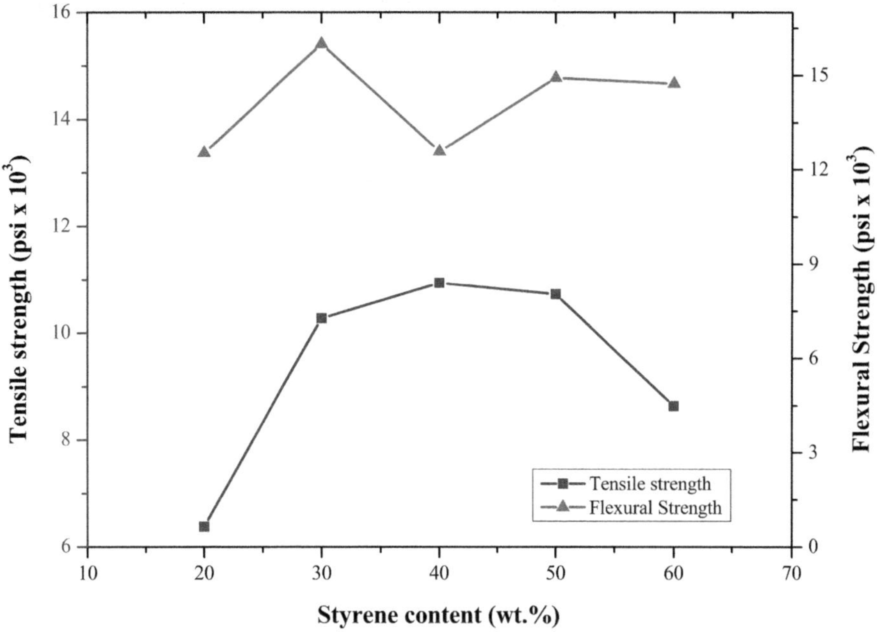

FIGURE 1.10 Change in the tensile and flexural strength of VERs with the styrene content in the resin.

and brominated VERs with styrene (Siva et al., 1994). The effect of these structures on the tensile and flexural properties of the resultant resins is depicted in Figure 1.11. Similarly, with the addition of rubber into VER matrices, the fracture toughness of the blend is seen to improve at lower loading rates despite the incompatible nature of VERs and rubber which was not the case under impact loads (Dreerman et al., 1999). Nevertheless, the tensile and the flexural strengths of the blends decrease with increasing rubber content. The compatibility between VERs and rubber phases can be enhanced by the use of reactive rubber compounds such as vinyl-terminated poly(butadiene-co-acrylonitrile), giving rise to more miscible blends with smaller rubber regions (Auad et al., 2001). In the same context, the structure of the epoxy hardener used to cure VER blends is also seen to affect the mechanical properties of VERs as seen in Cryshchuk & Karger-Kocsis (2004). Vinyl esters derived from the varying content of diacrylate and dimethacrylate oligomers of glycidyl ether from glycolysed polyethylene terephthalate (PET) were seen to affect the compressive strength of the blended vinyl ester-unsaturated polyester resins (Atta et al., 2005). In addition to the internal structure, the cure temperature of these resins also affected the compressive strength of the resins. The effect of these parameters on the compressive characteristics of the resin is illustrated in Figures 1.12 and 1.13. Similarly, the addition of ATLAC 363E polyester binders to VERs improved the tensile properties of the resin with its increasing content while the fracture toughness of the resin remained unaffected as a result of modifications in the cure kinetics (Brody & Gillespie, 2005). Thus,

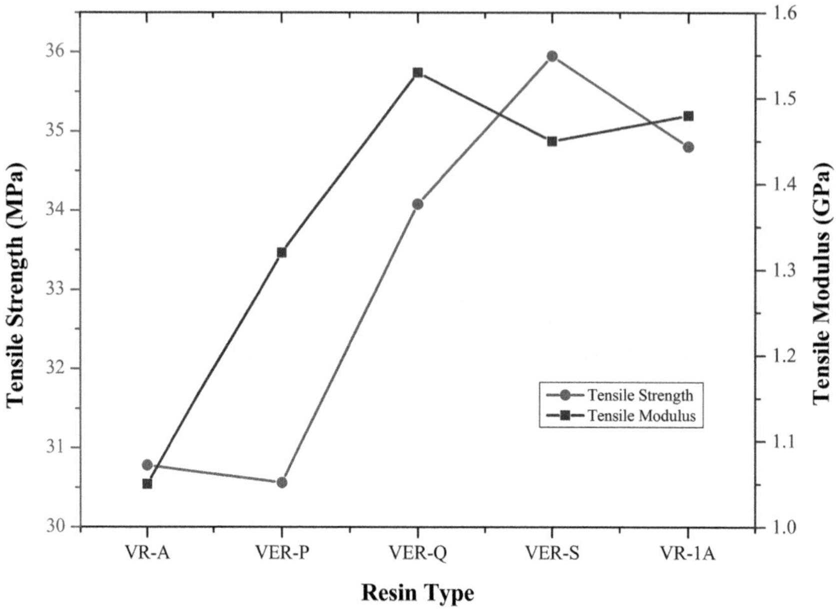

(a)

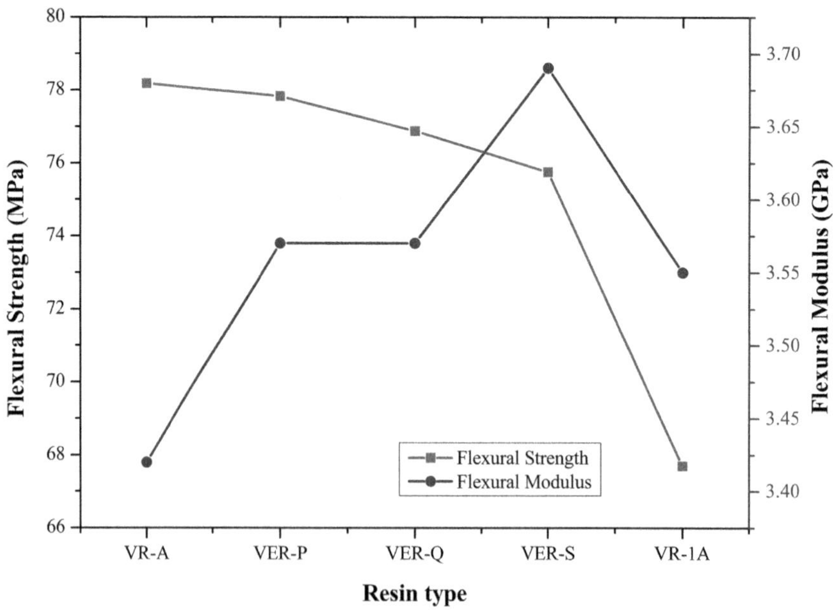

(b)

FIGURE 1.11 Variation of the (a) tensile properties and (b) flexural properties of VERs with different structural configurations. VR-A = 50 wt.% VER + 50 wt.% styrene; VER-P = 40 wt.% VER + 10 wt.% brominated VER + 50 wt.% styrene; VER-Q = 30 wt.% VER + 20 wt.% brominated VER + 50 wt.% styrene; VER-S = 20 wt.% VER + 30 wt.% brominated VER + 50 wt.% styrene; VR-1A = 50 wt.% brominated VER + 50 wt.% styrene.

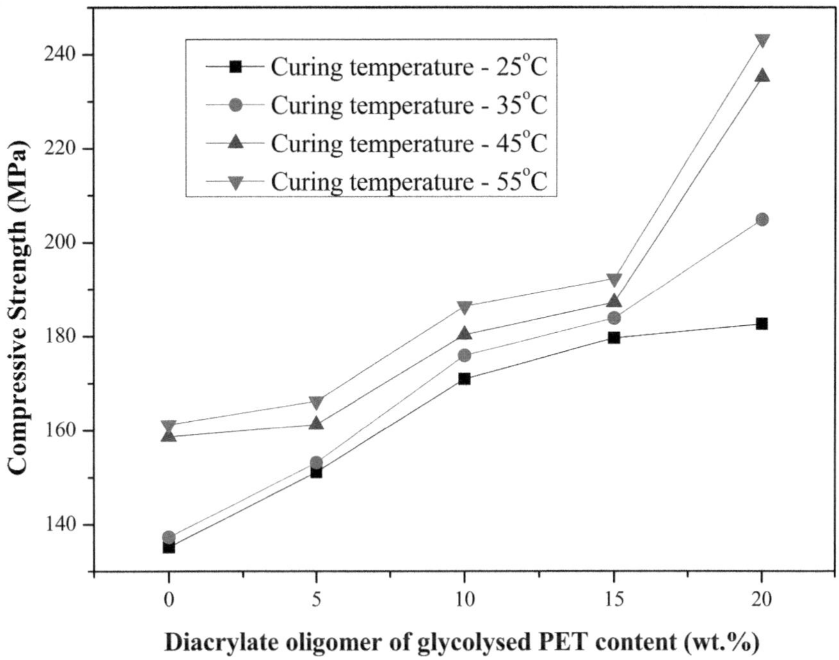

FIGURE 1.12 Effect of the content of diacrylate oligomer of glycolysed PET on the compressive strength of VERs at different cure temperatures.

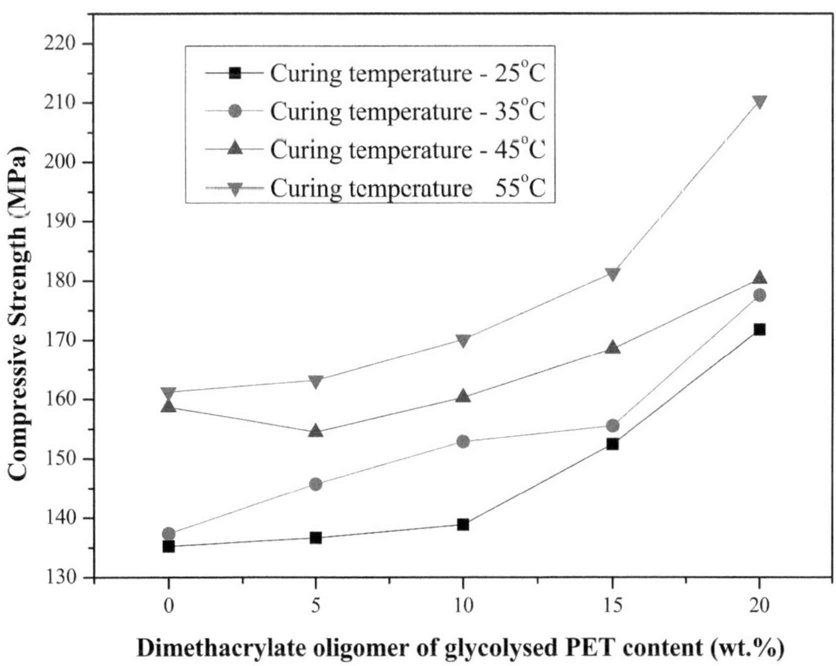

FIGURE 1.13 Effect of the content of dimethacrylate oligomer of glycolysed PET on the compressive strength of VERs at different cure temperatures.

the mechanical properties of VERs can be tailored in multiple ways such as the introduction of different constituent materials, altering the cure parameters, thermal and chemical treatments, among many others.

1.5.3 Thermal Properties

One of the major reasons for the widespread usage of vinyl esters is their seemingly appreciable thermal properties that combine well with their mechanical properties and corrosion resistance. Therefore, it is only natural to define the thermal behaviour of VERs and the factors that affect the same. The thermal properties are largely dependent on the curing process and the optimization of the cure parameters (Flores et al., 2002). Evaluation of the thermal degradation kinetics of VERs shows that several reactions happen simultaneously which usually commence with the C–C scission as a result of the presence of unreacted vinyl and epoxy groups at the ends (Vimalathithan et al., 2018). As a result, the resin exhibits lower activation energy at the initial stages of degradation. Additionally, the thermal oxidation behaviour of resins made from vinyl ester and unsaturated polyesters reveals the build-up of carbonyl products as a result of the oxidation of CH_2 and ester groups (Arrieta et al., 2016). However, in comparison of unsaturated polyester and VERs, the latter was more oxidizable while the former yielded more volatile compounds as a result of the distance between the oxidizable sites. This is an indication that VERs are less stable while providing better neutron shielding performances.

One of the most significant thermal properties of VERs is their flame retardancy. This property is usually imparted by a halogen phase within the VER molecule (Dev et al., 2017). However, studies by Malik et al. (2002) focused on the use of non-halogenic compounds for the synthesis of flame-retardant VERs using diglycidyl ether of bisphenol A with methacrylic acid at varying concentrations. The resins were diluted using reactive diluents such as glycidyl methacrylate and styrene or their combination. The properties of these resins were further exemplified with the use of flame-retardant additives that were a combination of ammonium polyphosphate (APP), azobiscarbonamide, tris(2-hydro-xyethyl) isocyanurate and hydrated alumina. Each of these additives has a different effect on the smoke density and the limiting oxygen index (LOI) of the final resin. In addition to this, the addition of glass fibres in a laminar fashion further increases the LOI and smoke density depending on the concentration of the reactive diluent and flame-retardant additive. For VERs modified by octaphenyl polyhedral oligomeric silsesquioxane (OPS), a gradual decrease in the peak heat release rate, total heat release and smoke release was observed as a result of the formation of carbon/silica layers that served as mass and heat transfer barriers (Zhang et al., 2019). Figure 1.14 depicts the variation of the key thermal parameters obtained from the thermogravimetric analysis and DSC of these composite materials. The effect of OPS on the flame-retardant properties of the resin is illustrated in Figure 1.15. Similar observations were made for VERs introduced with a combination of APP, 1-oxo-4-hydroxymethyl-2,6,7-trioxa-1-phosphabicyclo[2,2,2] octane (PEPA) and molybdenum trioxide (MoO_3) as seen in Figure 1.16 (Zeng et al., 2020).

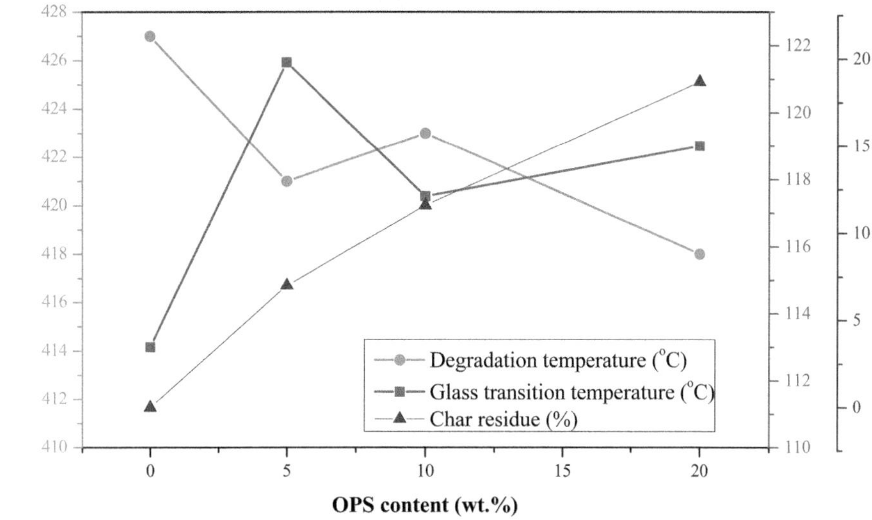

FIGURE 1.14 Effect of the OPS content on the thermal properties of the modified VER.

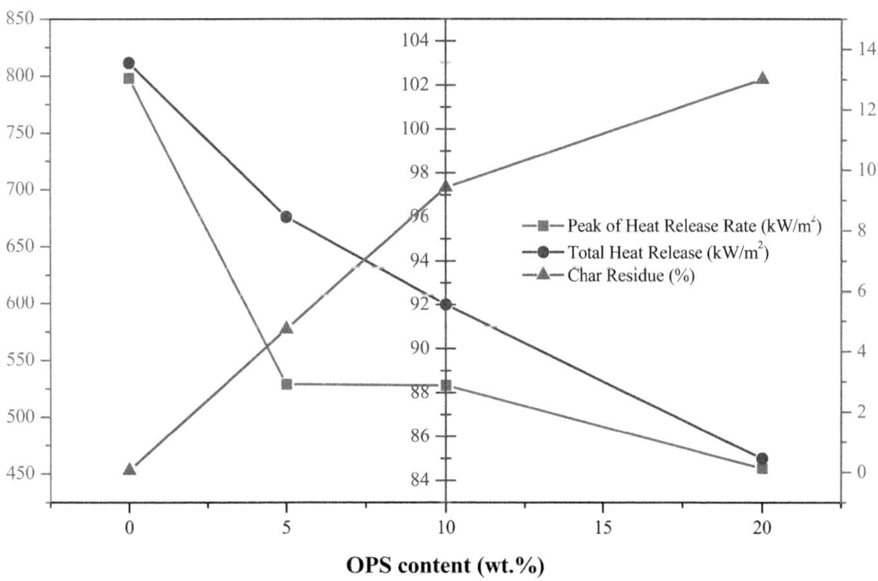

FIGURE 1.15 Effect of the OPS content on the flame-retardant properties of the modified VER.

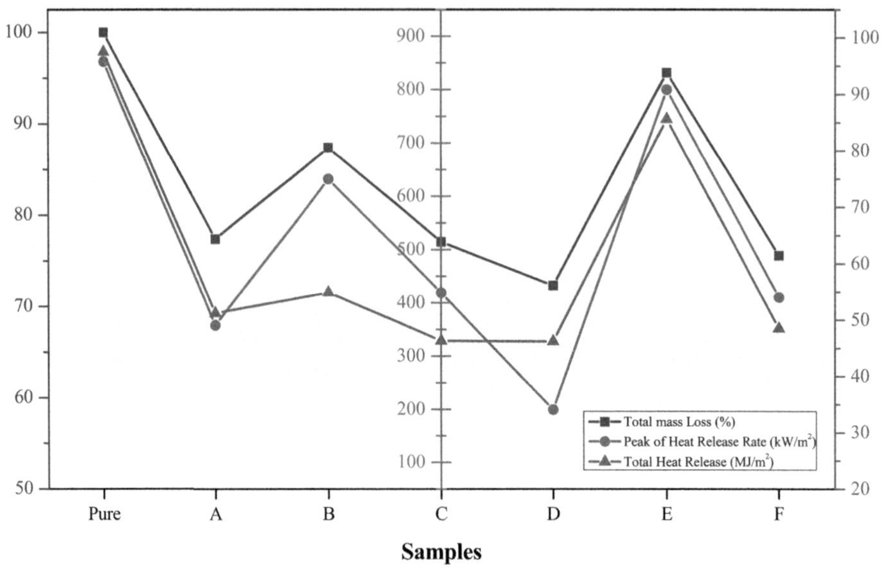

FIGURE 1.16 Flame-retardant properties of VERs modified with different flame-retardant additives. A = 75 wt.% VER + 20 wt.% APP + 5 wt.% MoO_3; B = 75 wt.% VER + 20 wt.% PEPA + 5 wt.% MoO_3; C = 80 wt.% VER + 10 wt.% APP + 10 wt.% PEPA; D = 75 wt.% VER + 10 wt.% APP + 10 wt.% PEPA + 5 wt.% MoO_3; E = 95 wt.% VER + 5 wt.% MoO_3; F = 75 wt.% VER + 15 wt.% APP + 10 wt.% PEPA.

1.6 CONCLUSIONS

In this chapter, a primary understanding and introduction to vinyl esters and their resins are provided. The chapter focuses on establishing the synthesis, curing and properties of the commonly found VERs. It is observed that the synthesis of VERs depends on the starting material and is usually influenced by the initiator and promoter material configuration along with their concentrations. The addition of external agents to impart specific characteristics and to toughen the resin also alters the synthesis process and the time required for the completion of the synthesis process. The synthesis is followed by curing of the resins where the final modifications to alter the physical, chemical, mechanical, thermal and other properties are carried out. Curing can be a physical, mechanical or chemical process depending on the chemistry of the participating compounds and the requirements of the final product. Lastly, an extensive discussion is conducted to establish and study the properties of VERs and their dependence on the synthesis and cure parameters apart from the molecular configuration and concentration of the materials.

REFERENCES

Abadie, M. J. M., Mekhissi, K., & Burchill, P. J. (2002). Effects of processing conditions on the curing of a vinyl ester resin. *Journal of Applied Polymer Science*, *84*(6), 1146–1154. https://doi.org/10.1002/app.10403.

Alia, C., Jofre-Reche, J. A., Suárez, J. C., Arenas, J. M., & Martín-Martínez, J. M. (2018). Characterization of the chemical structure of vinyl ester resin in a climate chamber under different conditions of degradation. *Polymer Degradation and Stability*, *153*, 88–99. https://doi.org/10.1016/j.polymdegradstab.2018.04.014.

Ambrogi, V., Carfagna, C., Giamberini, M., Amendola, E., & Douglas, E. P. (2002). Liquid crystalline vinyl ester resins for structural adhesives. *Journal of Adhesion Science and Technology*, *16*(1), 15–32. https://doi.org/10.1163/156856102527711130.

Anderson, T. F., & Messick, V. B. (1980). Vinyl ester resins. In *Developments in Reinforced Plastics-1: Resin Matrix Aspects* (Vol. 1, pp. 29–58). https://doi.org/10.1016/0141-3910(81)90028-8.

Ardhyananta, H., Sari, E. N., Wicaksono, S. T., Ismail, H., Tuswan, & Ismail, A. (2019). Characterization of vinyl ester bio-resin for core material sandwich panel construction of ship structure application: Effect of palm oil and sesame oil. *AIP Conference Proceedings: International Conference on Science and Applied Science (ICSAS) 2019*, *2202*, 020051-1-020051–020056. https://doi.org/10.1063/1.5141664.

Arrieta, J. S., Richaud, E., Fayolle, B., & Nizeyimana, F. (2016). Thermal oxidation of vinyl ester and unsaturated polyester resins. *Polymer Degradation and Stability*, *129*, 142–155. https://doi.org/10.1016/j.polymdegradstab.2016.04.003.

Atta, A. M., Abdel-Raouf, M. E., Elsaeed, S. M., & Abdel-Azim, A. A. A. (2006). Curable resins based on recycled poly(ethylene terephthalate) for coating applications. *Progress in Organic Coatings*, *55*(1), 50–59. https://doi.org/10.1016/j.porgcoat.2005.11.004.

Atta, A. M., El-Saeed, S. M., & Farag, R. K. (2006). New vinyl ester resins based on rosin for coating applications. *Reactive and Functional Polymers*, *66*(12), 1596–1608. https://doi.org/10.1016/j.reactfunctpolym.2006.06.002.

Atta, A. M., Elnagdy, S. I., Abdel-Raouf, M. E., Elsaeed, S. M., & Abdel-Azim, A. A. A. (2005). Compressive properties and curing behaviour of unsaturated polyester resins in the presence of vinyl ester resins derived from recycled poly(ethylene terephthalate). *Journal of Polymer Research*, *12*, 373–383. https://doi.org/10.1007/s10965-005-1638-3.

Auad, M. L., Frontini, P. M., Borrajo, J., & Aranguren, M. I. (2001). Liquid rubber modified vinyl ester resins: Fracture and mechanical behavior. *Polymer*, *42*, 3723–3730. https://doi.org/10.1016/S0032-3861(00)00773-4.

Bassett, A. W., Rogers, D. P., Sadler, J. M., La Scala, J. J., Wool, R. P., & Stanzione III, J. F. (2016). The effect of impurities in reactive diluents prepared from lignin model compounds on the properties of vinyl ester resins. *Journal of Applied Polymer Science*, *43817*, 1–10. https://doi.org/10.1002/app.44212.

Bearden, C. R. (1968). 2-Hydroxyalkyl acrylate and methacrylate dicarboxylic acid partial esters and the oxyalkylated derivatives there. In *United States Patent Office - No. 3367992*. https://doi.org/10.2307/1190003.

Benmokrane, B., Ali, A. H., Mohamed, H. M., ElSafty, A., & Manalo, A. (2017). Laboratory assessment and durability performance of vinyl-ester, polyester, and epoxy glass-FRP bars for concrete structures. *Composites Part B: Engineering*, *114*, 163–174. https://doi.org/10.1016/j.compositesb.2017.02.002.

Boinard, E., Pethrick, R. A., Dalzel-Job, J., & Macfarlane, C. J. (2000). Influence of resin chemistry on water uptake and environmental ageing in glass fibre reinforced composites-polyester and vinyl ester laminates. *Journal of Materials Science*, *35*, 1931–1937. https://doi.org/10.1023/A:1004766418966.

Brill, R. P., & Palmese, G. R. (2000). Investigation of vinyl-ester - styrene bulk copolymerization cure kinetics using Fourier transform infrared spectroscopy. *Journal of Applied Polymer Science*, *76*(10), 1572–1582. https://doi.org/10.1002/(SICI)1097-4628(20000606)76:10<1572::AID-APP12>3.0.CO;2-C.

Brinkman, W. H., Damen, L. W., & Salvatore, M. (1968). Accelerators for the organic peroxide curing of polyesters and factors influencing their behaviour. *23rd SPI Reinforced Plastics/Composites Conference*, Paper 19-D. https://doi.org/10.1093/nq/s8-IX.234.497-a.

Brody, J. C., & Gillespie, J. W. (2005). The effects of a thermoplastic polyester preform binder on vinyl ester resin. *Journal of Thermoplastic Composite Materials*, *18*, 157–179. https://doi.org/10.1177/0892705705043535.

Burchill, P. J., Kootsookos, A., & Lau, M. (2001). Benefits of toughening a vinyl ester resin matrix on structural materials. *Journal of Materials Science*, *36*(17), 4239–4247. https://doi.org/10.1023/A:1017937509700.

Can, E., Kinaci, E., & Palmese, G. R. (2015). Preparation and characterization of novel vinyl ester formulations derived from cardanol. *European Polymer Journal*, *72*, 129–147. https://doi.org/10.1016/j.eurpolymj.2015.09.010.

Cassis, F. A., & Talbot, R. C. (1998). Polyester and vinyl ester resins. In *Handbook of Composites* (pp. 34–47). https://doi.org/10.1007/978-1-4615-6389-1_3.

Cassoni, J. P., Harpell, G. A., Wang, P. C., & Zupa, A. (1977). Use of ketone peroxides for room temperature cure of thermoset resins. *32nd SPI Reinforced Plastics/Composites Conference*, Paper 3-E, Washington, DC.

Cook, W. D., Simon, G. P., Burchill, P. J., Lau, M., & Fitch, T. J. (1997). Curing kinetics and thermal properties of vinyl ester resins. *Journal of Applied Polymer Science*, *64*(4), 769–781. https://doi.org/10.1002/(SICI)1097-4628(19970425)64:4<769::AID-APP16>3.0.CO;2-P.

Cravens, T. (1972). DERAKANE 470-45, a new high temperature corrosion resistant resin. *27th SPI Reinforced Plastics/Composites Conference*, 3-B. https://doi.org/10.1016/s0047-2484(77)80016-x.

Cryshchuk, O., & Karger-Kocsis, J. (2004). Influence of the type of epoxy hardener on the structure and properties of interpenetrated vinyl ester/epoxy resins. *Journal of Polymer Science, Part A: Polymer Chemistry*, *42*, 5471–5481. https://doi.org/10.1002/pola.20371.

Cui, H. W., Jiu, J. T., Nagao, S., Sugahara, T., Suganuma, K., Uchida, H., & Schroder, K. A. (2014). Ultra-fast photonic curing of electrically conductive adhesives fabricated from vinyl ester resin and silver micro-flakes for printed electronics. *RSC Advances*, *4*(31), 15914–15922. https://doi.org/10.1039/c4ra00292j.

Dean, K., Cook, W. D., Rey, L., Galy, J., & Sautereau, H. (2001). Near-infrared and rheological investigations of epoxy-vinyl ester interpenetrating polymer networks. *Macromolecules*, *34*, 6623–6630. https://doi.org/10.1021/ma010438z.

Dev, S., Shah, P. N., Zhang, Y., Ryan, D., Hansen, C. J., & Lee, Y. (2017). Synthesis and mechanical properties of flame retardant vinyl ester resin for structural composites. *Polymer*, *133*, 20–29. https://doi.org/10.1016/j.polymer.2017.11.017.

Dey, T. (2007). Properties of vinyl ester resins containing methacrylated fatty acid comonomer: The effect of fatty acid chain length. *Polymer International*, *56*, 853–859. https://doi.org/10.1002/pi.

Dreerman, E., Narkis, M., Siegmann, A., Joseph, R., Dodiuk, H., & Dibenedetto, A. T. (1999). Mechanical behavior and structure of rubber modified vinyl ester resins. *Journal of Applied Polymer Science*, *72*, 647–657. https://doi.org/10.1002/(SICI)1097-4628(19990502)72:5<647::AID-APP5>3.0.CO;2-M.

Dua, S., McCullough, R. L., & Palmese, G. R. (1999). Copolymerization kinetics of styrene/vinyl-ester systems: Low temperature reactions. *Polymer Composites*, *20*(3), 379–391. https://doi.org/10.1002/pc.10364.

Fekete, F., Keenan, P. J., & Plant, W. J. (1965). *Ethylenically unsaturated dihydroxy diesters*. US patent, Bearing number: US3221043A.

Fekete, F., Keenan, P. J., & Plant, W. J. (1966). *Hydroxy polyether polyesters having terminal ethylenically unsaturated groups*. US patent, Bearing number: US3256226A.

Flores, F., Gillespie, J. W., & Bogetti, T. A. (2002). Experimental investigation of the cure-dependent response of vinyl ester resin. *Polymer Engineering and Science*, *42*(3), 582–590. https://doi.org/10.1002/pen.10973.

Garg, M. S., Srivastava, K., & Srivastava, D. (2015). Physical and chemical toughening of cardanol-based vinyl ester resin using CTBN: A study on spectral, thermal and morphological characteristics. *Progress in Organic Coatings*, *78*, 307–317. https://doi.org/10.1016/j.porgcoat.2014.08.004.

Gaur, B., & Rai, J. S. P. (1993). Rheological and thermal behaviour of vinyl ester resin. *European Polymer Journal*, *29*(8), 1149–1153.

Grishchuk, S., & Karger-Kocsis, J. (2011). Hybrid thermosets from vinyl ester resin and acrylated epoxidized soybean oil (AESO). *Express Polymer Letters*, *5*(1), 2–11. https://doi.org/10.3144/expresspolymlett.2011.2.

Han, C. D., & Lem, K. -W. (1984). Chemorheology of thermosetting resins. IV. The chemorheology and curing kinetics of vinyl ester resin. *Journal of Applied Polymer Science*, *29*, 1879–1902. https://doi.org/10.1002/app.1984.070290538.

Jang, C., Lacy, T. E., Gwaltney, S. R., Toghiani, H., & Pittman Jr., C. U. (2012). Erratum: Relative reactivity volume criterion for cross-linking: Application to vinyl ester resin molecular dynamics simulations. *Macromolecules*, *45*, 4876–4885. https://doi.org/10.1021/ma301156u.

Jaswal, S., & Gaur, B. (2014). New trends in vinyl ester resins. *Reviews in Chemical Engineering*, *30*(6), 567–581. https://doi.org/10.1515/revce-2014-0012.

Kandola, B. K., Ebdon, J. R., & Zhou, C. (2018). Development of vinyl ester resins with improved flame retardant properties for structural marine applications. *Reactive and Functional Polymers*, *129*, 111–122. https://doi.org/10.1016/j.reactfunctpolym.2017.08.006.

Kant, K., Mishra, A., & Rai, J. S. P. (1992). Curing studies on vinyl ester resin using acrylates as reactive diluents. *Polymer International*, *28*(3), 189–192. https://doi.org/10.1002/pi.4990280303.

Kuppusamy, R. R. P., Zade, A., & Kumar, K. (2020). Time-temperature-cure process window of epoxy-vinyl ester resin for applications in liquid composite moulding processes. *Materials Today: Proceedings*, *39*(January 2021), 1407–1411. https://doi.org/10.1016/j.matpr.2020.05.048.

La Scala, J. J., Orlicki, J. A., Winston, C., Robinette, E. J., Sands, J. M., & Palmese, G. R. (2005). The use of bimodal blends of vinyl ester monomers to improve resin processing and toughen polymer properties. *Polymer*, *46*, 2908–2921. https://doi.org/10.1016/j.polymer.2005.02.011.

Launikitis, M. B. (1982). Vinyl ester resins. In *Handbook of Composites* (pp. 38–49). https://doi.org/10.1007/978-94-009-8724-1_2.

Lee, J. H., & Lee, J. W. (1994). Kinetic parameters estimation for cure reaction of epoxy based vinyl ester resin. *Polymer Engineering & Science*, *34*(9), 742–749. https://doi.org/10.1002/pen.760340907.

Lee, S. B., Rockett, T. J., & Hoffman, R. D. (1992). Interactions of water with unsaturated polyester, vinyl ester and acrylic resins. *Polymer*, *33*(17), 3691–3697. https://doi.org/10.1016/0032-3861(92)90657-I.

Lewandowski, R. J., Ford, E. C., Longeneeker, D. M., Restaino, A. J., & Burns, J. (1975). New high performance corrosion resistant resin. *30th SPI Reinforced Plastics/Composites Conference*, 6-B, Washington, DC.

Li, H. (1998). *Synthesis, characterization and properties of vinyl ester matrix resins*. http://vtechworks.lib.vt.edu/handle/10919/30521.

Li, L., Sun, X., & Lee, L. J. (1999). Low temperature cure of vinyl ester resins. *Polymer Engineering and Science*, *39*(4), 646–661. https://doi.org/10.1002/pen.11454.

Li, P., Yu, Y., & Yang, X. (2008). Effects of initiators on the cure kinetics and mechanical properties of vinyl ester resins. *Journal of Applied Polymer Science*, *109*(4), 2539–2545. https://doi.org/10.1002/app.28234.

Li, S., Xia, J., Li, M., & Huang, K. (2013). New vinyl ester bio-copolymers derived from dimer fatty acids: Preparation, characterization and properties. *Journal of the American Oil Chemists' Society*, *90*, 695–706. https://doi.org/10.1007/s11746-013-2206-3.

Linow, W. H., Bearden, C. R., & Neuendorf, W. R. (1966). The DERAKANE resins-a valuable addition to thermosetting technology. *21st SPI Reinforced Plastics/Composites Conference*, *21*, 1-D, Chicago, IL.

Malik, M., Choudhary, V., & Varma, I. K. (2001). Effect of oxirane groups on curing behavior and thermal stability of vinyl ester resins. *Journal of Applied Polymer Science*, *82*(2), 416–423. https://doi.org/10.1002/app.1866.

Malik, M., Choudhary, V., & Varma, I. K. (2002). Effect of non-halogen flame retardant additives on the properties of vinyl ester resins and their composites. *Journal of Fire Sciences*, *20*(4), 329–342. https://doi.org/10.1177/073490402762574767.

Marsh, G. (2007). Vinyl ester -the midway boat building resin. *Reinforced Plastics*, *51*(8), 20–23. https://doi.org/10.1016/S0034-3617(07)70248-5.

Martin, J. S., Laza, J. M., Morraás, M. L., Rodriíguez, M., & Leoón, L. M. (2000). Study of the curing process of a vinyl ester resin by means of TSR and DMTA. *Polymer*, *41*(11), 4203–4211. https://doi.org/10.1016/S0032-3861(99)00631-X.

Mouritz, A. P., Kootsookos, A., & Mathys, G. (2004). Stability of polyester- and vinyl ester-based composites in seawater. *Journal of Materials Science*, *39*, 6073–6077. https://doi.org/10.1023/B:JMSC.0000041704.71226.ee.

Nodehi, M. (2022). Epoxy, polyester and vinyl ester based polymer concrete: A review. *Innovative Infrastructure Solutions*, *7*(64), 1–24. https://doi.org/10.1007/s41062-021-00661-3.

Nouranian, S., Jang, C., Lacy, T. E., Gwaltney, S. R., Toghiani, H., & Pittman, C. U. (2011). Molecular dynamics simulations of vinyl ester resin monomer interactions with a pristine vapor-grown carbon nanofiber and their implications for composite interphase formation. *Carbon*, *49*(10), 3219–3232. https://doi.org/10.1016/j.carbon.2011.03.047.

Padma, G., Varma, I. K., Sinha, T. J. M., & Patel, D. M. (1993). Effect of α-methyl styrene on properties of epoxy-novolac based vinyl ester resins. *Die Angewandte Makromolekulare Chemie*, *211*(1), 157–164. https://doi.org/10.1002/apmc.1993.052110114.

Pham, S., & Burchill, P. J. (1995). Toughening of vinyl ester resins with modified polybutadienes. *Polymer*, *36*(17), 3279–3285. https://doi.org/10.1016/0032-3861(95)99426-U.

Pratap, A. (2002). Vinyl ester and acrylic based polymer concrete for electrical applications. *Progress in Crystal Growth and Characterization of Materials*, *45*(1–2), 117–125. https://doi.org/10.1016/S0960-8974(02)00036-0.

Qin, C. L., Zhao, D. Y., Bai, X. D., Zhang, X. G., Zhang, B., Jin, Z., & Niu, H. J. (2006). Vibration damping properties of gradient polyurethane/vinyl ester resin interpenetrating polymer network. *Materials Chemistry and Physics*, *97*, 517–524. https://doi.org/10.1016/j.matchemphys.2005.10.022.

Robinette, E. J., Ziaee, S., & Palmese, G. R. (2004). Toughening of vinyl ester resin using butadiene-acrylonitrile rubber modifiers. *Polymer*, *45*, 6143–6154. https://doi.org/10.1016/j.polymer.2004.07.003.

Rodriguez, E., Larrañaga, M., Mondragón, I., & Vázquez, A. (2006). Relationship between the network morphology and properties of commercial vinyl ester resins. *Journal of Applied Polymer Science*, *100*, 3895–3903. https://doi.org/10.1002/app.22732.

Rosu, L., Cascaval, C. N., & Rosu, D. (2006). Curing of vinyl ester resins. Rheological behaviour. *Journal of Optoelectronics and Advanced Materials*, *8*(2), 690–693.

Rosu, L., Cascaval, C. N., & Rosu, D. (2009). Effect of UV radiation on some polymeric networks based on vinyl ester resin and modified lignin. *Polymer Testing*, *28*, 296–300. https://doi.org/10.1016/j.polymertesting.2009.01.004.

Scott, T. F., Cook, W. D., & Forsythe, J. S. (2002). Kinetics and network structure of thermally cured vinyl ester resins. *European Polymer Journal*, *38*(4), 705–716. https://doi.org/10.1016/S0014-3057(01)00244-0.

Scott, T. F., Cook, W. D., & Forsythe, J. S. (2008). Effect of the degree of cure on the viscoelastic properties of vinyl ester resins. *European Polymer Journal*, *44*, 3200–3212. https://doi.org/10.1016/j.eurpolymj.2008.07.009.

Shah, P. N., Kim, N., Huang, Z., Jayamanna, M., Kokil, A., Pine, A., Kaltsas, J., Jahngen, E., Ryan, D. K., Yoon, S., Kovar, R. F., & Lee, Y. (2015). Environmentally benign synthesis of vinyl ester resin from biowaste glycerin. *RSC Advances*, *5*, 38673–38679. https://doi.org/10.1039/c5ra03254g.

Shivakumar, K. N., Swaminathan, G., & Sharpe, M. (2006). Carbon/vinyl ester composites for enhanced performance in marine applications. *Journal of Reinforced Plastics and Composites*, *25*(10), 1101–1116. https://doi.org/10.1177/0731684406065194.

Signor, A. W., VanLandingham, M. R., & Chin, J. W. (2002). Effects of ultraviolet radiation exposure on vinyl ester resins: Characterization of chemical, physical and mechanical damage. *Polymer Degradation and Stability*, *79*(2), 359–368. https://doi.org/10.1016/S0141-3910(02)00300-2.

Siva, P., Varma, I. K., Patel, D. M., & Sinha, T. J. M. (1994). Effect of structure on properties of vinyl ester resins. *Bulletin of Materials Science*, *17*(6), 1095–1101. https://doi.org/10.1007/BF02757587.

Slama, W. R. (1996). Polyester and vinyl ester coatings. *Journal of Protective Coatings and Linings*, *13*(5), 88–109. https://doi.org/10.31399/asm.hb.v05b.a0006009.

Stanzione, J. F., Giangiulio, P. A., Sadler, J. M., La Scala, J. J., & Wool, R. P. (2013). Lignin-based bio-oil mimic as biobased resin for composite applications. *ACS Sustainable Chemistry and Engineering*, *1*(4), 419–426. https://doi.org/10.1021/sc3001492.

Sullivan, J. L., Kao, B. G., & Van Oene, H. (1984). Shear properties and a stress analysis obtained from vinyl-ester losipescu specimens. *Experimental Mechanics*, *24*(3), 223–232. https://doi.org/10.1007/BF02323169.

Sultania, M., Rai, J. S. P., & Srivastava, D. (2010). Studies on the synthesis and curing of epoxidized novolac vinyl ester resin from renewable resource material. *European Polymer Journal*, *46*, 2019–2032. https://doi.org/10.1016/j.eurpolymj.2010.07.014.

Sultania, M., Rai, J. S. P., & Srivastava, D. (2012). Modeling and simulation of curing kinetics for the cardanol-based vinyl ester resin by means of non-isothermal DSC measurements. *Materials Chemistry and Physics*, *132*(1), 180–186. https://doi.org/10.1016/j.matchemphys.2011.11.022.

Taillemite, S., & Pauer, R. (2009). Bright future for vinyl ester resins in corrosion applications. *Reinforced Plastics*, *53*(4), 34–37. https://doi.org/10.1016/s0034-3617(09)70151-1.

Tao, X., Duan, H., Dong, W., Wang, X., & Yang, S. (2018). Synthesis of an acrylate constructed by phosphaphenanthrene and triazine-trione and its application in intrinsic flame retardant vinyl ester resin. *Polymer Degradation and Stability*, *154*, 285–294. https://doi.org/10.1016/j.polymdegradstab.2018.06.015.

Thomas, A., Jacyszn, O., Schmitt, W., & Kolczynski, J. (1977). Methyl ethyl ketone perox ides, the relationship of reactivity to chemical structure. *32nd SPI Reinforced Plastics/Composites Conference*, Paper 3-B. https://doi.org/10.1093/mnras/stab2796.

Tu, R., & Sodano, H. A. (2021). Additive manufacturing of high-performance vinyl ester resin via direct ink writing with UV-thermal dual curing. *Additive Manufacturing*, *46*(May), 102180. https://doi.org/10.1016/j.addma.2021.102180.

Ullett, J. S., & Chartoff, R. P. (1995). Toughening of unsaturated polyester and vinyl ester resins with liquid rubbers. *Polymer Engineering & Science*, *35*(13), 1086–1097. https://doi.org/10.1002/pen.760351304.

Valea, A., Martinez, I., Gonzalez, M. L., Eceiza, A., & Mondragon, I. (1998). Influence of cure schedule and solvent exposure on the dynamic mechanical behavior of a vinyl ester resin containing glass fibers. *Journal of Applied Polymer Science*, *70*(13), 2595–2602. https://doi.org/10.1002/(SICI)1097-4628(19981226)70:13<2595::AID-APP5>3.0.CO;2-W.

Varco, P. (1975). Important cure criteria for chemical resistance and food handling applications of reinforced plastics. *30th SPI Reinforced Plastics/Composites Conference*, Paper 6-C, Washington, DC.

Varma, I. K., Rao, B. S., Choudhary, M. S., Choudhary, V., & Varma, D. S. (1985). Effect of styrene on properties of vinyl ester resins, I. *Die Angewandte Makromolekulare Chemie*, *130*, 191–199. https://doi.org/10.1002/apmc.1985.051300116.

Vimalathithan, P. K., Barile, C., & Vijayakumar, C. T. (2018). Investigation of kinetic triplets for thermal degradation of thermally cured vinyl ester resin systems and lifetime predictions. *Journal of Thermal Analysis and Calorimetry*, *133*(2), 881–891. https://doi.org/10.1007/s10973-018-7154-6.

Visco, A. M., Brancato, V., & Campo, N. (2012). Degradation effects in polyester and vinyl ester resins induced by accelerated aging in seawater. *Journal of Composite Materials*, *46*(17), 2025–2040. https://doi.org/10.1177/0021998311428533.

Vu, C. M., Nguyen, V. H., & Van, T. N. (2021). Phenyl/vinyl polysilsesquioxane based ladder polymer as novel additive for vinyl ester resin: Fabrication, mechanical and flame-retardant properties. *Polymer Testing*, *93*, 106987. https://doi.org/10.1016/j.polymertesting.2020.106987.

Yadav, S. K., Schmalbach, K. M., Kinaci, E., Stanzione, J. F., & Palmese, G. R. (2018). Recent advances in plant-based vinyl ester resins and reactive diluents. *European Polymer Journal*, *98*, 199–215. https://doi.org/10.1016/j.eurpolymj.2017.11.002.

Yang, G., Liu, H., Bai, L., Jiang, M., & Zhu, T. (2008). Preparation and characterization of novel poly(vinyl ester resin) monoliths. *Microporous and Mesoporous Materials*, *112*, 351–356. https://doi.org/10.1016/j.micromeso.2007.10.009.

Young, R. (1976). Vinyl ester resins. In *Unsaturated Polyester Technology* (pp. 315–319), Eds. Paul F. Bruins, New York: Gordon and Breach Science Publishers, Inc.

Yurdakul, H., Seyhan, A. T., Turan, S., Tanoĝlu, M., Bauhofer, W., & Schulte, K. (2010). Electric field effects on CNTs/vinyl ester suspensions and the resulting electrical and thermal composite properties. *Composites Science and Technology*, *70*(14), 2102–2110. https://doi.org/10.1016/j.compscitech.2010.08.007.

Zeng, G., Zhang, W., Zhang, X., Zhang, W., Du, J., He, J., & Yang, R. (2020). Study on flame retardancy of APP/PEPA/MoO3 synergism in vinyl ester resins. *Journal of Applied Polymer Science*, *137*(35), 1–11. https://doi.org/10.1002/app.49026.

Zhang, W., Zhang, X., Zeng, G., Wang, K., Zhang, W., & Yang, R. (2019). Flame retardant and mechanism of vinyl ester resin modified by octaphenyl polyhedral oligomeric silsesquioxane. *Polymers for Advanced Technologies*, *30*(12), 3061–3072. https://doi.org/10.1002/pat.4737.

Zhang, Y., Li, Y., Thakur, V. K., Gao, Z., Gu, J., & Kessler, M. R. (2018). High-performance thermosets with tailored properties derived from methacrylated eugenol and epoxy-based vinyl ester. *Polymer International*, *67*(5), 544–549. https://doi.org/10.1002/pi.5542.

2 Tensile, Flexural and Interfacial Properties of the Vinyl Ester-Based Bio-composites

Hossein Ebrahimnezhad-Khaljiri
University of Zanjan

CONTENTS

2.1 INTRODUCTION

Using composite materials as one of the advanced materials has been practiced in various structural applications. These materials can be fabricated from at least two distinct constituents. The matrix phase, which is usually more ductile, is the main phase into the polymeric composites. The other important phase is the reinforcement, which can be in the form of fibers, fabrics, powders, fillers, whiskers or nanomaterials [1–3].

Up to now, various thermoset resins have been developed for use in the composite materials. The epoxy, polyester and vinyl ester resins are known as the most accessible polymers for producing the composite parts. Among them, the vinyl ester is introduced as one of the proper matrices with special features. Vinyl ester has excellent chemical and corrosion resistance, better processibility, higher heat resistance and better mechanical properties than conventional polymers like polyester and epoxy resins [4]. Vinyl ester polymers are produced by reaction between epoxy resins and unsaturated carboxylic acids. Therefore, their structures are similar to the unsaturated polyester resins. For this reason, many attempts have been made to use the vinyl ester resins instead of polyester resins [5].

Reducing the resources for producing synthetic fibers and environmental issues of these fibers are the important challenges in the composite industries. For this reason,

DOI: 10.1201/9781003270997-2

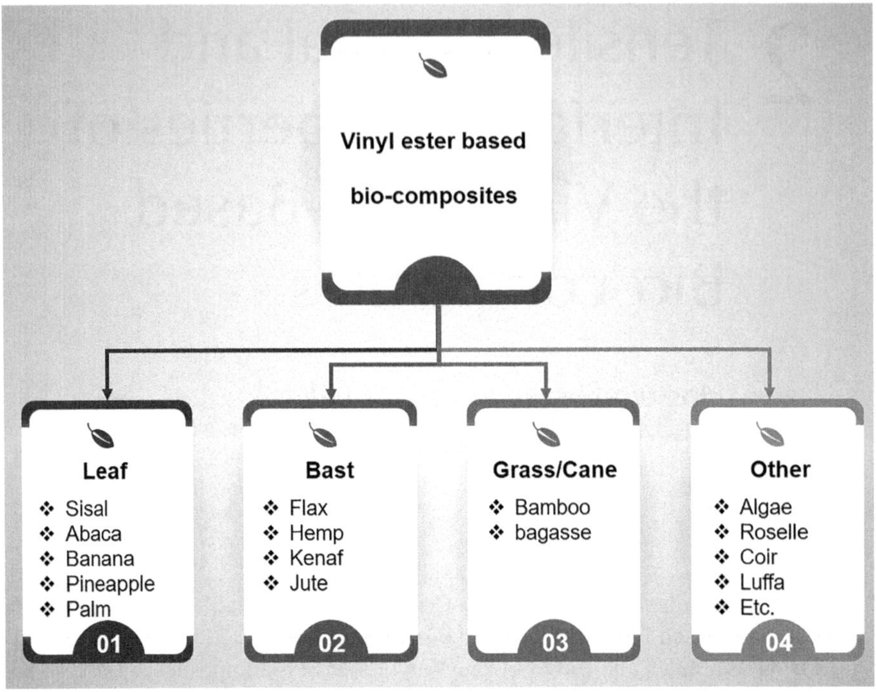

FIGURE 2.1 Categorizing the vinyl ester–based bio-composites.

during the last decade, many efforts have been made to develop natural fibers for using instead of synthetic fibers. According to their sources, the natural fibers can be extracted from plants, animals or mines.

The natural fibers/vinyl ester bio-composites are known as the new kind of bio-composites for use in structural applications. These bio-composites have their challenges like processibility, the interface between fibers/vinyl ester matrix, load transferring between fibers and matrix and affected features from environments. For this reason, a comprehensive cognition for introducing the mechanical properties of these bio-composites is very necessary. Therefore, this chapter aims to investigate the tensile, flexural and interfacial properties of these bio-composites. The vinyl ester bio-composites can be divided according to the used natural reinforcements, which can be seen in Figure 2.1.

2.2 LEAF-BASED FIBERS/VINYL ESTER BIO-COMPOSITES

One of the major sources for extracting bio-fibers for use in the composite parts is the leaf-based plant fibers. Sisal, abaca, henequen, banana, pineapple and palm plants have suitable leaves for extracting bio-fibers. Sisal is a structural fiber, which is extracted from stiff leaves of Agave Sisalana. This leaf-based fiber has lower production costs and weight than synthetic fibers, especially glass fibers. So, it can be said that it has proper potential for reinforcing the vinyl ester polymers. The length of sisal fibers, the content of sisal fibers in the matrix, type of surface modification of sisal fibers, and the fraction of lignin into the sisal fibers are the parameters, which can influence on the mechanical properties of sisal fibers/vinyl ester bio-composites. Li [6] focused on the production methods for fabrication of sisal/vinyl ester bio-composites. He fabricated this bio-composite with the resin transfer molding (RTM) and hot pressing methods. The used sisal fibers were in the form of plain weave fabrics, wherein the volume fraction of these fabrics into the vinyl ester polymer was about 30%. It should be mentioned that before fabricating these bio-composites, the sisal fabrics were surface modified by permanganate solution. The obtained tensile and flexural strength of fabricated bio-composite with the RTM method were about 31.8 and 72.7 MPa, respectively. These strengths were about 30.38 and 57.49 MPa, when the fabrication method was hot pressing.

In another work, Li and Mai [7] investigated the effect of silanization and oxidation surface treatments on the interfacial shear strength (IFFSS) of sisal fibers/vinyl ester bio-composites. For doing this, they silanized the sisal fibers with the 3-aminopropyltriethoxy and gamma-methacryloxy propyl trimethoxy silane agents. Also, the oxidation process was done by using the permanganate and dicumyl peroxide (DCP) agents. The shear strength of untreated sisal fibers/vinyl ester bio-composite was 8 MPa. By applying the silane treatments, this strength reached to 8.5 and 14.4 MPa, which was obtained with the 3-aminopropyltriethoxy and gamma-methacryloxy propyl trimethoxy silane agents, respectively. Given the obtained results, the oxidation treatments had the higher influence on the shear strength than silane agents. So, by using DCP and permanganate modification methods, the shear strength of bio-composite was about 20.5 and 26.6 MPa, respectively.

Mahato et al. [8] studied the effect of various subjecting times of sisal fibers into the 2 wt.% NaOH solution on the tensile and flexural properties of sisal fibers/vinyl ester bio-composite. The obtained results from this research work can be summarized in Figure 2.2. Based on this figure, it can be found that immersing sisal fibers into the NaOH solution for 8 h had the maximum improvement on the tensile strength and elastic modulus of this bio-composite. Whereas, the alkali treatment had not any especial effect on the flexural modulus and strength of sisal fibers-vinyl ester biocomposite. The effect of fibers' length on the mechanical performance of sisal fibers/vinyl ester bio composites was investigated by Navaneethakrishnan and Athijayamani [9]. They added fibers with the length of 3, 8 and 13 mm into the vinyl ester resin. In these bio-composites, the weight fraction of sisal fibers into the vinyl ester matrix was about the 52%. As per the reported results, the tensile strengths of bio-composites were 32.7, 33.1 and 35.9 MPa, which were obtained by adding the fibers with the length of 3, 8 and 13 mm, respectively. According to this research work, it seems that the fibers' length parameter had the higher influence on the improvement of flexural properties than the tensile properties.

Abaca with the scientific name of Musa textillisnee and common name of Manila hemp has 56–64 wt.% cellulose in its structure, which is apt for using in the bio-composites. For this reason, Ahmed et al. [10] fabricated the abaca fibers/vinyl ester bio-composites. For improving the mechanical properties, they modified these fibers with the hexamethyldisiloxane (HMDS). The tensile strength and elastic modulus of unmodified abaca fibers/vinyl ester composite were about 39 MPa and 3 GPa, respectively. In the optimum modifying condition (2 wt.% HMDS with the dispersion

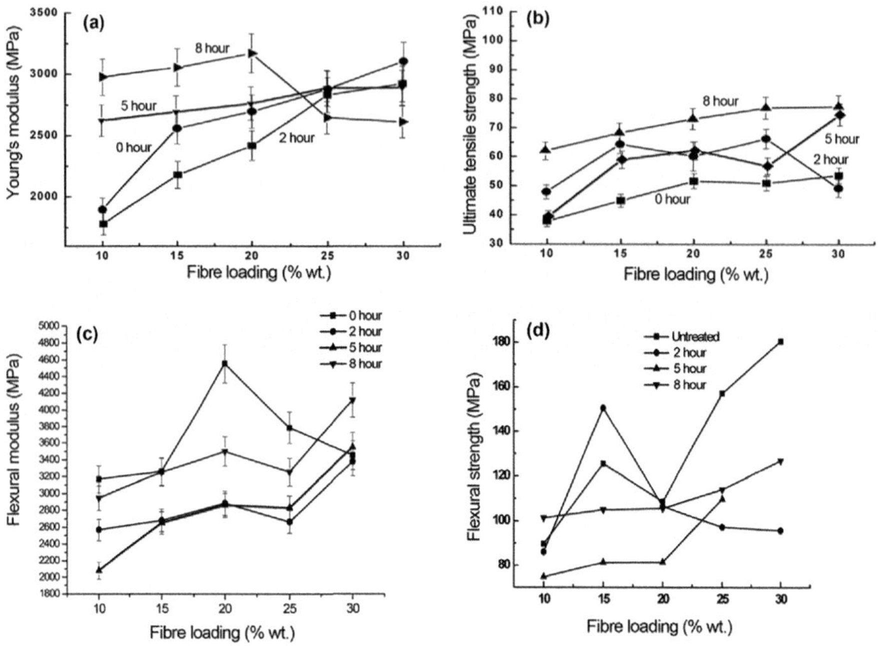

FIGURE 2.2 The effect of soaking times in alkali solution on the mechanical properties of sisal fibers/vinyl ester bio-composites (A: elastic modulus; B: tensile strength; C: flexural modulus; D: flexural strength).

Source: Adapted from Ref. [8] with the permission of Springer Nature.

time of 120 min), the obtained tensile strength and elastic modulus were 44 MPa and 6.65 GPa, respectively. Like tensile properties, the flexural properties had the maximum improvement in this condition. In other words, the flexural strength and modulus of unmodified abaca fibers/vinyl ester composite were 61 MPa and 1.75 GPa, respectively. By applying the silane treatment, these properties reached to 84.66 MPa and 4.9 GPa, respectively.

Banana fibers are another leaf-based bio-fiber, which has great potential for adding as reinforcement into the composite parts. Actually, these fibers are agricultural waste. Therefore, using them in the vinyl ester bio-composites reduces the production costs. Ghosh et al. [11] added 35 vol.% banana fibers into the vinyl ester resin to characterize the mechanical properties of this bio-composite. The tensile strength, elastic modulus, flexural strength and flexural modulus of this bio-composite were 114 MPa, 5.3 GPa, 124 MPa and 3.16 GPa, respectively.

Pineapple leaf fibers can be another source for fabricating the vinyl ester–based bio-composites. Like banana fibers, the pineapple fibers are known as agricultural waste, most of which is burnt by farmers after harvesting the pineapple. So, it can be said that developing pineapple fibers for substituting those with the synthetic fibers can help to reduce the fabrication costs of vinyl ester composites. Mohamed et al. [12] surveyed the interfacial features of pineapple fibers/vinyl ester bio-composites by the

IFSS method. To do so, they added nontreated, alkali treated and bleached pineapple fibers into the vinyl ester matrix. The shear strength of these bio-composites were 1.44, 1.90 and 1.77 MPa, respectively. It means that the alkali treatment had the higher influence than the bleaching treatment for improving the interfacial properties of pineapple fibers/vinyl ester composite. These research groups in another work [13] focused on the various extraction and modification methods of pineapple fibers for improving the flexural properties of pineapple fibers/vinyl ester bio-composite. The outcome of this research work has been illustrated in Figure 2.3. The tensile properties of this bio-composite were characterized by Yogesh and Hari Rao [14]. As per the reported results, the tensile strength and elastic modulus of bio-composite containing 30 wt.% pineapple fibers were 133 MPa and 3.64 GPa, respectively.

One of the most favorable sources for extracting the leaf-based bio-fibers is the palm fiber. Up to know, the different palm fibers have been introduced by various researchers. Oil palm, date palm, sugar palm, kernel shell, palmyra, toddy, peach, areca sheath, piassava and mauritiella armata trees are known as palm fibers sources. Unfortunately, the limited numbers of these palm sources have been developed for using as reinforcement into the polymeric composites. According to the researches, only sugar and date palm fibers have been added into the vinyl ester resins. For example, Ammar et al. [15] investigated the effect of various sugar palm fabric configurations on the tensile and flexural properties of palm fibers/vinyl ester bio-composites. These configurations were unidirectional, 0/90° and +45°. The obtained results showed that the unidirectional configuration with the elastic modulus of 2501 MPa and flexural modulus of 3328 MPa had the maximum elastic and flexural modulus, as compared with the other configurations. Also, it was found that all configurations had the lower tensile and flexural strength, as compared with neat vinyl ester. These reduction trends in these strengths can be due to the hollow structure of sugar palm fibers and poor interfacial bonding between fiber/matrix into these bio-composites. A similar trend has been

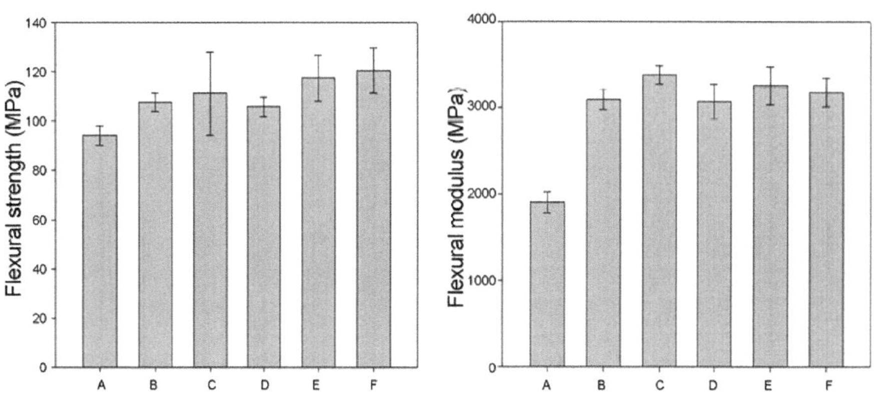

FIGURE 2.3 The effect of various pineapple leaf fiber extractions on the flexural properties of pineapple/vinyl ester composites (A: neat vinyl ester; B: extracted fibers from middle portion of the leaves; C: extracted fibers from whole portion of the leaves; D: extracted fibers by water soaking (24 hours); E: extracted fibers by abrasive-combed method; F: extracted fine fiber strands).

Source: Reprinted from Ref. [13] with the permission of Springer Nature.

reported by Huzaifah et al. [16]. Therefore, it can be concluded that the sugar palm fibers cannot be a good choice for use as reinforcement into the vinyl ester composites, especially when these composites are subjected to flexural and tensile loadings.

Nagaprasad et al. [17] characterized the mechanical features of date palm fibers/vinyl ester bio-composites. These researchers added the various weight fractions of date palm fibers into the vinyl ester resin. Based on the obtained results, the flexural and tensile properties can be maximum values, when the weight fraction of these fibers is 25%–30%. For having a better view of the mechanical properties of vinyl ester bio-composites containing the various weight fractions of date palm fibers, the summary of data in this research work has been depicted in Figure 2.4.

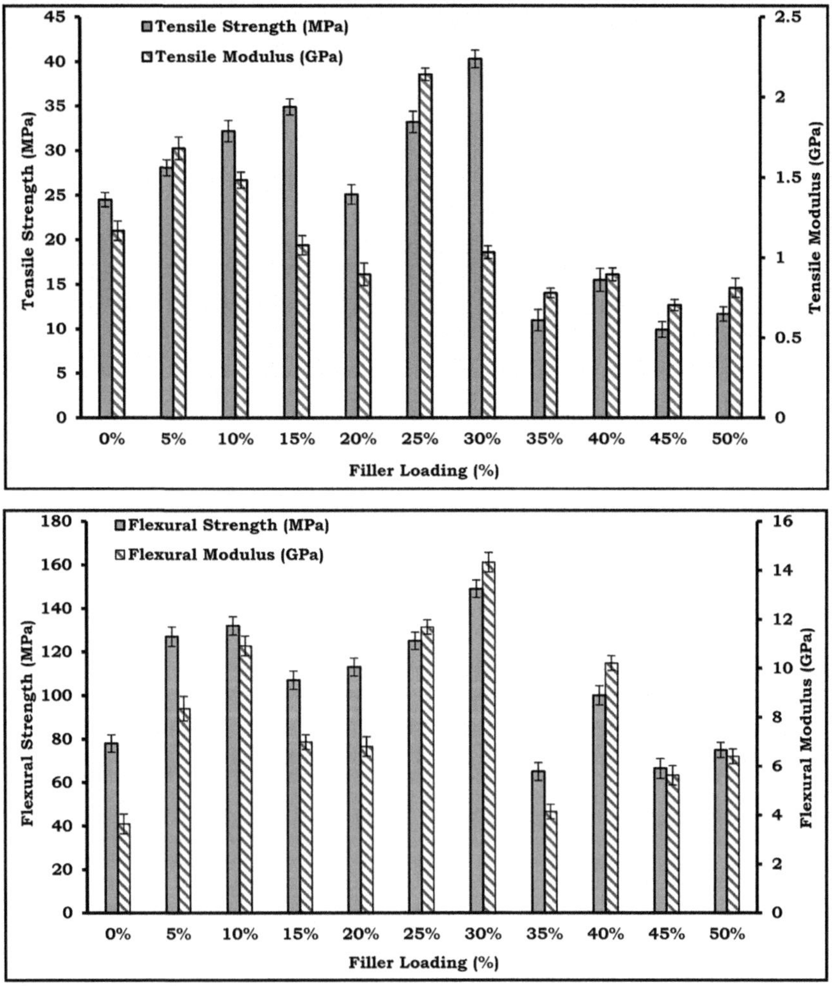

FIGURE 2.4 The effect of date filler incorporating the tensile and flexural properties of date palm/vinyl ester bio-composites.

Source: Adapted from Ref. [17] with the permission of Elsevier.

2.3 THE BAST-BASED FIBERS/VINYL ESTER BIO-COMPOSITES

The second proper bio-source to extract the fibers for reinforcing the vinyl ester polymers can be bast-based fibers. Flax, jute, hemp, kenaf and ramie are the examples of this major group. Among them, the flax fibers had the most attention for use in the vinyl ester bio-based composites. One of the interesting works for improving the mechanical properties of flax fibers/ vinyl ester bio-composites was done by Huo et al. [18]. This research group for improving the interfacial strength between fibers and matrix simultaneously investigated the effect of modifying the surface of fibers and adding the acrylic resin into the vinyl ester matrix. The optimum mechanical results obtained in the bio-composite contained 1 wt.% acrylic resin and NaOH-treated flax fibers. The interlaminar shear strength (ILSS), flexural strength, flexural modulus, tensile strength and elastic modulus of this bio-composite were about 28.9 MPa, 334.5 MPa, 34.1 GPa, 216.1 MPa and 13.23 GPa, respectively.

Seghini et al. [19] assessed the interfacial adhesion of flax fibers/vinyl ester bio-composites at the yarn scale with the fragmentation test and high resolution tomography methods. It was found that at the yarn scale, the breakage of flax yarn has been concentrated in the peripheral zone. One of the methods for improving the mechanical properties of flax fibers/vinyl ester bio-composites can be surface modification of these fibers. For this reason, many scientific groups have focused on this subject to develop using this bio-composite as structural parts. Some of these researches have interesting novelty, which can be used for other vinyl ester–based bio-composites. Seghini et al. [20] applied the surface treatment on the flax fibers with the supercritical carbon dioxide. By this method, the critical fragment length and debonding values of this bio-composite were reduced, which were 15.1 and 18.9%, respectively. In another interesting work, Whitacre et al. [21] studied the effect of corn zein protein as a new coupling agent on the mechanical performance of flax fibers/vinyl ester bio-composites. Given the obtained results, it was realized that adding 2.5% of this coupling agent improved the tensile, flexural and shear strength of this bio composite by 2.5%, 8% and 17%, respectively. By using the scanning electron microscope (SEM) and observing the surface of flax fibers, the role of this coupling agent on the interfacial bonding between flax fibers and vinyl ester matrix was confirmed, which can be seen in Figure 2.5.

One of the proper choices of bast-based fibers for reinforcing the vinyl ester polymers can be hemp fibers. These fibers have high specific strength and stiffness, which help them in developing the vinyl ester bio-composites. Like other natural fibers, the hemp fibers have a hydrophilic nature. To solve this problem, a surface modification process should be done on these fibers. Misnon et al. [22,23] added (approximately 37 vol.%) untreated and NaOH-treated hemp fibers into the vinyl ester resin. According to the reported results, the tensile strength, elastic modulus, flexural strength and flexural modulus of untreated hemp fibers/vinyl ester composite were 61.68 MPa, 6.2 GPa, 93.65 MPa and 5.62 GPa, respectively. By applying the NaOH treatment, the reducing trend in all results was seen, which was an unexpected trend. It seems that NaOH treatment was not proper for hemp fibers, because this treatment reduces the mechanical properties of hemp fibers/vinyl ester bio-composite.

The other renewable and biodegradable lignocellulose source for reinforcing the vinyl ester matrices is the kenaf fiber. These fibers are known as one of the light

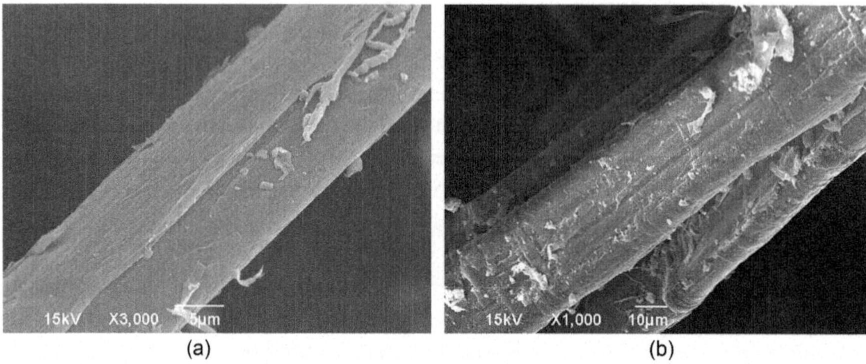

FIGURE 2.5 Comparing the surface morphology of flax fibers from flax/vinyl ester bio-composites (A: untreated flax fibers; B: treated by corn zein protein coupling agent).

Source: Reprinted from Ref. [21] with the permission of Elsevier.

cellulose fibers; these can be used in the bio-composites in the secondary or tertiary applications. The only interesting work on the kenaf fibers/vinyl ester bio-composite was done by the Fairuz et al. [24] research group, which investigated the effect of production parameters on the mechanical properties of this kenaf bio-composite. Based on this research work, it can be mentioned that the flexural and tensile properties can be the highest values, when the gelation and curing temperature of this bio-composite were set at 100°C and 180°C, respectively.

One of the bast-based fibers, which has great potential for use in the polymeric materials, is the corchorus fiber with the common name of jute fiber. The bundle of this fiber has high strength and initial modulus, as compared with other natural fibers. Like other natural fibers, the jute fibers need the surface modification for increasing the interfacial adhesion by creating the mechanical locks in the interface of fiber/matrix and reducing the hydrophilicity of jute fibers. This viewpoint resulted in the introduction of new methods for improving the mechanical properties of vinyl ester–based bio-composites by increasing the interfacial strength. As can be expected, the alkali treatment as the simplest method can be applied for improving the interfacial bonding, which was reported by Ray et al. [25] in the year of 2001. For doing this, these researchers soaked the jute fibers in 5 wt.% NaOH solution for 2, 4, 6 and 8 h at 30°C. Then, they added 8, 15, 23, 30 and 35 vol.% of these jute fibers into the vinyl ester resin. The maximum elastic modulus and flexural strength obtained in the bio-composites with 35 vol.% jute fibers. In this composite, the elastic modulus and flexural strength were 11.89 GPa and 199 MPa, respectively, when the untreated jute fibers were used. The highest flexural strength in this composite was obtained, when the jute fibers were soaked for 2 h in the NaOH solution. By immersing the jute fibers for 4 h in the NaOH solution, the elastic modulus reached the highest value.

Ray et al. [26] coated the jute fibers with the styrene butadiene rubber (SBR) latex and then fabricated the vinyl ester composites by using these coated fibers. Although, applying this coat reduced the elastic modulus and tensile strength, the breaking energy improved. The reason for this reduction trend was the lack of covalent bonding between hydrophilic jute fibers and the hydrophobic latex layer. By using and developing this method, the interfacial bonding between vinyl ester, the latex layer, and jute fibers can be improved, so that, this improvement can enhance the tensile strength and fracture toughness of bio-composite at the same time. In another work, Ray et al. [27] treated jute fibers with shellac. Actually, the shellac resin has acidic structure and is capable of solving alkaline solutions. For this treatment, these researchers dissolved the shellac resin with the various weight percentages of 1, 2 and 3 wt.% into the water containing 0.5 wt.% diethanolamine. Then, the vinyl ester polymer was reinforced by these treated jute fibers. As per the reported results, the composite containing untreated jute fibers had 10.67 GPa flexural modulus, with the flexural strength of about 136 MPa. By applying surface treatment with 1 wt.% shellac solution, the flexural modulus and strength reached 11.65 GPa and 139 MPa, respectively.

In another work, Ray et al. [28] used guar gam as the surface treatment agent of jute fibers. Characterizing the mechanical properties showed that by applying this treatment with the solution containing 0.2 wt.% guar gam, the flexural modulus of composite reached from 761 to 9.43 GPa. Also, the flexural strength increased from 148 to 159.5 MPa. Treating jute fibers with 2-hydroxyethyl methacrylate was performed by Khan and Bhattacharia [29] for improving the mechanical properties of vinyl ester composites. By performing the surface treatment with this coupling agent, the 70 increment in flexural strength and 45% improvement in the flexural modulus of this composite were seen.

2.4 THE GRASS- AND CANE-BASED FIBERS/ VINYL ESTER BIO-COMPOSITES

The other developed plant source for extracting fibers to compose with the vinyl ester polymers can be grass- and cane-based fibers. The bamboo and bagasse fibers are known as the most famous grass and cane fibers, respectively. Chin et al. [30] firstly modified the bamboo fibers with the NaOH solution. After that, they added the various weight percentages (10%, 20%, 30% and 40%) of these fibers into the vinyl ester polymer. The tensile strength of composite containing 40 wt.% bamboo fibers was 96 MPa, which showed 839% increment, as compared with neat vinyl ester. Like tensile strength, the maximum flexural strength obtained in this weight percentage was 149 MPa. So, when compared with neat vinyl ester, this flexural strength showed 294% improvement. In the study by Chen et al. [31], the effect of fibers' surface modifications on the shear strength of bamboo fibers/vinyl ester bio-composites was assessed. The applied surface treatments were alkali treatment, acetylation, silanization and oxidization ($KMnO_4$) methods. By applying the silane surface treatment, the highest improvement (36.5%) was obtained in the shear strength of this bio-composite.

The study on the mechanical properties of bagasse fibers/vinyl ester bio-composites was performed by Athijayamani et al. [32]. Therefore, they used the mat fibers for fabricating these composites. Also, the bagasse contents into the vinyl ester were 17, 36, 44, 53 and 60 wt.%. Given the obtained results, the optimum weight fraction was 53%. In this weight percentage, the tensile strength and elastic modulus of this bio-composite were 79.4 and 1449 MPa, respectively. The shear strength of neat vinyl ester was 5.21, whereas by adding 53 wt.% bagasse fibers, this strength reached to 10.23 MPa. Similarly, the flexural strength was enhanced from 29.6 to 112.5 MPa, by adding 53 wt.% bagasse fibers.

2.5 OTHER BIO-REINFORCEMENT/VINYL ESTER BIO-COMPOSITES

In the previous sections, three major groups of vinyl ester–based bio-composites were introduced. As we know, there are many sources for extracting the bio-reinforcement, especially bio-fibers, which can be added into the vinyl ester polymers. The research and development of these sources for reinforcing the vinyl ester polymers are so limited. Therefore, the aim of this section is to introduce these sources, which have great potential for use in the composite industries in the near future. The first of this bio-reinforcement can be algae, which is biodegradable, aquatic organism abundantly available in the ocean and ponds. This filler with the volume fraction of 5, 10, 15, 20 and 25% was composed with the vinyl ester by Bharathiraja et al. [33]. The obtained results showed that the vinyl ester bio-composite with 25 vol.% algae filler had the highest fracture strength, which was about 43 Pa.m,$^{1/2}$ whereas the neat vinyl ester had the fracture strength of 26.6 Pa.m$^{1/2}$. This means that this filler has strong interfacial bonding with the vinyl ester polymers.

Roselle with the scientific name of Hibiscus sabdariffa L. is one of the plants, the stem of which can be suitable for use in the production of bio-fibers. These fibers have a weak interface with the polymers. Therefore, before use in the polymers (especially vinyl ester families), surface modification process is necessary, which was performed by Nadlene et al. [34]. As per this research work, it seems that the most suitable treatment for these fibers is the silane surface modification. By this modification method, the tensile strength of this bio-composite increased from 17.53 to 24.81 MPa.

Limonia acidissima or wood apple is known as biowaste, which has proper mechanical and thermal properties, as compared with other biowastes. Therefore, it can be a good choice for use as bio-filler into the vinyl ester polymers. This filler in the shape of thin powder was firstly modified by Shravanabelagola et al. [35] via alkali treatment for investigating the mechanical properties of Limonia acidissima powder/vinyl ester bio-composite. The tensile and flexural properties of this bio-composite were the highest values, when the weight fraction of that was 15%. Betal nut husk (BNH) is another agro-waste fiber, which is obtained from betal nut by the water retting method. Its tensile strength is comparable with the kenaf fiber with the significantly higher elongation at break. In the study by Yusriah et al. [36], the flexural strength of vinyl ester composite containing the different weight percentages of these fibers was investigated. According to the results, by adding and increasing the weight fraction of BNH into the vinyl ester, the flexural strength exhibited a significant loss. This trend can be due to incompatibility between BNH/vinyl ester and degradation of fibers during the manufacturing process.

Alhuthali et al. [37] investigated the mechanical and thermal properties of recycled cellulose fibers (RCFs)/vinyl ester bio-composites containing the various percentages of nanoclay. In this scientific work, the flexural properties had the highest improvement, when the 3 wt.% nanoclay was added into the RCFs/vinyl ester bio-composite. Improving the interfacial bonding between RCFs and vinyl ester by nanoclay particles caused to reduce the pulling out and debonding of fibers.

Silk is known as one of the animal-based sources for extracting bio-fibers. As we know, these fibers are extensively used in the textile industries. These fibers have a lot of waste, which can be used in novel applications. One of the applications can be as a reinforcing agent in the polymeric materials like vinyl ester family, which was performed by Ravindra Rama and Rai [38]. To investigate the physical and mechanical properties of silk fibers/vinyl ester bio-composites, these researchers added 5, 10 and 15 wt.% silk fibers into the vinyl ester resin. As per the obtained results, the composite with 15 wt.% silk fibers had the maximum tensile strength and elastic modulus, which were more than 70 and 1300 MPa, respectively. It seems that adding the higher percentage of silk fibers into the vinyl ester resin cannot improve the mechanical properties, because of increasing the void content into this bio-composite.

Generally, the polyalthia longifolia tree is used to prepare herbal concoctions, and the light trunks are used for sailing and pencil boxes. According to the new researches, it was found that the seed of this tree can be used as filler in the polymeric composites. Having features like renewability, sustainability, nontoxicity and biodegradability has caused that this filler to be one of the new environment-friendly reinforcing materials. Characterizing the vinyl ester–based composites containing this filler was done by Stalin et al. [39]. The recorded data showed that by incorporating 25 wt.% of this filler seed, the tensile strength and elastic modulus had the highest values, which were 32 MPa and 1.23 GPa, respectively. In this weight fraction, the flexural strength of this bio-composite reached 125 MPa, which is 60.26% higher than the flexural strength of the neat vinyl ester. The luffa sponge is known as vegetable sponge, which is cultivated in the tropical countries. The luffa is categorized as fruit-based natural fibers, which has features such as biomass degradation, nontoxicity and chemical/physical stability. Having hollow structures can enable this bio-fiber to fabricate the bio-composites with the sound or energy absorption features. For adding 15 wt.% of this sponge into the vinyl ester resin, Siqueira and Botaro [40] used the ethanol/cyclohexane extracting, mercerizing and 1,2,4,5-benzenetetracarboxylic dianhydride (PMDA) modification methods. The obtained results showed that by applying PMDA surface modification method, the elastic modulus of composite reached 3906 MPa, whereas the elastic modulus of untreated luffa/vinyl ester composite and neat vinyl ester were 2870 and 2541 MPa, respectively. Also, the tensile strengths of neat vinyl ester, untreated bio-composite and PMDA-treated composite were 14, 15.9 and 18.6 MPa, respectively.

One of the interesting sources for extracting the bio-reinforcement is coconut. The coconut coir can be used as bio-fiber, whereas the coconut shell can be powdered and used as bio-filler in the polymeric composites. In the study by Udaya Kumar et al. [41], the coconut coir fiber and shell powder were added into the vinyl ester polymer, with the content of these bio-reinforcements being 30 and 5 wt.%, respectively. The tensile strength of neat vinyl ester was about 26 MPa. By incorporating this bio-fiber and bio-filler, the tensile strength improved and reached 55 and 38 MPa, respectively. Also, the elastic modulus of neat vinyl ester was 19 GPa, which after adding these bio-materials reached 1.9 and 1.2 GPa, respectively. In an interesting work, Sundarababu et al. [42] simultaneously added coconut shell powder (CSP) and rice husk powder

(RHP) into the vinyl ester resin. The ratios of CSP and RHP and vinyl ester into the bio-composites were 20:20:60, 15:15:70 and 10:10:80 wt.%. Based on the obtained results, the flexural and tensile strength had the maximum values, when the ratio of those was 10:10:80 wt.%. These strengths were 17.79 and 38.45 MPa, respectively.

2.6 FUTURE PERSPECTIVE

Although, there have been many efforts to characterize the tensile, flexural and interfacial properties of vinyl ester bio-composites, the documentation about these features is limited. Therefore, it can be predicted that in the near future, many researchers will focus on characterizing these properties. Using other bio-fibers as reinforcement in the vinyl ester polymers can be another trend. Modifying the interface between natural fibers and vinyl ester polymers can be another progressive trend, which can result in fabricating the vinyl ester bio-composites with the higher mechanical properties. The progressive view about this kind of bio-composites shows that they will have proper capability to use instead of conventional composite parts in the near future.

2.7 CONCLUSION

Easier curing than epoxy resins and having higher mechanical properties than polyester families are the key factors for developing the vinyl ester–based composites. Composing vinyl ester polymers with natural reinforcements has been done to fabricate the composite structures with the acceptable mechanical properties and lower environmental issues. According to the used bio-reinforcements, these composites can be categorized into four separate parts, which are leaf fibers, bast fibers, grass/cane fibers and other bio-fillers vinyl ester composites. To enhance the mechanical properties of these bio-composites such as tensile, flexural and interfacial features, many efforts have been made by different science groups. Surface modifying the natural fibers, creating the chemical bonding between the natural fibers/vinyl ester matrix, adding the nanoparticles especially nanocellulose as second reinforcement and incorporating the second polymer for tuning the mechanical properties are the examples of these efforts.

REFERENCES

1. Mohammadi, M. A., R. Eslami-Farsani, and H. Ebrahimnezhad-Khaljiri. 2020. Experimental Investigation of the Healing Properties of the Microvascular Channels-Based Self-Healing Glass Fibers/Epoxy Composites Containing the Three-Part Healant. *Polymer Testing* 91:106862.
2. Eslami-Farsani, R., and H. Ebrahimnezhad-khaljiri. 2021. Smart Epoxy Composites. In *Epoxy Composites: Preparation, Characterization and Applications*, eds. J. Parameswaranpillai, H. Pulikkalparambil, S. M. Rangappa, and S. Siengchin, pp. 349–94. Weinheim: Wiley- VCH Verlag GmbH & Co.
3. Shamohammadi Maryan, M., H. Ebrahimnezhad-Khaljiri, and R. Eslami-Farsani. 2022. The Experimental Assessment of the Various Surface Modifications on the Tensile and Fatigue Behaviors of Laminated Aluminum/Aramid Fibers-Epoxy Composites. *International Journal of Fatigue* 154:106560.

4. Jaswal, S., and B. Gaur. 2014. New Trends in Vinyl Ester Resins. *Reviews in Chemical Engineering* 30(6):567–81.
5. Johnson, R. D. J., V. Arumugaprabu, and T. J. Ko. 2019. Mechanical Property, Wear Characteristics, Machining and Moisture Absorption Studies on Vinyl Ester Composites – A Review. *Silicon* 11(5):2455–70.
6. Li, Y. 2006. Processing of Sisal Fiber Reinforced Composites by Resin Transfer Molding. *Materials and Manufacturing Processes* 21(2):181–90.
7. Li, Y., and Y. W. Mai. 2006. Interfacial Characteristics of Sisal Fiber and Polymeric Matrices. *Journal of Adhesion* 82(5):527–54.
8. Mahato, K., S. Goswami, and A. Ambarkar. 2014. Morphology and Mechanical Properties of Sisal Fibre/Vinyl Ester Composites. *Fibers and Polymers* 15(6):1310–20.
9. Navaneethakrishnan, S., and A. Athijayamani. 2017. Taguchi Method for Optimization of Fabrication Parameters with Mechanical Properties in Sisal Fibre–Vinyl Ester Composites. *Australian Journal of Mechanical Engineering* 15(2):74–83.
10. Ahmed, S. N., M. N. Prabhakar, Siddaramaiah, and J. I. Song. 2018. Influence of Silane-Modified Vinyl Ester on the Properties of Abaca Fiber Reinforced Composites. *Advances in Polymer Technology* 37(6):1970–78.
11. Ghosh R., K.V. Narasimham, and M. P. Kalyan. 2019. Study of Mechanical Properties of Banana-Fiber-Reinforced Vinyl Ester Resin Composites. In *Recent Advances in Material Sciences, Lecture Notes on Multidisciplinary Industrial Engineering*, eds. S. Pujari, S. Srikiran, and S. Subramonian, pp. 375–82. Singapore: Springer Nature.
12. Mohamed, A. R., S. M. Sapuan, M. Shahjahan, and A. Khalina. 2010. Effects of Simple Abrasive Combing and Pretreatments on the Properties of Pineapple Leaf Fibers (Palf) and Palf-Vinyl Ester Composite Adhesion. *Polymer - Plastics Technology and Engineering* 49(10):972–78.
13. Mohamed, A. R., S. M. Sapuan, and A. Khalina. 2014. Mechanical and Thermal Properties of Josapine Pineapple Leaf Fiber (PALF) and PALF-Reinforced Vinyl Ester Composites. *Fibers and Polymers* 15(5):1035–41.
14. Yogesh, M., and A. N. Hari Rao. 2018. Fabrication, Mechanical Characterization of Pineapple Leaf Fiber (PALF) Reinforced Vinylester Hybrid Composites. *AIP Conference Proceedings* 1943:020062.
15. Ammar, I. M., M. R. M. Huzaifah, S. M. Sapuan, Z. Leman, and M. R. Ishak. 2019. Mechanical Properties of Environment-Friendly Sugar Palm Fibre Reinforced Vinyl Ester Composites at Different Fibre Arrangements. *EnvironmentAsia* 12(1):25–35.
16. Huzaifah, M. R.M., S. M. Sapuan, Z. Leman, and M. R. Ishak. 2019. Effect of Fibre Loading on the Physical, Mechanical and Thermal Properties of Sugar Palm Fibre Reinforced Vinyl Ester Composites. *Fibers and Polymers* 20(5):1077–84.
17. Nagaprasad, N., S. Balasubramaniam, V. Venkataraman, R. Manickam, R. Nagarajan, and I. S. Oluwarotimi. 2020. Effect of Cellulosic Filler Loading on Mechanical and Thermal Properties of Date Palm Seed/Vinyl Ester Composites. *International Journal of Biological Macromolecules* 147:53–66.
18. Huo, S., V. S. Chevali, and C. A. Ulven. 2013. Study on Interfacial Properties of Unidirectional Flax/Vinyl Ester Composites: Resin Manipulation on Vinyl Ester System. *Journal of Applied Polymer Science* 128(5):3490–3500.
19. Seghini, M. C., F. Touchard, F. Sarasini, L. Chocinski-Arnault, D. Mellier, and J. Tirillò. 2018. Interfacial Adhesion Assessment in Flax/Epoxy and in Flax/Vinylester Composites by Single Yarn Fragmentation Test: Correlation with Micro-CT Analysis. *Composites Part A: Applied Science and Manufacturing* 113:66–75.
20. Seghini, M. C., F. Touchard, L. Chocinski-arnault, V. Placet, C. François, L. Plasseraud, M. P. Bracciale, J. Tirill, and F. Sarasini. 2020. Environmentally Friendly Surface Modification Carbon Dioxide. *Molecules* 25(428):1–16.

21. Whitacre, R., A. Amiri, and C. Ulven. 2015. The Effects of Corn Zein Protein Coupling Agent on Mechanical Properties of Flax Fiber Reinforced Composites. *Industrial Crops and Products* 77:232–38.
22. Misnon, M. I., M. M. Islam, J. A. Epaarachchi, H. Chen, K. Goda, and M. T. I. Khan. 2018. Flammability Characteristics of Chemical Treated Woven Hemp Fabric Reinforced Vinyl Ester Composites. *Science and Technology of Materials* 30(3):174–88.
23. Misnon, M. I., M. M. Islam, J.A. Epaarachchi, N. Da. N. Affandi, and H. Wang. 2018. Water Exposure, Tensile and Fatigue Properties of Treated Hemp Reinforced Vinyl Ester Composites. *AIP Conference Proceedings* 1985:030006.
24. Fairuz, A. M., S. M. Sapuan, E. S. Zainudin, and C. N.A. Jaafar. 2015. The Effect of Gelation and Curing Temperatures on Mechanical Properties of Pultruded Kenaf Fibre Reinforced Vinyl Ester Composites. *Fibers and Polymers* 16(12):2645–51.
25. Ray, D., B. K. Sarkar, A. K. Rana, and N. R. Bose. 2001. Mechanical Properties of Vinylester Resin Matrix Composites Reinforced with Alkali-Treated Jute Fibres. *Composites Part A: Applied Science and Manufacturing* 32(1):119–27.
26. Ray, D., N. R. Bose, A. K. Mohanty, and M. Misra. 2007. Modification of the Dynamic Damping Behaviour of Jute/Vinylester Composites with Latex Interlayer. *Composites Part B: Engineering* 38(3):380–85.
27. Ray, D., D. Bhattacharya, A. K. Mohanty, L. T. Drzal, and M. Mishra. 2006. Static and Dynamic Mechanical Properties of Vinylester Resin Matrix Composites Filled with Fly Ash. *Macromolecular Materials and Engineering* 291(7):784–92.
28. Ray, D., A. K. Rana, N. R. Bose, and S. P. Sengupta. 2005. Effect of Guar-Gum Treatment on Mechanical Properties of Vinylester Resin Matrix Composites Reinforced with Jute Yarns. *Journal of Applied Polymer Science* 98(2):557–63.
29. Khan, M. A., and S. K. Bhattacharia. 2007. Effect of Novel Coupling Agent on the Mechanical and Thermal Properties of Unidirectional Jute-Vinyl Ester Composites. *Journal of Reinforced Plastics and Composites* 26(6):617–27.
30. Chin, S. C., K. F. Tee, F. S. Tong, H. R. Ong, and J. Gimbun. 2020. Thermal and Mechanical Properties of Bamboo Fiber Reinforced Composites. *Materials Today Communications* 23:100876.
31. Chen, H., M. Miao, and X. Ding. 2011. Chemical Treatments of Bamboo to Modify Its Moisture Absorption and Adhesion to Vinyl Ester Resin in Humid Environment. *Journal of Composite Materials* 45(14):1533–42.
32. Athijayamani, A., B. Stalin, S. Sidhardhan, and A. B. Alavudeen. 2016. Mechanical Properties of Unidirectional Aligned Bagasse Fibers/Vinyl Ester Composite. *Journal of Polymer Engineering* 36(2):157–63.
33. Bharathiraja, G., N. Karunagaran, V. Jayakumar, and S. Ganesh. 2020. Investigation on Fracture Toughness of Algae Filler Vinyl Ester Composite. *Materials Today: Proceedings* 22:1233–35.
34. Nadlene, R., S. M. Sapuan, M. Jawaid, M. R. Ishak, and L. Yusriah. 2018. The effects of chemical treatment on the structural and thermal, physical, and mechanical and morphological properties of roselle fiber-reinforced vinyl ester composites. *Polymer Composites* 39(1):274–87.
35. Shravanabelagola N. S., V. Kumar, G. Govardhan, S. M. Rangappa, and S. Siengchin. 2021. Raw and Chemically Treated Bio-Waste Filled (Limonia Acidissima Shell Powder) Vinyl Ester Composites: Physical, Mechanical, Moisture Absorption Properties, and Microstructure Analysis. *Journal of Vinyl and Additive Technology* 27(1):97–107.
36. Yusriah, L., S. M. Sapuan, E. S. Zainudin, M. Mariatti, and M. Jawaid. 2016. Thermo-Physical, Thermal Degradation, and Flexural Properties of Betel Nut Husk Fiber-Reinforced Vinyl Ester Composites. *Polymer Composites* 37(7):2008–17.

37. Alhuthali, A., I. M. Low, and C. Dong. 2012. Characterisation of the Water Absorption, Mechanical and Thermal Properties of Recycled Cellulose Fibre Reinforced Vinyl-Ester Eco-Nanocomposites. *Composites Part B: Engineering* 43(7):2772–81.
38. Ravindra Rama, S., and S. K. Rai. 2011. Performance Analysis of Waste Silk Fabric-Reinforced Vinyl Ester Resin Laminates. *Journal of Composite Materials* 45(23):2475–80.
39. Stalin, B., N. Nagaprasad, V. Vignesh, M. Ravichandran, N. Rajini, Si. O. Ismail, and F. Mohammad. 2020. Evaluation of Mechanical, Thermal and Water Absorption Behaviors of Polyalthia Longifolia Seed Reinforced Vinyl Ester Composites. *Carbohydrate Polymers* 248:116748.
40. Siqueira, É. J., and V. R. Botaro. 2013. Luffa Cylindrica Fibres/Vinylester Matrix Composites: Effects of 1,2,4,5-Benzenetetracarboxylic Dianhydride Surface Modification of the Fibres and Aluminum Hydroxide Addition on the Properties of the Composites. *Composites Science and Technology* 82:76–83.
41. Udaya Kumar, P. A., B. Suresha, N. Rajini, and K.G. Satyanarayana. 2018. Effect of Treated Coir Fiber/Coconut Shell Powder and Aramid Fiber on Mechanical Properties of Vinyl Ester. *Polymer Composites* 39(12):4542–50.
42. Sundarababu, J., S. S. Anandan, and P. Griskevicius. 2020. Evaluation of Mechanical Properties of Biodegradable Coconut Shell/Rice Husk Powder Polymer Composites for Light Weight Applications. *Materials Today: Proceedings* 39:1241–47.

3 Compression and Impact Properties of Vinyl Ester-Based Bio-Composites

M. Meena
S.T. Hindu College

Senthil Muthu Kumar Thiagamani
Kalasalingam Academy of Research and Education

CONTENTS

3.1 INTRODUCTION

Recent research has specialized in the production of bio-composites based on natural fillers in conjunction with synthetic thermoplastic polymers [1–4]. A composite material is a blend of two materials with unique qualities that is often formed by the fortification of a matrix structure. The thermoset or thermoplastic resins such as polyester, epoxy, and vinyl ester (VE) are the most regularly deployed matrix materials. Carbon, aramid, and glass fibers are the most often used reinforcing agents. Natural fiber-reinforced composites have attracted a lot of interest in recent years because of its lightweight, nonabrasive, flammable, nontoxic, cheap cost, and biodegradable qualities [5–8]. Expanding worldwide environmental and social issue, high percentage of petroleum resource exhaustion and new environmental legislation have compelled the quest for environmentally friendly composites. Recently, bio-composites evolved as a trend of research, due to its variety of applications like automobiles, railway, sports equipment, etc. [9–12].

DOI: 10.1201/9781003270997-3

3.1.1 Thermoset Composites

A composite material is made up of two or more separate phases (matrix region and dispersion part) with considerably differing bulk characteristics from one another. The matrix's task is to hold the dispersion medium together in an organized array and to safeguard them from the environment. The matrix transmits loads to the dispersed medium, which is crucial in compression loading because it prevents brittle fracture. The matrix also adds hardness, fracture toughness, and impact and abrasion resistance to the composite. The matrix's properties also set the optimum operating temperature, moisture and fluid resistance, and thermal and oxidative sustainability. The matrix could be made of metal, ceramic, or polymeric material [13–17]. The polymer matrix might be thermoplastic or thermosetting, but the latter is considerably preferred. [18–20].

Superior features of the thermosetting resin include increased mechanical strength, great chemical resistance, amazing dimensional stability, and exceptional weatherability [21–24]. Thermosets are low-molecular-weight, low-viscosity monomers (2000 cP) that cure into three-dimensional (3D) cross-linked structures that are infusible. Thermoset composite matrices include polyesters, VEs, epoxies, bismaleimides, cyanate esters, polyimides, and phenolics. When thermoset is subjected to heat, its chemical properties are irrevocably intensified. Once thermoset plastics have been modified by heat, they cannot be remolded [25–27].

3.1.2 Vinyl Ester and Its Properties

Vinyl ester (VE) is a thermoset plastic that is widely used for developing advanced composites due to its good stiffness, dimensional stability, chemical resistance, higher strength than polyester resins, and lower cost than epoxy resin [28]. VEs were created to combine the comfort of use and low cost of polyesters with the thermal and mechanical properties of epoxies. So, their property acts the halfway between the polyesters and epoxy [29]. VE resins have a similar molecular structure to polyesters, but they differ primarily in terms of their chemical groups, which are only found at the ends of the chains. VE resins are more long-lasting and adaptable than polyesters because the length of the chain is available to absorb impact loads. The VE molecule has very few ester groups as well. Because such ester groups are susceptible to degradation by water, VE has greater water resistance and many other chemicals than its polyester equivalents and, as a result, is frequently used as a transmission fluid [29].

3.1.2.1 Properties of VE

Because the cross-chain link molecules absorb the impact vibration, it has great chemical resistance, is less absorbent to water, has outstanding thermal resistance, and has less fracture initiation and propagation. Polyester and epoxy resins have intermediate properties, strength, and cost. VE resins are more robust than epoxy resins and more lasting than polyester resins. VE resin has a viscosity of around 200 cps, which is lower than polyester and epoxy, which have viscosities of 500 and 900 cps, respectively. Environmental elements such as temperature and moisture are more vulnerable to VE resins [30]. Figure 3.1 shows the properties of VE.

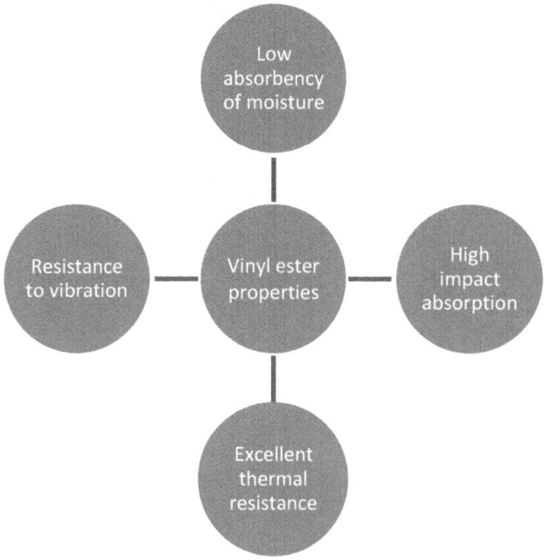

FIGURE 3.1 Salient properties of VE.

Overall, the long length of the cross-linked molecular chain absorbs loading shocks, making VE resins tougher and more resilient than polyester resins. The VE molecule's ester groups are resistant to water and many other chemicals. As a result, it is primarily used in marine applications, as well as in the chemical industry to store and transport chemicals [31]. It is also used in the production of polymer composite parts for automobiles and other surface transportation, bridge reinforcements, infrastructure construction, military and aerospace applications, and so on [32–35].

In order to use the VE resin in various industries, with excellent properties based on environmental concern, VE is combined with natural or synthetic fiber to form a bio-composite hybrid material for future applications. However, the interaction of matrix materials with the fillers highly influences the mechanical, thermal, and chemical properties of prepared composites. So, in order to use the prepared bio-composites for various applications one must carefully investigate the changes in their mechanical and other properties. Studies on compression and impact properties of thermoset bio-composites suggest their use not only in automotive industry, but also for other industry applications as well [36–38].

3.1.3 Vinyl Ester (VE)-Based Bio-Composites

VE resins are a sort of polyester resin used in industrial applications that require enhanced strength and chemical resistance. These materials are created by combining an epoxy resin with methacrylic acid to get the desired structure shown in Figure 3.2.

Since the active dual bonds occur at the endpoints of rather long chains, the cross-link density of these polyester resins is lower than that of conventional polyester resins. At a higher cost, this results in materials with higher failure strain, superior mechanical and impact features, and improved chemical resistance.

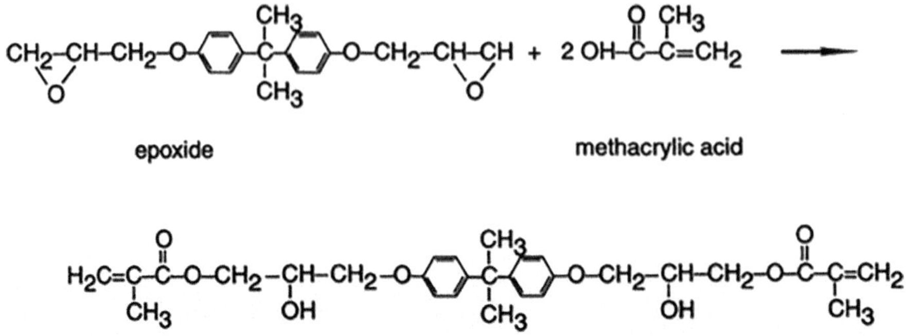

FIGURE 3.2 Formation of VE by the reaction between epoxy and methacrylic acid [28].

A composite is highly influenced by its matrix. It can be customized for a particular use by selecting the appropriate polymer. For example, the use of VE may improve a composite's stiffness, dimensional stability, chemical resistance, and strength. In addition, VE costs less than epoxy resins [39]. VE possesses mechanical properties relatively similar to those of epoxy resins, particularly in terms of hydrolytic stability. It is not difficult to handle this polymer at room temperature, so it offers greater control over cure rate and reaction conditions compared to epoxy resins. However, VE resin is brittle and reinforcement of the fillers is necessary to improve performance and minimize the cost of the resin [40]. Particulate fillers are normally mixed with VE during shape casting of the product to reinforce the resin.

3.2 COMPRESSION PROPERTIES OF VE-BASED BIO-COMPOSITES

For applications involving compressive loads on structures, compressive strength may be a critical design parameter. Composites are subjected to significant compressive stress in electrical applications such as power transformer coil supports or core support blocks. All-composite component designs must carefully evaluate the compressive strength requirements for a compressively loaded application, whether it is a composite mold platen or insulating plate. The matrix strength has a greater influence on compressive strength in composites over tensile strength (TS), especially in lower fiber-volume composites. The compressive strength of a composite can also be affected by matrix filler. In practice, injecting filler to the matrix reduces compressive strength. Nevertheless, some fillers can be incorporated to resins at rather high quantities (up to 50% of the total weight of the composite) without having a significant impact on compressive strength. With adequate loadings, platy fillers can improve compressive strength, but lubricating fillers (i.e., carbon black) can dramatically reduce compressive strength at low loading levels. When fillers that do not attach properly to the resin are put into the matrix, their compressive strength is always reduced. Figure 3.3 represents the Scanning Electron Microscope (SEM) image of the samples after the compression test.

Composite flaws will always have a negative impact on mechanical performance. Flaws have a greater detrimental impact on compressive strength, especially if they are related to the matrix. Voids, inclusions, fissures, and other matrix faults are examples of matrix flaws.

FIGURE 3.3 SEM image for after compression test: (a) 1% addition of coir fiber 80× magnification and (b) 1% addition of coir fiber 350× magnification [41].

When natural fiber is reinforced with VE matrix, it forms a bio-composite which improves its strength. Many of the researchers reported the formation of VE-based bio-composites by adding VE with jute [42], roselle [43], flax [44], pineapple [45,46], kenaf [47], sisal [48], bagasse fiber [49], unidirectional aligned bagasse fiber [48], betel nut husk (BNH) [43], and rice husk impregnate coir [50].

Razavi et al. [51] fabricated kenaf fiber–vinyl ester composite with 10%, 30%, and 40% fiber-volume fraction. The fiber was treated with 5% NaOH for 3 h by the immersion method. This process of treatment of the fiber was carried out in order to achieve improved mechanical properties of the resulting kenaf fiber bio-composites product. The compression testing was carried out using a universal testing machine with 500 kN capacity. 40% fiber-volume-fraction VE composite exhibits 31% more compressive strength than the other two volume fractions.

Jahagirdar and Kulkarni [52] developed two biodegradable composites using the hand lay-up technique: coconut fiber–vinyl ester composite and coconut fiber–rubber particles–vinyl ester composite. The TS and tensile modulus of both these composites are higher than plain VE, and both properties increase as the composition of coconut fibers increases.

Compressive strength and compressive moduli of different concentration micro-balloon-reinforced VE syntactic foam was reported by Gupta et al. [54]. The pure VE resin and syntactic foams have comparable stress–strain profiles, with a linear elastic zone accompanied by a strain softening region with a modest drop in stress. If the compression is increased further, the tension level begins to rise once more. In the case of plain resin, the increase in tension is faster and substantially greater than in the case of syntactic resin. It is dependent on the type and volume percentage of micro-balloons in foams. Although stress–strain behavior may be strain rate-dependent, the compression rate is remained constant.

The effects of methyl methacrylate (MMA) concentrations and curing temperatures on compressive strengths, flexural strengths, the coefficient of thermal expansion, and the modulus of elasticity are investigated in polymer concrete utilizing MMA monomer-modified VE resin as a binder. The compressive strength test was carried out in accordance with ASTM C 579 (Standard Test Method for the Compressive Strength of Chemical-Resistant Mortars, Grouts, Monolithic Surfacing's, and Polymer Concretes). The compressive strength rose rapidly until 24 h, and then slowly until 168 h. This tendency varied with MMA concentration and curing temperature, with 168-h compressive strength ranging from 43.8 to 77.2 MPa. These compressive strength values are lower than those of other polymer concrete kinds [55,56].

Under impact, the compressive failure of carbon fiber-reinforced unidirectional vinyl ester composites (CFRVE) is investigated experimentally and analytically. For CFRP with VE ranging from 10% to 60%, impact and quasi-static compressive tests are performed.

3.3 IMPACT PROPERTIES OF VE-BASED BIO-COMPOSITES

The impact energy of a material is the amount of energy required to fracture a given volume of the material [4]. Therefore, the impact strength of a material is the energy required to initiate and propagate a crack through the material. The crack propagation energy is related to the toughness of the material and the length of that crack tip that must travel in order to fracture a component. This means the lower the value of the impact energy, the more brittle the material behaves. The standard tests for impact strength of a material include Charpy test, Izod test, drop weight impact test, chip impact test, and compression-after-impact and tension-after-impact tests.

Shivakumar et al. [31] reported the compressive test analysis of VE/FLYASH (33%) composites, an analysis of the average energy needed to cause the crack within these groups, can provide a useful indicator of the first failure of the specimens. The average energy necessary to induce the crack between the specimens cured under ambience and microwave conditions with a power level of 180 W is shown in Table 3.1.

Microwave-cured samples appeared to demand lower energy to induce the crack. It uses 0.62 J less energy than those treated in ambient conditions. Furthermore, the distribution of this group was the smallest when compared to others. The average energy required to induce the crack in specimens cured with microwaves for 40 s was found to be nearly equal to those cured for 35 s. The average energy required to break the specimens differed by only 0.15 J, while the difference in the group cured under ambient conditions was 0.18 J. The dispersion was less than that of those microwaved for 35 s.

Instrumented impact tests using Charpy V-notch specimens are used to determine the strength and toughness qualities of a glass fiber-reinforced VE. The large values of energy absorbed by the specimens at high temperatures, on the one side, and the decreasing values of maximum load with increasing temperature, on the other side, must be correlated with the presence of delamination of the specimen at these high temperatures and facilitating non-negligible load values at this delamination [57].

The low-velocity falling weight impact damage behavior of flax-basalt/VE hybrid composites was investigated [58]. Incident impact energies of 50, 60, and 70 J were used to achieve total perforation in order to characterize distinct impact damage parameters such as energy absorption properties and damage modes and mechanisms. The testing results demonstrated that the basalt hybrid system had a higher impact energy and peak

TABLE 3.1
Average Energy Required to Fracture Specimens Cured with a Power Level of 180 W [31]

Curing condition	Microwave with a power level of 180 W			
Specimen type	Fractured specimens			
Symbols representation	σ = standard deviation; m = meter			
Specimens' materials	VE/FLYASH (33%)			
	Ambient Condition	**Microwave Cured**		
Exposure time	(0 s)	30 s	35 s	40 s
Energy used to initiate the crack	8.84 J (σ = 0.893)	8.22 J (σ = 0.803)	8.81 J (σ = 1.160)	8.66 J (σ = 1.6064)
Energy used to propagate the crack	2.16 J (σ = 0.590)	2.62 J (σ = 0.818)	2.16 J (σ = 0.878)	2.16 J (σ = 0.574)
Total energy dissipated	11.00 J (σ = 0.3555)	10.84 J (σ = 0.297)	10.97 J (σ = 0.386)	10.82 J (σ = 0.528)
Displacement at peak force	0.0018 m (σ = 0.0001)	0.0017 m (σ = 0.0001)	0.0018 m (σ = 0.0001)	0.0017 m (σ = 0.0001)

load than the flax/VE composite without hybridization, showing that a hybrid approach is a promising strategy for improving the toughness attributes of natural fiber composites. Obtained impact parameters by the authors are depicted in Table 3.2.

Furthermore, the authors predicted that, from the time vs load graph of the proposed samples as shown in Figure 3.4, basalt fiber hybridization on the front face of the laminates greatly improved the impact load and total energy of flax/VE composites, demonstrating the potential of flax/VE bio-based composites for semi-structural or structural applications.

TABLE 3.2
Important Impact Parameters and Corresponding Values Obtained from Impact Testing [59]

Specimen	Peak Load, F_m (kN)	Incipient Damage Load, F_i (kN)	Maximum Energy, E_m (J)	Incipient Energy, E_i (J)	Total Energy, E_t (J)
Energy at 50 J Flax/VE	5.39	4.09	16.81	2.22	43.03
Flax/VE/basalt	8.40	2.36	31.67	1.67	59.88
Energy at 60 J Flax/VE	5.42	4.11	14.67	2.70	40.44
Flax/VE/basalt	8.27	2.36	31.67	1.67	59.88
Energy at 70 J Flax/VE	5.51	4.21	15.73	2.82	41.44
Flax/VE/basalt	8.30	3.90	28.58	2.34	62.89

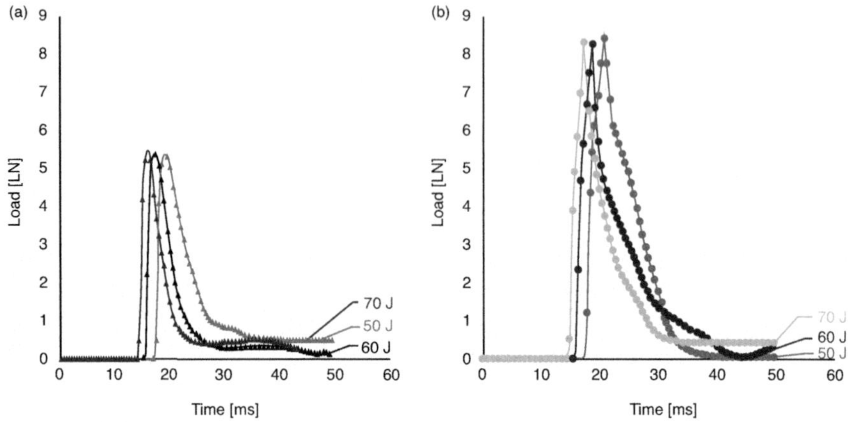

FIGURE 3.4 Typical load vs time: (a) flax and (b) flax/basalt hybrid composites [59].

The impact damage properties of glass fiber-reinforced poly(vinyl ester) beams are discovered to be significantly temperature-sensitive. The glass transition temperature influences the failure mode of the damaged beams significantly. Interlaminar fracture on the neutral plane was the primary mechanism at test temperatures below T_g. Interlaminar fracture on the neutral plane cannot occur at test temperatures over T_g due to matrix phase relaxation [60].

Motru et al. [61] investigated the toughness and notched sensitivity of the algal filler VE composite, and the composite was cut into pieces according to ASTM D256 specifications for impact testing. The dimensions of the impact testing specimens are (64 × 12.5 × 3.2) mm. It is an interesting discovery that the inclusion of filler up to 10% (v/v%) of algae filler boosts the impact strength of the composite material. But, injecting excess filler weakens the impact strength of the composite material just little. The fracture toughness of the composite material is too responsible. However, the value is still more than that of pure resin. Table 3.3 shows the impact parameters of few VE-based bio-composites.

The performance of sugar palm fiber-reinforced VE composites with various fiber configurations was investigated by Ishak [65]. There were three fiber layouts chosen: unidirectional, weaved, and roving. The tensile modulus, flexural strength, flexural modulus, and impact strength of the unidirectional fiber composites were good, with values of 2501 MPa, 93.08 MPa, 3328 MPa, and 33.66 kJ/m^2, respectively.

Stalin et al. [74] used 65 × 13 × 3 mm dimension Tamarind seed filler (TSF) VE composite to carry out impact testing. The value of impact strength increases as the weight fraction of the TSF-VE composite increases. The composite with 15% filler has 14.37 kJ/m^2, which is larger than the pure VE matrix. The impact strength of the TSF-reinforced VE composite increases by 21.5%. It is found that the inclusion of TSF reinforced with VE raises the strength by 15%, while the other weight percentages result in low impact values due to inadequate TSF dispersion in the matrix. Furthermore, decreased VE content in the highest TSF-filled composites has resulted in a decline in the composites' ability to absorb impact energy. TSF-VE composite

TABLE 3.3
Impact Parameters of VE-Based Bio-Composites

Composite Name	Impact Parameter: Impact Strength (KJ/m²)	Impact Parameter: Hardness	Reference
VE Sheets		-	
BS	42.11		
ES	48.37		[62]
FS	48.11		
GS	38.49		
VE	5.438	-	[63]
Roselle fiber-reinforced VE (10, 20, 30, 40 and 50 wt.% filler)	0.79–1.34	-	[64]
Sugar palm fiber-reinforced VE composites	33.66	-	[65]
Dry leaf fiber-reinforced polymer matrix composites		-	
VE resin fiber			
VERF5	0.1		
VERF10	0.25		[66]
VERF15	0.36		
	0.28		
Chopped roselle fiber-reinforced commercial-grade VE resin	1.59	-	[67]
Roselle plant fiber-reinforced commercial-grade VE resin	5.438	-	[68]
Roselle fiber-reinforced commercial-grade VE resin	6.60	-	[69]
Roselle fiber-reinforced commercial-grade VE resin	2.39	-	[70]
Unidirectional aligned bagasse fiber-reinforced commercial-grade VE resin	5.89	-	[48]
Polyalthia longifolia seed filler loading with 5–50 wt.% /VE	10–31.09	23–36.5	[71]
Date seed filler loading between 5% and 50%/VE	9.43–17.03	20.33–51	[72]
Tamarind seed filler reinforced with 5–50 wt.%/VE	7–14	23–42.33	[71]
Sugarcane bagasse fillers in the range of 5–50 vol.%/ unsaturated polyester resin	3.5–7	8–14	[73]
Pineapple chaff fillers in the range of 5–50 vol.%/ unsaturated polyester resin	3.5–5.2	9–13	[73]

has a higher impact strength than fly ash-reinforced polyester composite and red mud-reinforced polyester composite. The TSF-VE composites have a 44% increase in impact strength. In comparison to other fly-ash-reinforced polyester composites, it is 31% for red mud-reinforced polyester composite. Few TSF-filled VE composite products used in our daily life are reported by Stalin et al. [74].

Impact strength of dates seed filler/TSF-VE composites (DSF/TSF-VE) was investigated with different filler weight percentages by Stalin et al. [74]. Resultant impact strength enhanced from 10 to 22 kJ/m^2 when the filler content increased from

0 to 5 wt.%. 5 wt.% filler-loaded VE composites exhibit 1.18 times enhanced impact energy when compared with pure resin. Increasing the filler content up to 10 wt.% favors the enhancement in impact strength and energy, whereas further increase in filler to 20 wt.% leads to lowering of impact strength from 22 to 19 kJ/m^2.

An attempt was made to reinforce vinyl ester matrix (VE) with *Polyalthia longifolia* seed filler (PLSF), with varying filler content ranging from 5 to 50 wt.% loadings, by Stalin et al. [71]. Pure VE resin has an impact strength of 11.83 kJ/m^2. However, it was raised to 13.50, 23.91, and 26.24 kJ/m^2 following reinforcement with filler components of 5%, 10%, and 15% by weight. The experimental results demonstrated that an improvement was possible. Impact strength was measured up to 45 wt.% PLSF content. There was, however, a significant reduction with 50% PLSF-VE composite. The improved interfacial property of the PLSF-VE composite endorsed its increased impact strength of 1.83, 2.22, 2.35, 5.45, and 9.42 times over date seed filler/VE, TSF/VE, sisal fiber/epoxy, BNH fiber/VE, and thyme herbs/polyethylene composites, respectively.

The impact strength of the neat VE has been modified by the addition of unripe, ripe, and matured BNH fiber, with the unripe and ripe BNH fiber-reinforced VE composites showing a significant improvement in impact strength. Incorporation of developed BNH fiber in VE resin, on the other hand, has resulted in a significant reduction in the composites' impact strength [43].

With reinforcement of dry leaf fillers in VE matrix, the compression stress decreases abruptly because of gaps between the fibers, and then increases by 1.822 times after the addition of 10 wt.% dry leaf in the polymer matrix composites due to higher compatibility and lower voids. Finally, for 15 wt.% dry leaves fiber polymer matrix composites, it is reduced by 0.74 times. Due to the effective reinforcement of fiber, the impact energy of the VE polymers tends to increase by 2.5 and 3.6 times with the addition of fibers content of 5 and 10 wt.%, respectively. Due to incompatibility between fibers and resin material, this energy decreases by 0.778 times in 15 wt.% of dry leaves fiber reinforcement in the polymer content [66].

3.4 CONCLUSIONS

Bio-fiber-reinforced polymer composites have advantages such as low density, lower cost, and lower solidity, allowing them to be used in commercial applications in sectors such as automotive, construction, and aerospace. Mechanical properties and environmental endurance are the desirable characteristics in such high-performance applications. This research examines the mechanical properties, including compression and impact properties of bio-fiber-based VE composites. Further, the formation of VE thermoset composites and its properties are also discussed in this chapter. The use of natural fibers as reinforcement for VE-based composites has a positive impact on polymer mechanical behavior. The chemical treatment of natural fiber-reinforced composites has shown better impact and compression properties than the untreated composites. This may be owing to interfacial bonding between the fiber and polymer matrix. And also, the higher weight percentage of natural fibers has enhanced compression and impact strength of the bio-composites.

REFERENCES

1. R. Campilho, "Recent innovations in biocomposite products," in *Biocomposites for High-Performance Applications*, D. Ray (ed.), Sawston: Woodhead publishing, 2017, pp. 275–306.
2. M. R. Sanjay, S. Siengchin, C. I. Pruncu, M. Jawaid, T. S. M. Kumar, and N. Rajini, "Biomedical applications of polymer/layered double hydroxide bionanocomposites," in *Nanostructured Polymer Composites for Biomedical Applications*, S. K. Swain and M. Jawaid (eds.), Amsterdam: Elsevier, 2019, pp. 315–322.
3. A. Atiqah, M. Chandrasekar, T. S. M. Kumar, K. Senthilkumar, and M. N. M. Ansari, "Characterization and interface of natural and synthetic hybrid composites," in *Reference Module in Materials Science and Materials Engineering - Encyclopedia of Renewable and Sustainable Materials*, Amsterdam: Elsevier, 2019, vol. 4, pp. 380–400. https://doi.org/10.1016/B978-0-12-803581-8.10805-7 2019.
4. T. S. M. Kumar, N. Rajini, H. Tian, A. V. Rajulu, J. T. Winowlin Jappes, and S. Siengchin, "Development and analysis of biodegradable poly (propylene carbonate)/ tamarind nut powder composite films," *Int. J. Polym. Anal. Charact.*, vol. 22, no. 5, pp. 415–423, 2017.
5. K. Rohit and S. Dixit, "A review-future aspect of natural fiber reinforced composite," *Polym. from Renew. Resour.*, vol. 7, no. 2, pp. 43–59, 2016.
6. S. M. K. Thiagamani, R. Nagarajan, M. Jawaid, V. Anumakonda, and S. Siengchin, "Utilization of chemically treated municipal solid waste (spent coffee bean powder) as reinforcement in cellulose matrix for packaging applications," *Waste Manag.*, vol. 69, pp. 445–454, 2017, doi: 10.1016/j.wasman.2017.07.035.
7. T. Senthil Muthu Kumar, N. Rajini, K. Obi Reddy, A. Varada Rajulu, S. Siengchin, and N. Ayrilmis, "All-cellulose composite films with cellulose matrix and Napier grass cellulose fibril fillers," *Int. J. Biol. Macromol.*, 2018, doi: 10.1016/j.ijbiomac.2018.01.167.
8. T. S. M. Kumar, N. Rajini, M. Jawaid, A. V. Rajulu, and J. T. W. Jappes, "Preparation and properties of cellulose/tamarind nut powder green composites," *J. Nat. Fibers*, vol. 15, pp. 1–10, 2017.
9. S. M. K. Thiagamani, N. Rajini, S. Siengchin, A. V. Rajulu, N. Hariram, and N. Ayrilmis, "Influence of silver nanoparticles on the mechanical, thermal and antimicrobial properties of cellulose-based hybrid nanocomposites," *Compos. Part B Eng.*, vol. 165, pp. 516–525, 2019.
10. T. S. M. Kumar, K. S. Kumar, N. Rajini, S. Siengchin, N. Ayrilmis, and A. V. Rajulu, "A comprehensive review of electrospun nanofibers: Food and packaging perspective," *Compos. Part B Eng.*, vol. 175, p. 107074, 2019.
11. S. M. K. Thiagamani et al., "Investigation into mechanical, absorption and swelling behaviour of hemp/sisal fibre reinforced bioepoxy hybrid composites: Effects of stacking sequences," *Int. J. Biol. Macromol.*, vol. 140, pp. 637–646, 2019.
12. S. Krishnasamy et al., "Recent advances in thermal properties of hybrid cellulosic fiber reinforced polymer composites," *Int. J. Biol. Macromol.*, vol. 141, pp. 1–13, 2019.
13. A. K. Sharma, R. Bhandari, A. Aherwar, and R. Rimašauskienė, "Matrix materials used in composites: A comprehensive study," *Mater. Today Proc.*, vol. 21, pp. 1559–1562, 2020.
14. M. Rimašauskas, T. Kuncius, and R. Rimašauskienė, "Processing of carbon fiber for 3D printed continuous composite structures," *Mater. Manuf. Process.*, vol. 34, no. 13, pp. 1528–1536, 2019.
15. T. Senthil Muthu Kumar, N. Rajini, S. Siengchin, A. Varada Rajulu, and N. Ayrilmis, "Influence of Musa acuminate bio-filler on the thermal, mechanical and visco-elastic behavior of poly (propylene) carbonate biocomposites," *Int. J. Polym. Anal. Charact.*, vol. 24, no. 5, pp. 439–446, 2019.

16. I. D. MP and S. Siengchin, "Antimicrobial properties of poly (propylene) carbonate/Ag nanoparticle-modified tamarind seed polysaccharide with composite films," *Ionics (Kiel).*, vol. 25, no. 7, pp. 3461–3471, 2019.
17. M. P. I. Devi et al., "Biodegradable poly (propylene) carbonate using in-situ generated CuNPs coated Tamarindus indica filler for biomedical applications," *Mater. Today Commun.*, vol. 19, pp. 106–113, 2019.
18. R. M. Kumar, N. Rajini, T. S. M. Kumar, K. Mayandi, S. Siengchin, and S. O. Ismail, "Thermal and structural characterization of acrylonitrile butadiene styrene (ABS) copolymer blended with polytetrafluoroethylene (PTFE) particulate composite," *Mater. Res. Express*, vol. 6, no. 8, p. 85330, 2019.
19. S. Krishnasamy et al., "Effect of fibre loading and $Ca(OH)_2$ treatment on thermal, mechanical, and physical properties of pineapple leaf fibre/polyester reinforced composites," *Mater. Res. Express*, vol. 6, no. 8, p. 85545, 2019.
20. M. Chandrasekar et al., "Flax and sugar palm reinforced epoxy composites: Effect of hybridization on physical, mechanical, morphological and dynamic mechanical properties," *Mater. Res. Express*, vol. 6, no. 10, p. 105331, 2019.
21. F. Yeasmin et al., "Remarkable enhancement of thermal stability of epoxy resin through the incorporation of mesoporous silica micro-filler," *Heliyon*, vol. 7, no. 1, p. e05959, 2021.
22. K. Senthilkumar et al., "Tribological characterization of cellulose fiber-reinforced polymer composites," in *Tribology of Polymer Composites*, S. M. Rangappa, S. Siengchin, J. Parameswaranpillai, and K. Friedrich (eds.), Amsterdam: Elsevier, 2021, pp. 95–113.
23. M. Chandrasekar et al., "Effect of adding sisal fiber on the sliding wear behavior of the coconut sheath fiber-reinforced composite," in *Tribology of Polymer Composites*, S. M. Rangappa, S. Siengchin, J. Parameswaranpillai, and K. Friedrich (eds.), Amsterdam: Elsevier, 2021, pp. 115–125.
24. K. Senthilkumar et al., "Free vibration analysis of bamboo fiber-based polymer composite," in *Bamboo Fiber Composites*, M. Jawaid, S. M. Rangappa, and S. Siengchin (eds.), Singapore: Springer, 2021, pp. 97–110.
25. S. Krishnasamy et al., "Effects of stacking sequences on static, dynamic mechanical and thermal properties of completely biodegradable green epoxy hybrid composites," *Mater. Res. Express*, vol. 6, no. 10, p. 105351, 2019.
26. K. Senthilkumar, M. Chandrasekar, N. Rajini, S. Siengchin, and V. Rajulu, "Characterization, thermal and dynamic mechanical properties of poly (propylene carbonate) lignocellulosic *Cocos nucifera* shell particulate biocomposites," *Mater. Res. Express*, vol. 6, no. 9, p. 96426, 2019.
27. T. S. M. Kumar, N. Rajini, T. Huafeng, A. V. Rajulu, N. Ayrilmis, and S. Siengchin, "Improved mechanical and thermal properties of spent coffee bean particulate reinforced poly (propylene carbonate) composites," *Part. Sci. Technol.*, vol. 37, no. 5, pp. 643–650, 2019.
28. J. Hodgkin, "Thermosets: Epoxies and polyesters," *Encycl. Mater. Sci. Technol.*, vol. 1, pp. 9215–9221, 2011.
29. K. Józefiak and R. Michalczyk, "Prediction of structural performance of vinyl ester polymer concrete using FEM elasto-plastic model," *Materials (Basel).*, vol. 13, no. 18, p. 4034, 2020.
30. K. Yorseng, N. Rajini, S. Siengchin, N. Ayrilmis, and V. Rajulu, "Mechanical and thermal properties of spent coffee bean filler/poly (3-hydroxybutyrate-co-3-hydroxyvalerate) biocomposites: Effect of recycling," *Process Saf. Environ. Prot.*, vol. 124, pp. 187–195, 2019.
31. K. N. Shivakumar, G. Swaminathan, and M. Sharpe, "Carbon/vinyl ester composites for enhanced performance in marine applications," *J. Reinf. Plast. Compos.*, vol. 25, no. 10, pp. 1101–1116, 2006.

32. A. T. Bhatt, P. P. Gohil, and V. Chaudhary, "Primary manufacturing processes for fiber reinforced composites: History, development & future research trends," in *IOP Conf. Ser.: Mater. Sci. Eng.*, 2018, vol. 330, p. 12107.
33. I. O. Oladele, T. F. Omotosho, and A. A. Adediran, "Polymer-based composites: An indispensable material for present and future applications," *Int. J. Polym. Sci.*, vol. 2020, pp. 1–12, 2020.
34. K. Senthilkumar et al., "Performance of sisal/hemp bio-based epoxy composites under accelerated weathering," *J. Polym. Environ.*, vol. 29, no. 2, pp. 624–636, 2021.
35. N. Rajini, A. Alavudeen, S. Siengchin, V. Rajulu, and N. Ayrilmis, "Development and Analysis of Completely Biodegradable Cellulose/Banana Peel Powder Composite Films," *J. Nat. Fibers*, vol. 18, no. 1, pp. 1–10, 2019, doi: 10.1080/15440478.2019.1612811.
36. T. Senthil Muthu Kumar et al., "Influence of titanium dioxide particles on the filtration of 1, 4-dioxane and antibacterial properties of electrospun cellulose acetate and polyvinylidene fluoride nanofibrous membranes," *J. Polym. Environ.*, vol. 29, no. 3, pp. 775–784, 2021.
37. K. Senthilkumar et al., "Dual cantilever creep and recovery behavior of sisal/hemp fibre reinforced hybrid biocomposites: Effects of layering sequence, accelerated weathering and temperature," *J. Ind. Text.*, vol. 51, 2_suppl, pp. 2372S–2390S, 2022. doi:10.1177/1528083720961416.
38. T. Kumar et al., "Characterization, thermal and antimicrobial properties of hybrid cellulose nanocomposite films with in-situ generated copper nanoparticles in *Tamarindus indica* Nut Powder," *J. Polym. Environ.*, vol. 29, no. 4, pp. 1134–1142, 2021.
39. N. A. S. Aprilia, H. P. S. A. Khalil, A. H. Bhat, R. Dungani, and M. S. Hossain, "Exploring material properties of vinyl ester biocomposites filled carbonized jatropha seed shell," *BioResources*, vol. 9, no. 3, pp. 4888–4898, 2014.
40. V. N. P. Naidu, G. R. Reddy, M. A. Kumar, M. M. Reddy, P. N. Khanam, and S. V. Naidu, "Compressive & impact properties of sisal/glass fiber reinforced hybrid composites," *Int. J. Fibre Text. Res.*, vol. 1, no. 1, pp. 11–14, 2011.
41. K. Korniejenko, E. Frączek, E. Pytlak, and M. Adamski, "Mechanical properties of geopolymer composites reinforced with natural fibers," *Procedia Eng.*, vol. 151, pp. 388–393, 2016.
42. D. Ray, B. K. Sarkar, A. K. Rana, and N. R. Bose, "The mechanical properties of vinylester resin matrix composites reinforced with alkali-treated jute fibres," *Compos. Part A Appl. Sci. Manuf.*, vol. 32, no. 1, pp. 119–127, 2001.
43. R. Nadlene, S. M. Sapuan, M. Jawaid, M. R. Ishak, and L. Yusriah, "The effects of chemical treatment on the structural and thermal, physical, and mechanical and morphological properties of roselle fiber-reinforced vinyl ester composites," *Polym. Compos.*, vol. 39, no. 1, pp. 274–287, 2018.
44. S. Huo, V. S. Chevali, and C. A. Ulven, "Study on interfacial properties of unidirectional flax/vinyl ester composites: Resin manipulation on vinyl ester system," *J. Appl. Polym. Sci.*, vol. 128, no. 5, pp. 3490–3500, 2013.
45. A. R. Mohamed, S. M. Sapuan, and A. Khalina, "Mechanical and thermal properties of josapine pineapple leaf fiber (PALF) and PALF-reinforced vinyl ester composites," *Fibers Polym.*, vol. 15, no. 5, pp. 1035–1041, 2014.
46. J. Sangilimuthukumar, T. S. M. Kumar, C. Santulli, M. Chandrasekar, K. Senthilkumar, and S. Siengchin, "The use of pineapple fiber composites for automotive applications: A short review," *J. Mater. Sci. Res. Rev.*, vol. 6, no. 3, pp. 39–45, 2020.
47. N. A. Nasimudeen et al., "Mechanical, absorption and swelling properties of vinyl ester based natural fibre hybrid composites," *Appl. Sci. Eng. Prog.*, vol. 14, no. 4, pp. 680–688, 2021.
48. S. Navaneethakrishnan and A. Athijayamani, "Taguchi method for optimization of fabrication parameters with mechanical properties in fiber and particulate reinforced composites," *Int. J. Plast. Technol.*, vol. 19, no. 2, pp. 227–240, 2015.

49. C. Manickam, J. Kumar, A. Athijayamani, and N. Diwahar, "Mechanical and wear behaviors of untreated and alkali treated roselle fiber-reinforced vinyl ester composite," *J. Eng. Res.*, vol. 3, no. 3, pp. 1–13, 2015.
50. M. V. Ramana and S. Ramprasad, "Experimental investigation on jute/carbon fibre reinforced epoxy based hybrid composites," *Mater. Today Proc.*, vol. 4, no. 8, pp. 8654–8664, 2017.
51. M. Razavi, O. E. Babatunde, J. M. Yatim, M. Razavi, and N. M. Azzmi, "Compressive properties of Kenaf/Vinylester composite with different fibre volume," *Adv. Sci. Lett.*, vol. 24, no. 6, pp. 3894–3897, 2018.
52. M. S. Jahagirdar and S. R. Kulkarni, "Biodegradable composites: Vinyl ester reinforced with coconut fibers and vinyl ester reinforced with coconut fibers and rubber particles," *Int. J. Innov. Res. Sci. Eng. Technol.*, vol. 3, no. 8, pp. 15486–15494, 2014.
53. P. A. Udaya Kumar, B. Suresha, N. Rajini, and K. G. Satyanarayana, "Effect of treated coir fiber/coconut shell powder and aramid fiber on mechanical properties of vinyl ester," *Polym. Compos.*, vol. 39, no. 12, pp. 4542–4550, 2018.
54. A. Gupta, A. Kumar, A. Patnaik, and S. Biswas, "Effect of different parameters on mechanical and erosion wear behavior of bamboo fiber reinforced epoxy composites," *Int. J. Polym. Sci.*, vol. 2011, 837875, 35 pp, 2011.
55. D. Liu, J. Song, D. P. Anderson, P. R. Chang, and Y. Hua, "Bamboo fiber and its reinforced composites: Structure and properties," *Cellulose*, vol. 19, no. 5, pp. 1449–1480, 2012.
56. J. G. Gwon, S. Y. Lee, G. H. Doh, and J. H. Kim, "Characterization of chemically modified wood fibers using FTIR spectroscopy for biocomposites," *J. Appl. Polym. Sci.*, vol. 116, no. 6, pp. 3212–3219, 2010.
57. L. C. Ferreira, W. G. Trindade, E. Frollini, and Y. Kawano, "Raman and infrared spectra of natural fibers," in *Fifth International Symposium on Natural Polymers and Composites (ISNaPol 2004) Proceedings*. Sao Pedro, Brasil, 2004, pp. 12–15.
58. H. N. Dhakal, Z. Y. and Zhang, and M. O. W. Richardson, "Effect of water absorption on the mechanical properties of hemp fibre reinforced unsaturated polyester composites," *Compos. Sci. Technol.*, vol. 67, no. 7–8, pp. 1674–1683, 2007.
59. H. N. Dhakal, E. Le Méner, M. Feldner, C. Jiang, and Z. Zhang, "Falling weight impact damage characterisation of flax and flax basalt vinyl ester hybrid composites," *Polymers (Basel).*, vol. 12, no. 4, p. 806, 2020.
60. A. Sangregorio, N. Guigo, J. C. van der Waal, and N. Sbirrazzuoli, "All 'green' composites comprising flax fibres and humins' resins," *Compos. Sci. Technol.*, vol. 171, pp. 70–77, 2019.
61. S. Motru, V. H. Adithyakrishna, J. Bharath, and R. Guruprasad, "Development and evaluation of mechanical properties of biodegradable PLA/flax fiber green composite laminates," *Mater. Today Proc.*, vol. 24, pp. 641–649, 2020.
62. N. Agarwal, A. Singh, I. K. Varma, and V. Choudhary, "Effect of structure on mechanical properties of vinyl ester resins and their glass fiber-reinforced composites," *J. Appl. Polym. Sci.*, vol. 108, no. 3, pp. 1942–1948, 2008.
63. S. D. Malingam, N. L. Feng, N. C. Sean, K. Subramaniam, N. Razali, and Z. Mustafa, "Mechanical properties of hybrid kenaf/kevlar fibre reinforced thermoplastic composites," *Sci. Technol. Res. Inst. Def.*, vol. 11, no. 2, pp. 181–209, 2018.
64. M. Ramakrishnan, S. Ramasubramanian, V. Subbarayalu, and A. Ayyanar, "Study of mechanical properties of roselle fiber reinforced vinyl ester biocomposite based on the length and content of fiber: Mechanical properties of roselle fiber reinforced vinyl ester biocomposite," *Mechanics*, vol. 27, no. 3, pp. 265–269, 2021.
65. M. A. B. Ishak, "Performance of sugar palm fibre-reinforced vinyl ester composites at different fibre arrangements," Thesis Submitted to the School of Graduate Studies, Universiti Putra Malaysia, 2017.

66. H. Mohit, "Studies on structural properties of dry leaf fiber reinforced vinyl ester polymer matrix composites," *Stress*, vol. 40, no. 50, p. 60.
67. J. Chandradass, M. Ramesh Kumar, and R. Velmurugan, "Effect of clay dispersion on mechanical, thermal and vibration properties of glass fiber-reinforced vinyl ester composites," *J. Reinf. Plast. Compos.*, vol. 27, no. 15, pp. 1585–1601, 2008.
68. P. Compston, J. Schiemer, and A. Cvetanovska, "Mechanical properties and styrene emission levels of a UV-cured glass-fibre/vinylester composite," *Compos. Struct.*, vol. 86, no. 1–3, pp. 22–26, 2008.
69. X. Dirand, B. Hilaire, J. P. Soulier, and M. Nardin, "Interfacial shear strength in glass-fiber/vinylester-resin composites," *Compos. Sci. Technol.*, vol. 56, no. 5, pp. 533–539, 1996.
70. M. A. Fuqua, S. Huo, and C. A. Ulven, "Natural fiber reinforced composites," *Polym. Rev.*, vol. 52, no. 3, pp. 259–320, 2012.
71. B. Stalin et al., "Evaluation of mechanical, thermal and water absorption behaviors of *Polyalthia longifolia* seed reinforced vinyl ester composites," *Carbohydr. Polym.*, vol. 248, p. 116748, 2020.
72. N. Nagaraj, S. Balasubramaniam, V. Venkataraman, R. Manickam, R. Nagarajan, and I. S. Oluwarotimi, "Effect of cellulosic filler loading on mechanical and thermal properties of date palm seed/vinyl ester composites," *Int. J. Biol. Macromol.*, vol. 147, pp. 53–66, 2020.
73. S. O. Adeosun, O. P. Gbenebor, E. I. Akpan, and F. A. Udeme, "Influence of organic fillers on physicochemical and mechanical properties of unsaturated polyester composites," *Arab. J. Sci. Eng.*, vol. 41, no. 10, pp. 4153–4159, 2016.
74. B. Stalin, N. Nagaprasad, V. Vignesh, and M. Ravichandran, "Evaluation of mechanical and thermal properties of tamarind seed filler reinforced vinyl ester composites," *J. Vinyl Addit. Technol.*, vol. 25, no. s2, pp. E114–E128, 2019.

4 Thermal Properties of Vinyl Ester-Based Biocomposites

Tarkan Akderya
İzmir Bakırçay University

Buket Okutan Baba
Izmir Katip Çelebi University

CONTENTS

4.1 INTRODUCTION

The effect of situations such as the increase in the frequency and severity of natural disasters and the disruption of ecological balance on the rise of environmental awareness, the emergence of concerns about the depletion of fossil fuels, and the re-industrialization, which has entered a transformation with the bio-economy fiction and desire that emerged in developed countries, is related to the interest and orientation towards sustainability phenomenon and sustainable technologies (Pandey et al. 2015; Ray and Sain 2017; Chang et al. 2020). The finding of bio-based products in the market, which can be seen in almost every industrial area in daily life, emerges as a result of increased sensitivity and the search for environmentally friendly products (Saba et al. 2017a, b). This search has opened up a new research area for the possibility and usability of bio-based alternatives of synthetic fibre and synthetic matrix-based composite materials.

The use of fibre-reinforced composite materials in daily life has reached enormous levels, with global production reaching 5.9 million tons in 1999 and 8.7 million tons in 2011 (Reux 2012; Ray and Sain 2017). While the global need for carbon fibre was approximately 33 kilotons (kt) in 2010, it has been stated that this figure reached 98 kt in 2020 and will increase to 120 kt in 2022 (Toray 2012; Zhang et al. 2020; Statista 2022). It has been declared that the global need for glass fibres in 2021 is approximately 6 million tons (GlobalNewsWire 2021). In addition to excellent

DOI: 10.1201/9781003270997-4

mechanical properties, superior properties such as high corrosion resistance and ultra-lightness have made glass and carbon fibre-reinforced composites preferred for use in energy, transportation, aircraft and marine industries. The increase in the use of fibre-reinforced plastics (FRP) brings with it some problems such as non-recyclability (Palanikumar 2008; Sapuan et al. 2017; Santhosh Kumar and Hiremath 2020). In order to prevent this problem, bioresources have been introduced to the FRP industry and biofibre-reinforced biocomposite materials have been started to be used and researched instead of synthetic fibre-reinforced composites.

Within the scope of the search for sustainable materials and technologies, there is a serious increasing trend in the use and research of bio-based plastics and their composites. Biocomposites, in which fibres obtained from regionally different local sources are used as additives, contribute to an increase in ecological efficiency, while paving the way for green chemistry and green industry fiction to be realized and ensuring the continuation of sustainability. Thanks to their biodegradability, biocomposites contribute to a sustainable and healthy ecosystem, while their low cost, easy availability and high performance meet the economic interests of the industry. Biocomposite materials can be considered as a serious alternative in reducing the carbon footprint created by synthetic-based composite materials, in making waste management more effective, and in terms of sustainability (Chinga-Carrasco 2018; Holmes 2019).

Compared to other synthetic fibre-reinforced composite materials, natural fibre-reinforced composite materials that use natural additives as reinforcement have some superior properties such as high availability, affordability, renewability, biodegradability, low density, low carbon footprint, and abundant diversity (Chandramohan and Marimuthu 2011; Karkri 2017; Naushad et al. 2017). While forming the composite material, when natural fibres are selected as reinforcement material, characteristic features such as the thermal properties of fibres, the fibre volumetric ratio, the average dimensions of the fibres, the fibre orientation, the moisture content of the fibres, and their wettability are the parameters to be considered (Santhosh Kumar and Hiremath 2020; Praveena et al 2021). A schematic representation of the classification of natural fibres is given in Figure 4.1. In this classification, there are organic natural fibres, which are divided into three main subclasses as plant-based, animal-based, and non-organic mineral-based natural fibres. Within the subclassification of plant-based natural fibres, abaca (Bledzki et al. 2015), banana (Rodríguez et al. 2020), cantala (Raharjo et al. 2020), henequen (Choi et al. 2009), pineapple (Kumar et al. 2022), and sisal (Kittikorn et al. 2017) are examples of leaf fibres. Plant-based stem fibres' examples include flax (Stochioiu et al. 2021), hemp (Colomer-Romero et al. 2020), jute (Velasco-Parra et al. 2021), kenaf (Waghmare et al. 2020), and ramie (Jin et al. 2017). Coir (Ng et al. 2018) and oil palm (Warid et al. 2020) can be considered as plant-based natural fibres under fruit fibre category. Cotton (Ayan et al. 2020) and kapok (Sangalang 2021) are examples of plant-based seed fibre category, while bamboo (Noori et al. 2021), bagasse (Balaji et al. 2015), and totora (Wille et al. 2021) are examples of plant-based grass fibre category. For another plant-based fibre separation categorized as stalk, examples include barley (Serra-Parareda et al. 2019), rice (Singh and Singh 2019), and wheat (Mohammed et al. 2021), and broom (Nouar et al. 2020) can be considered as an example when the classification is based on root.

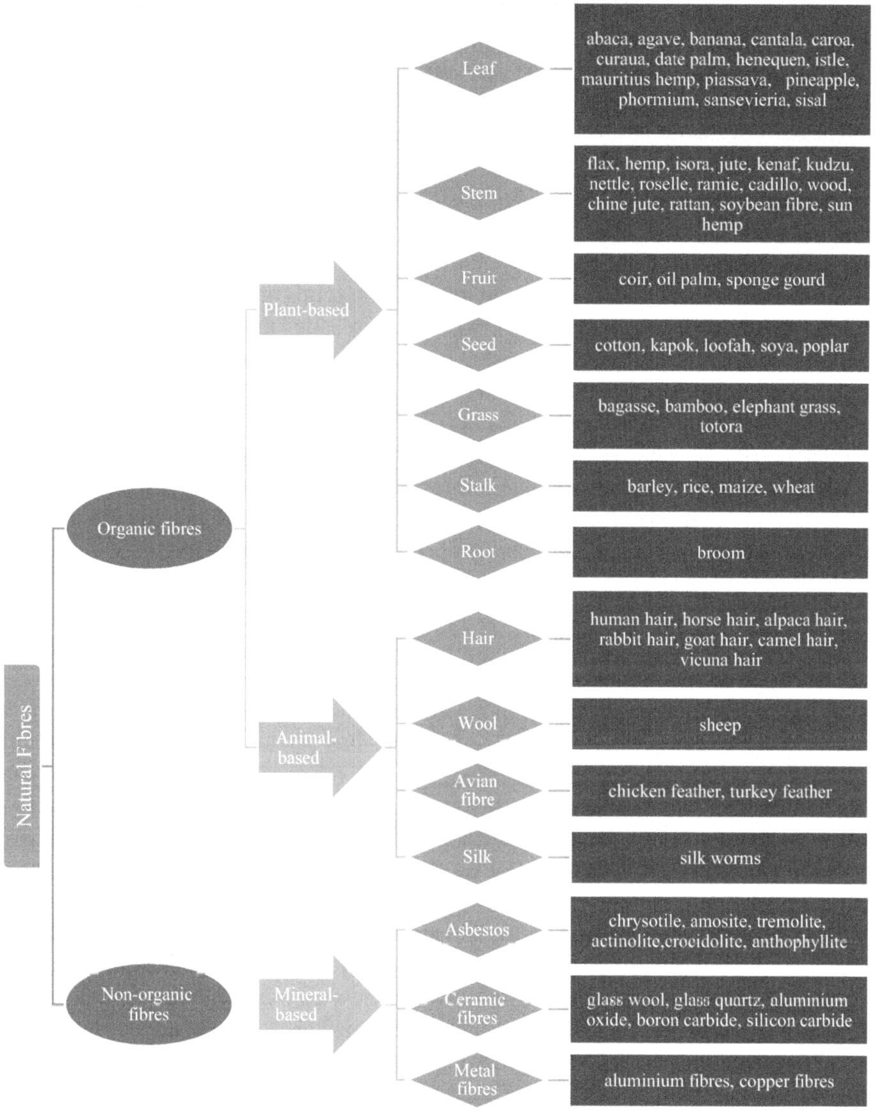

FIGURE 4.1 Classification of natural fibres and their origins (Riedel and Nickel 2002; John and Thomas 2008; Tridico 2009; Chandramohan and Marimuthu 2011; Akil et al. 2011; Azwa et al. 2013; Pandey et al. 2015; Bharath and Basavarajappa 2016; Sapuan et al. 2017; Mayandi et al. 2018; Santhosh Kumar and Hiremath 2020; Hiremath and Sridhar 2020).

In animal-based natural fibres classification, human hair (Yu et al. 2015), horsehair (Kumar et al. 2019), alpaca hair (Weber 2016), and goat hair (Jayaseelan et al. 2017) fall under hair fibre category, while sheep wool (Bharath et al. 2021) can be cited as wool. Examples of avian fibre category include chicken feather (Akderya et al. 2021) and turkey feather (Soykan 2022). In the silk category, silkworms (Cheung and

Lau 2007) can be considered. While chrysotile (Luo et al. 2018), amosite (Germine and Puffer 2020), tremolite (Bloise et al. 2008), actinolite (Bloise 2019), crocidolite (Colomban and Kremenović 2020), and anthophyllite (Pollastri et al. 2017) fibres can be examples of the category of asbestos fibres from non-organic mineral-based natural fibres, glass wool (Lund and Yue 2008), glass quartz (Sharafeddin et al. 2014), aluminium oxide (Grashchenkov et al. 2012), boron carbide (Wawner 2017), and silicon carbide (Choi et al. 2018) can be considered as ceramic fibres examples. In addition, aluminium fibres (Sabapathy et al. 2021) can be considered as mineral-based non-organic natural metal fibres.

Composite materials are formed by mixing the reinforcement materials with the matrix phase, which is a thermoset and thermoplastic polymer that performs the binding function. Examples of natural resins are wheat starch (Shah et al. 2019), corn starch (Santos and Spinacé 2021), potato starch (Jumaidin et al. 2019), biodegradable polyester (Sengupta 2020), and polyacids (Khan 2017). Epoxy (Militello et al. 2020), polyester (Ammayappan et al. 2016), and phenolic (Berni et al. 2019) can be counted as thermoset polymers. Among the thermoplastic polymers are polycarbonate (Fang et al. 2020), polyvinyl chloride (Jiang et al. 2020), nylon (Randhawa and Patel 2021), polystyrene (Adeniyi et al. 2021), and polyethylene (Bazan et al. 2020).

4.2 VINYL ESTER-BASED BIOCOMPOSITES

One of the most important types of epoxy resins is vinyl ester resins. In addition to providing thermal stability, thermal resistance, and toughness from the elemental properties of thermoset polymers, this resin also has outstanding hydrolysis, chemical resistance, and excellent mechanical properties (Suresha and Kumar 2009; Pham and Marks 2012; Sri Aprilia et al. 2014). Thanks to these features, it has found application as a construction material for tanks where corrosive chemical materials are stored, pipes used for the transport of various chemicals such as petroleum products, and auxiliary equipment in the processing of chemicals. Vinyl esters, which find use in the transportation industry, are used in automobile valve covers, oil pans, exterior panels, and truck boxes in the automotive industry, as well as structural parts in boats. In this context, considering the area of use, vinyl ester epoxies are considered to be a premium resin with outstanding features such as higher chemical resistance at higher costs. They are chosen for use at points where polyesters, which are cheaper than vinyl ester epoxies in terms of price, cannot meet the performance criteria that arise due to service conditions (Pham and Marks 2012; Sri Aprilia et al. 2014; Ranjan and Goswami 2020).

Vinyl ester composite materials reinforced with various synthetic fibres are used industrially; examples include glass fibre-reinforced vinyl ester composites (Chandradass et al. 2008; Shamsuddoha et al. 2017; Shaker et al. 2019), carbon fibre-reinforced vinyl ester composites (Afshar et al. 2016; Garcia et al. 2019; Prabhakar et al. 2020), aramid fibre-reinforced vinyl ester composites (Karahan et al. 2010; Pagnoncelli et al. 2018), and glass/carbon fibre-reinforced vinyl ester hybrid composites (Suresha et al. 2010; Guggari et al. 2021). Biocomposites with vinyl ester matrix have been produced by using alternative natural reinforcements such as pineapple leaf fibre (Abilash et al. 2019; Zin et al. 2019), abaca fibre (Kumar et al. 2017;

Lee et al. 2017), sugar palm fibre (Huzaifah et al. 2018; Ammar et al. 2019), and banana fibre (Kumar et al. 2017; Umachitra et al. 2019) within the scope of sustainability. In addition, it is also being studied to obtain hybrid biocomposites using fibre types such as vetiver/banana fibres (Stalin et al. 2020a), pineapple leaf/kenaf fibres (Mazlan et al. 2019), angustata/vetiver/banana fibre mat (Stalin et al. 2021), kenaf/coconut fibres (Khalil et al. 2017), and kenaf/pineapple fibres (Mazlan et al. 2019).

The manufacturing process of the resins consists of reacting epoxy resins with methacrylic acid and diluting them with styrene to solvent (Pham and Marks 2012; Alia et al. 2018). Important studies have been carried out to reduce the styrene levels in its structure in order to be more environmentally friendly. Considering the occupational health and safety criteria, vinyl ester epoxies containing less styrene have been developed to expose workers to lower levels of styrene (Jaswal and Gaur 2014; Yadav et al. 2018; Ranjan and Goswami 2020).

4.3 THERMAL PROPERTIES OF VINYL ESTER-BASED BIOCOMPOSITES

One of the most important disadvantages of various natural fibres used as reinforcement in biocomposites is their low thermal stability. For this purpose, thermogravimetric analysis (TGA) is used to analyze the thermal resistance and the thermal stability of natural fibre-reinforced biocomposites in a chamber where the temperature is raised in increments (Huzaifah et al. 2019c). Considering the data given in Table 4.1, the thermal decomposition values of the biocomposites were obtained using TGA curves.

In a typical TGA curve, in the first step of mass loss, it is seen that the volatile content and moisture in the test sample evaporate in proportion to the temperature. It can be determined from TGA thermographs that the ratio of this part, called volatile content, to the whole sample by mass is quite small. It is known that in the second mass loss step, biocomposite samples begin to degrade. In the third step, the sample starts to decompose structurally, and the effect of temperature is most severely seen in this step. With the increase in temperature, mass losses increase, and the most serious mass losses are experienced in this step. This means that the thermal stability values decrease drastically at this step where the degradation of the material occurs most rapidly (Akderya and Çevik 2018; Gheith et al. 2019; Akderya et al. 2020). In addition to that, by determining the peak point in the derivative thermogravimetry analysis (DTG), the temperature value at which the highest degradation occurs can be obtained. Due to their thermal stability, which is one of the good essential properties of composite materials, they can be used up to high temperatures without degradation. The thermal stability temperature, attributed as the highest temperature at which thermal resistance is maintained, is used as the onset temperature (T_{onset}) (Siakeng et al. 2018; Yang et al. 2019). Another data obtained by TGA analysis is T_{50}, which is dedicated as the structural destabilization point of the sample (Khalil et al. 2017). Another thermal characteristic of materials is the glass transition temperature (T_g), and it is important to determine the glass transition temperature to be able to comment on the thermal stability of a composite material. It is generally known that

TABLE 4.1
Thermal Properties of Vinyl Ester-Based Biocomposites

Study Done by	Material	Thermal Properties					
Sri Aprilia et al. (2014)	Carbonized Jatropha seed shell-reinforced vinyl ester biocomposite (CB/VE)	**Onset temperature (T_{onset}) (°C)**	**DTG peak temperature (°C)**	**Glass transition temperature (T_g) (°C)**	**First melting temperature (°C)**	**Heat capacity (ΔC_p) (J/g°C)**	**Enthalpy (ΔH) (J/g)**
	Neat vinyl ester	407	427	118	414	0.39	244.20
	CB/VE (%10)	382	400	127	408	0.28	237.25
	CB/VE (%20)	379	394	122	405	0.36	243.18
	CB/VE (%30)	373	392	108	401	0.49	232.89
	CB/VE (%40)	367	394	107	400	0.34	179.50
Pashaei et al. (2011)	Turmeric spent and organically modified layered silicates (o-montmorillonite, MMT) (organoclays)-reinforced vinyl ester composites (o-MMT/TS /VE)	**Onset temperature (T_{onset}) (°C)**	**DTG peak temperature (°C)**	**Glass transition temperature (T_g) (°C)**	**$T_{50\%}$**	**Residue at 600°C (%)**	
	o-MMT (2%)/TS (0%)/VE	393	440	103.1	436	12.3	
	o-MMT (2%)/TS (2.5%)/VE	401	452	93.7	445	13.0	
	o-MMT (2%)/TS (5%)/VE	387	445	95.6	438	16.0	
	o-MMT (2%)/TS (7.5%)/VE	389	96.8	445	439		
	o-MMT (2%)/TS (10%)/VE	386	97.7	444	438		
Nadlene et al. (2018)	Roselle fibre-reinforced vinyl ester composites (RF/VE)	**First degradation phase (°C)**	**First degradation phase DTG peak temperature (°C)**	**Second degradation phase (°C)**	**Second degradation phase DTG peak temperature (°C)**		
	RF/VE	200–307	312.76	307–400	348.49		

(Continued)

TABLE 4.1 (*Continued*)
Thermal Properties of Vinyl Ester-Based Biocomposites

Study Done by	Material	Thermal Properties			
Huzaifah et al. (2019a)	Sugar palm fibre-reinforced vinyl ester composites (SPF/VE)	**Onset temperature (T_{onset}) (°C)**	**DTG peak temperature (°C)**	**Endset temperature (T_{endset}) (°C)**	**Residue at 600°C (%)**
	Neat vinyl ester	329.20	430.56	480.33	2.19
	SPF/VE (10%)	270.83	428.66	459.50	7.68
	SPF/VE (20%)	239.63	422.66	456.83	12.77
	SPF/VE (30%)	240.33	423.33	455.16	12.08
	SPF/VE (40%)	196.67	420.58	458.52	18.14
Huzaifah et al. (2019b)	Sugar palm fibre-reinforced vinyl ester composites (SPF/VE)	**Onset temperature (T_{onset}) (°C)**	**DTG peak temperature (°C)**	**Endset temperature (T_{endset}) (°C)**	
	Neat vinyl ester	335.59	369.22	425.51	
	SPF/VE (10%)	365.13	395.42	425.04	
Khalil et al. (2017)	Coconut shell-based nanoparticles in kenaf/coconut fibres-reinforced vinyl ester hybrid composites (K-C-K)	**Onset temperature (T_{onset}) (°C)**	**$T_{50\%}$**	**Residue at 800°C**	
	K-C-K (0% nanoparticles)	306	409	32.44	
	K-C-K (1% nanoparticles)	308	410	32.61	
	K-C-K (3% nanoparticles)	311	412	35.62	
	K-C-K (5% nanoparticles)	314	415	37.01	

(*Continued*)

TABLE 4.1 (*Continued*)
Thermal Properties of Vinyl Ester-Based Biocomposites

Study Done by	Material	Thermal Properties	
Ranjan and Goswami (2020)	Kenaf fibre-reinforced vinyl ester (VE)/polyurethane (PU) interpenetrating polymer network (IPN)-based composites (VE:PU IPN)	**Onset temperature (T_{onset}) (°C)**	**Glass transition temperature (T_g) (°C)**
	Vinyl ester resin	380	-
	VE:PU IPN	320	59.85
Huzaifah et al. (2019c)	Sugar palm fibre-reinforced vinyl ester composites (SPF/VE)	**Onset temperature (T_{onset}) (°C)**	**DTG peak temperature (°C)**
	Neat vinyl ester	364.31	405.84
	SPF (from Tawau)/VE (10%)	376.38	410.84
	SPF (from Jempol)/VE	374.56	405.64
	SPF (from Indonesia)/VE	373.00	408.77
Shahedifar and Rezadoust (2013)	Cotton-reinforced vinyl ester composites (Co/VE)	**Onset temperature (T_{onset}) (°C)**	
	Co/VE (%30)	290	
Razali et al. (2016)	Roselle fibre-reinforced vinyl ester composites (RF/VE)	**DTG peak temperature (°C)**	
	Neat vinyl ester	440.17	
	RF/VE (10%)	431.25	
	RF/VE (20%)	428.33	
	RF/VE (30%)	427.83	
	RF/VE (40%)	428.84	

(*Continued*)

TABLE 4.1 (*Continued*)
Thermal Properties of Vinyl Ester-Based Biocomposites

Study Done by	Material	Thermal Properties		
Nagaraj et al. (2020)	Date seed filler-reinforced vinyl ester (DSF/VE)	**DTG peak temperature (°C)**		
	DSF/VE (5%)	382		
	DSF/VE (30%)	400.2		
Stalin et al. (2020b)	*Polyalthia longifolia* seed-reinforced vinyl ester composites (PLSF/VE)	**Onset temperature (T_{onset}) (°C)**	**Heat deflection temperature (HDT) (°C)**	
	Neat vinyl ester	385	53	
	PLSF/VE (25%)	410	66	
	PLSF/VE (50%)	430	47	
Ray et al. (2004)	Alkali-treated jute fibres-reinforced vinyl ester composite (JF/VE)	**DTG peak temperature (°C)**	**Residue at 600°C**	
	Neat vinyl ester	418.3	7.23	
	JF/VE	364.2	8.86	
Yusriah et al. (2016)	Betel nut husk fibre-reinforced vinyl ester composites (BNH/VE)	**Transition temperature range (°C)**	**DTG peak temperature (°C)**	**Residue at 700°C**
	Neat vinyl ester	287–425	378.48	4.77
	BNH/VE (10%)	325–425	373.78	3.27
	BNH/VE (40%)	325–500	379.88	5.03
Prabhakar et al. (2022)	Flax fabric-reinforced flame-retardant vinyl ester composites (FF/VE)	**Onset temperature (T_{onset}) (°C)**	**DTG peak temperature (°C)**	**Residue at 700°C**
	FF/VE	370.67	435.86	9.7

some characteristic physical properties of amorphous materials, such as viscosity, coefficient of thermal expansion, and thermal capacity, can vary considerably around T_g (Tokiwa and Calabia 2006; Xue et al. 2013; Yang et al. 2018). One of the thermal properties mentioned here is the temperature that causes a material to deviate by 0.25 mm when a load of 455 kPa is applied, technically called the heat deflection temperature, and this temperature is attributed as the softening temperature of the material (Stalin et al. 2019, 2020b).

The char formation obtained at the end of the TGA analysis temperature range in biocomposite materials can be attributed to the degradation of the vinyl ester as well as the fibres. The fact that the transition temperature range of some biocomposites is higher than that of neat vinyl ester is attributed to the effect of preventing the cell walls of natural fibres from pyrolysis at high temperatures and thus preventing the degradation of the fibre (Taj and Khan 2007; Yusriah et al. 2016). It is noticed that the amount of char accumulated at the end of TGA analysis is higher in these biocomposites with higher fibre content. In addition, when the TGA analysis of the biocomposites is examined, the presence of small peaks and one major peak is striking (Ray et al. 2004; Shahedifar et al. 2016; Ranjan and Goswami 2020). It is observed that the thermal stability value decreases as the mass fibre ratio increases in some biocomposites, and this is associated with the decomposition of hemicellulose, cellulose, and lignin in the structure of the fibres (Sri Aprilia et al. 2014; Razali et al. 2016; Huzaifah et al. 2019a).

4.4 CONCLUSION

The desire to implement a long-term sustainability strategy, developing technology, and increasing environmental awareness have led to the orientation towards natural resources that can be an alternative to synthetic materials in industrial and engineering applications. Biocomposite materials have unique advantages such as being non-toxicity, recyclability, biodegradability, having high chemical resistance, low cost, easy availability, high fatigue strength, and high mechanical, thermal, and physical properties. Vinyl ester, which is a thermoset polymer, has good mechanical properties such as high hardness, high strength, good thermal properties such as high thermal stability and thermal resistance, and good chemical properties such as high corrosive resistance and chemical consistency. Vinyl ester resin, which is a thermosetting polymer resulting from the esterification process of an epoxy resin with acrylic or methacrylic acids, offers high thermal stability and therefore safe use in service conditions at high temperatures. Neat vinyl ester, which offers thermal stability in the range of 329°C–407°C, serves in a wider thermal stability range of 196°C–401°C in biocomposites.

REFERENCES

Abilash MS, Arumugaprabu V, Rajini N, et al (2019) Moisture absorption and chemical resistance studies on pineapple fiber reinforced vinyl ester composite. *Int J Eng Adv Technol* 9. https://doi.org/10.35940/ijeat.a1081.1291s419.

Adeniyi AG, Adeoye AS, Ighalo JO, Onifade DV (2021) FEA of effective elastic properties of banana fiber-reinforced polystyrene composite. *Mech Adv Mater Struct* 28. https://doi.org/10.1080/15376494.2020.1712628.

Afshar A, Liao HT, Chiang FP, Korach CS (2016) Time-dependent changes in mechanical properties of carbon fiber vinyl ester composites exposed to marine environments. *Compos Struct* 144. https://doi.org/10.1016/j.compstruct.2016.02.053.

Akderya T, Çevik M (2018) Investigation of thermal-oil environmental ageing effect on mechanical and thermal behaviours of E-glass fibre/epoxy composites. *J Polym Res* 25:214. https://doi.org/10. 1007/s10965-018-1615-2.

Akderya T, Özmen U, Baba BO (2020) Investigation of long-term ageing effect on the thermal properties of chicken feather fibre/poly(lactic acid) biocomposites. *J Polym Res* 27. https://doi.org/10. 1007/s10965-020-02132-2.

Akderya T, Özmen U, Baba BO (2021) Revealing the long-term ageing effect on the mechanical properties of chicken feather fibre/poly(lactic acid) biocomposites. *Fibers Polym* 22:2602–2611. https://doi.org/10.1007/s12221-021-0304-7.

Akil HM, Omar MF, Mazuki AAM, et al (2011) Kenaf fiber reinforced composites: A review. *Mater. Des.* 32:4107–4121.

Alia C, Jofre-Reche JA, Suárez JC, et al (2018) Characterization of the chemical structure of vinyl ester resin in a climate chamber under different conditions of degradation. *Polym Degrad Stab* 153. https://doi.org/10.1016/j.polymdegradstab.2018.04.014.

Ammar IM, Muhammad Huzaifah MR, Sapuan SM, et al (2019) Mechanical properties of environment-friendly sugar palm fibre reinforced vinyl ester composites at different fibre arrangements. *EnvironmentAsia* 12. https://doi.org/10.14456/ea.2019.4.

Ammayappan L, Das S, Guruprasad R, et al (2016) Effect of lac treatment on mechanical properties of jute fabric/polyester resin based biocomposite. *Indian J Fibre Text Res* 41:312–317.

Ayan MÇ, Kiriş S, Yapici A, et al (2020) Investigation of cotton fabric composites as a natural radar-absorbing material. *Aircr Eng Aerosp Technol* 92. https://doi.org/10.1108/AEAT-01-2020-0018.

Azwa ZN, Yousif BF, Manalo AC, Karunasena W (2013) A review on the degradability of polymeric composites based on natural fibres. *Mater. Des.* 47:424–442.

Balaji A, Karthikeyan B, Sundar Raj C (2015) Bagasse fiber – The future biocomposite material: A review. *Int J ChemTech Res* 7:223–233.

Bazan P, Mierzwiński D, Bogucki R, Kuciel S (2020) Bio-based polyethylene composites with natural fiber: Mechanical, thermal, and ageing properties. *Materials (Basel)* 13. https://doi.org/10.3390/ma13112595.

Berni R, Cai G, Hausman JF, Guerriero G (2019) Plant fibers and phenolics: A review on their synthesis, analysis and combined use for biomaterials with new properties. *Fibers* 7.

Bharath KN, Basavarajappa S (2016) Applications of biocomposite materials based on natural fibers from renewable resources: A review. *Sci Eng Compos Mater.* 23:123–133.

Bharath KN, Madhu P, Yashas Gowda TG, et al (2021) Mechanical and chemical properties evaluation of sheep wool fiber-reinforced vinylester and polyester composites. *Mater Perform Charact* 10. https://doi.org/10.1520/MPC20200036.

Bledzki AK, Franciszczak P, Osman Z, Elbadawi M (2015) Polypropylene biocomposites reinforced with softwood, abaca, jute, and kenaf fibers. *Ind Crops Prod* 70. https://doi.org/10.1016/j.indcrop.2015.03.013.

Bloise A (2019) Thermal behaviour of actinolite asbestos. *J Mater Sci* 54. https://doi.org/10.1007/s10853-019-03738-8.

Bloise A, Fornero E, Belluso E, et al (2008) Synthesis and characterization of tremolite asbestos fibres. *Eur J Mineral* 20. https://doi.org/10.1127/0935-1221/2009/0021-1838.

Chandradass J, Ramesh Kumar M, Velmurugan R (2008) Effect of clay dispersion on mechanical, thermal and vibration properties of glass fiber-reinforced vinyl ester composites. *J Reinf Plast Compos* 27. https://doi.org/10.1177/0731684407081368.

Chandramohan D, Marimuthu K (2011) A review on natural fibers. *Int J Res Rev Appl Sci* 8:194–206.

Chang BP, Mohanty AK, Misra M (2020) Studies on durability of sustainable biobased composites: A review. *RSC Adv*. 10:17955–17999.

Cheung HY, Lau KT (2007) Study on a silkworm silk fiber/biodegradable polymer biocomposite. In: *ICCM International Conferences on Composite Materials*, Kyoto, Japan

Chinga-Carrasco G (2018) Novel biocomposite engineering and bio-applications. *Bioengineering* 5:1–3.

Choi HY, Han SO, Lee JS (2009) The effects of surface and pore characteristics of natural fiber on interfacial adhesion of henequen fiber/PP biocomposites. *Compos Interf* 16:359–376

Choi JH, Nam YW, Jang MS, Kim CG (2018) Characteristics of silicon carbide fiber-reinforced composite for microwave absorbing structures. *Compos Struct* 202. https://doi.org/10.1016/j.compstruct.2018.01.081.

Colomban P, Kremenović A (2020) Asbestos-Based pottery from corsica: The first fiber-reinforced ceramic matrix composite. *Materials (Basel)* 13. https://doi.org/10.3390/MA13163597.

Colomer-Romero V, Rogiest D, García-Manrique JA, Crespo JE (2020) Comparison of mechanical properties of hemp-fibre biocomposites fabricated with biobased and regular epoxy resins. *Materials (Basel)* 13. https://doi.org/10.3390/ma13245720.

Fang M, Zhang N, Huang M, et al (2020) Effects of hydrothermal aging of carbon fiber reinforced polycarbonate composites on mechanical performance and sand erosion resistance. *Polymers (Basel)* 12. https://doi.org/10.3390/polym12112453.

Garcia R, Castellanos AG, Prabhakar P (2019) Influence of Arctic seawater exposure on the flexural behavior of woven carbon/vinyl ester composites. *J Sandw Struct Mater* 21. https://doi.org/10.1177/1099636217710821.

Germine M, Puffer JH (2020) Analytical transmission electron microscopy of amosite asbestos from South Africa. *Arch Environ Occup Heal* 75. https://doi.org/10.1080/19338244.2018.1556201.

Gheith MH, Aziz MA, Ghori W, et al (2019) Flexural, thermal and dynamic mechanical properties of date palm fibres reinforced epoxy composites. *J Mater Res Technol* 8:853–860. https://doi.org/10.1016/j. jmrt.2018.06.013.

GlobalNewsWire (2021) Global Glass Fiber Reinforcements Market Forecasts 2021–2026: Global Market Forecast to Reach 8.2 Million Metric Tons by 2026. In: Res. Mark. https://www.globenewswire.com/news-release/2021/09/09/2293995/28124/en/Global-Glass-Fiber-Reinforcements-Market-Forecasts-2021-2026-Global-Market-Forecast-to-Reach-8-2-Million-Metric-Tons-by-2026.html#:~:text=According to statistics published by, at %249.7.

Grashchenkov D V., Balinova YA, Tinyakova E V. (2012) Aluminum oxide ceramic fibers and materials based on them. *Glas Ceram (English Transl Steklo i Keramika)* 69. https://doi.org/10.1007/s10717-012-9430-4.

Guggari GS, Shivakumar S, Manjunath GA, et al (2021) Thermal and mechanical properties of vinyl ester hybrid composites with carbon black and glass reinforcement. *Adv Mater Sci Eng* 2021. https://doi.org/10.1155/2021/6030096.

Hiremath A, Sridhar T (2020) Use of bio-fibers in various practical applications. In: *Encyclopedia of Renewable and Sustainable Materials*, Hashmi S, Choudhury IA (Ed.), Amsterdam: Elsevier, pp. 931–935.

Holmes M (2019) Biocomposites take natural step forward: Applications for biocomposites and the use of natural fiber reinforcements are increasing. Reinforced Plastics looks at a number of examples. *Reinf Plast* 63. https://doi.org/10.1016/j.repl.2019.04.069.

Huzaifah MRM, Sapuan SM, Leman Z, et al (2018) Effect of soil burial on water absorption of sugar palm fibre reinforced vinyl ester composites. In: *6th Postgraduate Seminar on Natural Fiber Reinforced Polymer Composites 2018*, Malaysia.

Huzaifah MRM, Sapuan SM, Leman Z, Ishak MR (2019a) Effect of fibre loading on the physical, mechanical and thermal properties of sugar palm fibre reinforced vinyl ester composites. *Fibers Polym* 20. https://doi.org/10.1007/s12221-019-1040-0.

Huzaifah MRM, Sapuan SM, Leman Z, Ishak MR (2019b) Effect of soil burial on physical, mechanical and thermal properties of sugar palm fibre reinforced vinyl ester composites. *Fibers Polym* 20. https://doi.org/10.1007/s12221-019-9159-6.

Huzaifah MRM, Sapuan SM, Leman Z, Ishak MR (2019c) Comparative study of physical, mechanical, and thermal properties on sugar palm fiber (*Arenga pinnata* (Wurmb) Merr.) reinforced vinyl ester composites obtained from different geographical locations. *BioResources* 14. https://doi.org/10.15376/biores.14.1.619-637.

Jaswal S, Gaur B (2014) New trends in vinyl ester resins. *Rev Chem Eng* 30. https://doi.org/10.1515/revce-2014-0012.

Jayaseelan J, Vijayakumar KR, Ethiraj N, et al (2017) The effect of fibre loading and graphene on the mechanical properties of goat hair fibre epoxy composite. *IOP Conf Ser: Mater Sci Eng* 282:012018.

Jiang L, Fu J, Liu L (2020) Seawater degradation resistance of straw fiber-reinforced polyvinyl chloride composites. *BioResources* 15. https://doi.org/10.15376/biores.15.3.5305-5315.

Jin X, Chen X, Cheng Q, et al (2017) Non-isothermal crystallization kinetics of ramie fiber-reinforced polylactic acid biocomposite. *RSC Adv* 7. https://doi.org/10.1039/c7ra09418c.

John MJ, Thomas S (2008) Biofibres and biocomposites. *Carbohydr Polym.* 71:343–364.

Jumaidin R, Adam NW, Ilyas RA, et al (2019) Water transport and physical properties of sugarcane bagasse fibre reinforced thermoplastic potato starch biocomposite. *J Adv Res Fluid Mech Therm Sci* 61:273–281.

Karahan M, Ulcay Y, Eren R, et al (2010) Investigation into the tensile properties of stitched and unstitched woven aramid/vinyl ester composites. *Text Res J* 80. https://doi.org/10.1177/0040517509346441.

Karkri M (2017) Thermal conductivity of biocomposite materials. In: *Biopolymer Composites in Electronics*, Sadasivuni KK, Ponnamma D, Kim J, Cabibihan JJ, AlMaadeed MA (Eds.), Elsevier, pp. 129–153.

Khalil AHPS, Masri M, Saurabh CK, et al (2017) Incorporation of coconut shell based nanoparticles in kenaf/coconut fibres reinforced vinyl ester composites. *Mater Res Express* 4. https://doi.org/10.1088/2053-1591/aa62ec.

Khan B (2017) Development of antibacterial hemp hurd/poly(lactic acid) biocomposite for food packaging.. Doctor of Philosophy Thesis, University of Southern Queensland Toowoomba, QLD 4350, Australia.

Kittikorn T, Kongsuwan S, Malaku R (2017) Investigation of the durability of sisal fiber/PLA biocomposite through evaluation of biodegradability by means of microbial growth. *J Met Mater Miner* 27:23–34.

Kumar AM, Parameshwaran R, Kumar PS, et al (2017) Effects of abaca fiber reinforcement on the dynamic mechanical behavior of vinyl ester composites. *Mater Test* 59. https://doi.org/10.3139/120.111044.

Kumar N, Singh A, Ranjan R (2019) Fabrication and mechanical characterization of horse hair (HH) reinforced polypropylene (PP) composites. *Mater Today: Proc* 19:622–625.

Kumar S, Saha A, Bhowmik S (2022) Accelerated weathering effects on mechanical, thermal and viscoelastic properties of kenaf/pineapple biocomposite laminates for load bearing structural applications. *J Appl Polym Sci* 139. https://doi.org/10.1002/app.51465.

Lee D-W, Park B-J, Song J-I (2017) A study on fire resistance of abaca/vinyl-ester composites. *Compos Res* 30. https://doi.org/10.7234/composres.2017.30.1.059.

Lund MD, Yue Y (2008) Fractography and tensile strength of glass wool fibres. *J Ceram Soc Japan* 116. https://doi.org/10.2109/jcersj2.116.841.

Luo J, Zhang C, Li L, et al (2018) Intrinsic sensing properties of chrysotile fiber reinforced piezoelectric cement-based composites. *Sensors (Switzerland)* 18. https://doi.org/10.3390/s18092999.

Mayandi K, Rajini N, Manojprabhakar M, et al (2018) Recent studies on durability of natural/synthetic fiber reinforced hybrid polymer composites. In: *Durability and Life Prediction in Biocomposites, Fibre-Reinforced Composites and Hybrid Composites*, Jawaid M, Thariq M and Saba N (Eds.), Amsterdam: Elsevier, pp. 1–13.

Mazlan AA, Sultan MTH, Shah AUM, Safri SNA (2019) Thermal properties of pineapple leaf/kenaf fibre reinforced vinyl ester hybrid composites. *IOP Conf Ser: Mater Sci Eng* 670:012030.

Militello C, Bongiorno F, Epasto G, Zuccarello B (2020) Low-velocity impact behaviour of green epoxy biocomposite laminates reinforced by sisal fibers. *Compos Struct* 253. https://doi.org/10.1016/j.compstruct.2020.112744.

Mohammed AABA, Borhana Omran AA, Hasan Z, et al (2021) Wheat biocomposite extraction, structure, properties and characterization: A review. *Polymers (Basel)* 13:1–27.

Nadlene R, Sapuan SM, Jawaid M, et al (2018) The effects of chemical treatment on the structural and thermal, physical, and mechanical and morphological properties of roselle fiber-reinforced vinyl ester composites. *Polym Compos* 39. https://doi.org/10.1002/pc.23927.

Nagaraj N, Balasubramaniam S, Venkataraman V, et al (2020) Effect of cellulosic filler loading on mechanical and thermal properties of date palm seed/vinyl ester composites. *Int J Biol Macromol* 147. https://doi.org/10.1016/j.ijbiomac.2019.11.247.

Naushad M, Alfadul SM, Al-Muhtaseb AH, et al (2017) Progress from composite materials to biocomposite materials and their applications. In: *Modified Biopolymers: Challenges and Opportunities*, Pathania D, Sharma G, Kumar A (Eds.), Nova Science Publishers, Inc., pp. 163–188.

Ng YR, Shahid SNAM, Nordin NIAA (2018) The effect of alkali treatment on tensile properties of coir/polypropylene biocomposite. *IOP Conf Ser: Mater Sci Eng* 368:012048.

Noori A, Lu Y, Saffari P, et al (2021) The effect of mercerization on thermal and mechanical properties of bamboo fibers as a biocomposite material: A review. *Constr Build Mater* 279. https://doi.org/10.1016/j.conbuildmat.2021.122519.

Nouar Y, Zouaoui F, Nekkaa S, et al (2020) Effect of chemical treatment on thermophysical behavior of Spanish broom flour-reinforced polypropylene biocomposite. *J Polym Eng*. https://doi.org/10.1515/polyeng-2020-0073.

Pagnoncelli M, Piroli V, Romanzini D, et al (2018) Mechanical and ballistic analysis of aramid/vinyl ester composites. *J Compos Mater* 52. https://doi.org/10.1177/0021998317705976.

Palanikumar K (2008) Application of Taguchi and response surface methodologies for surface roughness in machining glass fiber reinforced plastics by PCD tooling. *Int J Adv Manuf Technol* 36. https://doi.org/10.1007/s00170-006-0811-0.

Pandey JK, Nagarajan V, Mohanty AK, Misra M (2015) Commercial potential and competitiveness of natural fiber composites. In: *Biocomposites: Design and Mechanical Performance*, Misra M, Pandey J, Mohanty A (Eds.), Amsterdam: Elsevier Inc., pp. 1–15.

Pashaei S, Siddaramaiah, Syed AA (2011) Investigation on mechanical, thermal and morphological behaviors of turmeric spent incorporated vinyl ester green composites. *Polym - Plast Technol Eng* 50. https://doi.org/10.1080/03602559.2011.551975.

Pham HQ, Marks MJ (2012) Epoxy resins Ullmann's encylcopedia of industrial chemistry. *Ullmann's Encycl Ind Chem* 13:156–244.

Pollastri S, Perchiazzi N, Gigli L, et al (2017) The crystal structure of mineral fibres. 2. Amosite and fibrous anthophyllite. *Period di Mineral* 86:55–65. https://doi.org/10.2451/2017PM693.

Prabhakar MN, Naga Kumar C, Dong Woo L, Jung-IL S (2022) Hybrid approach to improve the flame-retardant and thermal properties of sustainable biocomposites used in outdoor engineering applications. *Compos Part A Appl Sci Manuf* 152. https://doi.org/10.1016/j.compositesa.2021.106674.

Prabhakar P, Garcia R, Imam MA, Damodaran V (2020) Flexural fatigue life of woven carbon/vinyl ester composites under sea water saturation. *Int J Fatigue* 137. https://doi.org/10.1016/j.ijfatigue.2020.105641.

Praveena BA, Buradi A, Santhosh N, et al (2021) Study on characterization of mechanical, thermal properties, machinability and biodegradability of natural fiber reinforced polymer composites and its Applications, recent developments and future potentials: A comprehensive review. *Mater Today Proc.* https://doi.org/10.1016/j.matpr.2021.11.049.

Raharjo WW, Kusharjanto B, Triyono T (2020) Alkali treatment effect on the tensile and impact properties of recycled high-density polyethylene composites reinforced with short cantala fiber. *J Southwest Jiaotong Univ* 55. https://doi.org/10.35741/issn.0258-2724.55.3.25.

Randhawa KS, Patel AD (2021) A review on tribo-mechanical properties of micro and nanoparticulate-filled nylon composites. *J Polym Eng* 41:339–355.

Ranjan JK, Goswami S (2020) Effect of surface treatment of natural reinforcement on thermal and mechanical properties of vinyl ester/polyurethane interpenetrating polymer network-based biocomposites. *J Elastomers Plast* 52. https://doi.org/10.1177/0095244318819214.

Ray D, Sain S (2017) Plant fibre reinforcements. In: *Biocomposites for High-Performance Applications: Current Barriers and Future Needs Towards Industrial Development*, Ray D (Ed.), Woodhead Publishing, pp. 1–s21.

Ray D, Sarkar BK, Basak RK, Rana AK (2004) Thermal behavior of vinyl ester resin matrix composites reinforced with alkali-treated jute fibers. *J Appl Polym Sci* 94. https://doi.org/10.1002/app.20754.

Razali N, Sapuan SM, Jawaid M, et al (2016) Mechanical and thermal properties of roselle fibre reinforced vinyl ester composites. *BioResources* 11. https://doi.org/10.15376/BIORES.11.4.9325-9339.

Reux F (2012) Worldwide composites market: Main trends of the composites industry. In: *5th Innovative Composites Summit-JEC ASIA*, pp. 26–28.

Riedel U, Nickel J (2002) Applications of natural fiber composites for constructive parts in aerospace, automobiles, and other areas. In: *Biopolymers*. Hoboken, NJ: Wiley Online Library, Vol. 10, pp. 1–28.

Rodríguez LJ, Fabbri S, Orrego CE, Owsianiak M (2020) Life cycle inventory data for banana-fiber-based biocomposite lids. *Data Br* 30. https://doi.org/10.1016/j.dib.2020.105605.

Saba N, Jawaid M, Sultan MTH, Alothman OY (2017a) *Green Energy and Technology*. Springer, Cham.

Saba N, Jawaid M, Sultan MTH, Alothman OY (2017b) Green biocomposites for structural applications. *Green Energy Technol.* https://doi.org/10.1007/978-3-319-49382-4_1.

Sabapathy YK, Sabarish S, Nithish CNA, et al (2021) Experimental study on strength properties of aluminium fibre reinforced concrete. *J King Saud Univ - Eng Sci* 33. https://doi.org/10.1016/j.jksues. 2019.12.004.

Sangalang RH (2021) Kapok fiber- structure, characteristics and applications: A review. *Orient J Chem* 37. https://doi.org/10.13005/ojc/370301.

Santhosh Kumar S, Hiremath SS (2020) Natural fiber reinforced composites in the context of biodegradability: A review. In: *Encyclopedia of Renewable and Sustainable Materials*, Hashmi S, Choudhury IA (Eds.), Amsterdam: Elsevier, pp. 160–178.

Santos TA, Spinacé MAS (2021) Sandwich panel biocomposite of thermoplastic corn starch and bacterial cellulose. *Int J Biol Macromol* 167. https://doi.org/10.1016/j.ijbiomac.2020.11.156.

Sapuan SM, Tamrin KF, Nukman Y, et al (2017) Natural fiber-reinforced composites: Types, development, manufacturing process, and measurement. In: *Comprehensive Materials Finishing*. Hashmi MSJ (Ed.), Amsterdam: Elsevier Inc., pp. 203–230.

Sengupta S (2020) Development of Jute fabric for Jute-polyester biocomposite considering structure–property relationship: Jute fabric structure for biocomposite. *J Nat Fibers*. https://doi.org/10.1080/15440478.2020.1788495.

Serra-Parareda F, Tarrés Q, Delgado-Aguilar M, et al (2019) Biobased composites from biobased-polyethylene and barley thermomechanical fibers: Micromechanics of composites. *Materials (Basel)* 12. https://doi.org/10.3390/ma12224182.

Shah Z, Fernandes C, Suares D (2019) Investigation of effect of anti-angiogenic green tea extract on the mechanical, physical and wound healing property of 2D wheat starch-sodium alginate biocomposite film. *J Drug Deliv Sci Technol* 52. https://doi.org/10.1016/j.jddst.2019.04.022.

Shahedifar V, Masoud Rezadoust A, Amiri Amraei I (2016) Comparison of physical, thermal, and thermomechanical properties of cotton/epoxy composite and cotton/vinyl ester composite inhibitors. *Propellants, Explos Pyrotech* 41. https://doi.org/10.1002/prep.201500005.

Shahedifar V, Rezadoust AM (2013) Thermal and mechanical behavior of cotton/vinyl ester composites: Effects of some flame retardants and fiber treatment. *J Reinf Plast Compos* 32. https://doi.org/10.1177/0731684413475911.

Shaker K, Nawab Y, Saouab A (2019) Influence of silica fillers on failure modes of glass/vinyl ester composites under different mechanical loadings. *Eng Fract Mech* 218. https://doi.org/10.1016/j.engfracmech.2019.106605.

Shamsuddoha M, Djukic LP, Islam MM, et al (2017) Mechanical and thermal properties of glass fiber–vinyl ester resin composite for pipeline repair exposed to hot-wet conditioning. *J Compos Mater* 51. https://doi.org/10.1177/0021998316661869.

Sharafeddin F, Alavi AA, Zare S (2014) Fracture resistance of structurally compromised premolar roots restored with single and accessory glass or quartz fiber posts. *Dent Res J (Isfahan)* 11:264–271.

Siakeng R, Jawaid M, Ariffin H, Salit MS (2018) Effects of surface treatments on tensile, thermal and fibre-matrix bond strength of coir and pineapple leaf fibres with poly lactic acid. *J Bionic Eng* 15:1035–1046. https://doi.org/10.1007/s42235-018-0091-z.

Singh P, Singh B (2019) Assessment of mechanical properties of biocomposite material by using sawdust and rice husk. *INCAS Bull* 11. https://doi.org/10.13111/2066-8201.2019.11.3.13.

Soykan U (2022) Development of turkey feather fiber-filled thermoplastic polyurethane composites: Thermal, mechanical, water-uptake, and morphological characterizations. *J Compos Mater* 56. https://doi.org/10.1177/00219983211056137.

Sri Aprilia NA, Abdul Khalil HPS, Bhat AH, et al (2014) Exploring material properties of vinyl ester biocomposites filled carbonized Jatropha seed shell. *BioResources* 9. https://doi.org/10.15376/biores.9.3.4888-4898.

Stalin A, Mothilal S, Vignesh V, et al (2020a) Mechanical properties of hybrid vetiver/banana fiber mat reinforced vinyl ester composites. *J Ind Text*. https://doi.org/10.1177/1528083720938161.

Stalin A, Mothilal S, Vignesh V, et al (2021) Mechanical properties of Typha Angustata/vetiver/banana fiber mat reinforced vinyl ester hybrid composites. *J Nat Fibers*. https://doi.org/10.1080/15440478.2021.1875366.

Stalin B, Nagaprasad N, Vignesh V, et al (2020b) Evaluation of mechanical, thermal and water absorption behaviors of Polyalthia longifolia seed reinforced vinyl ester composites. *Carbohydr Polym* 248. https://doi.org/10.1016/j.carbpol.2020.116748.

Stalin B, Nagaprasad N, Vignesh V, Ravichandran M (2019) Evaluation of mechanical and thermal properties of tamarind seed filler reinforced vinyl ester composites. *J Vinyl Addit Technol* 25. https://doi.org/10.1002/vnl.21701.

Statista (2022) Global demand for carbon fiber from 2010 to 2022. https://www.statista.com/statistics/380538/projection-demand-for-carbon-fiber-globally/.

Stochioiu C, Gheorghiu HM, Artimon FPG (2021) Visco-elastoplastic characterization of a flax-fiber reinforced biocomposite. *Mater Plast* 58. https://doi.org/10.37358/MP.21.1.5447.

Suresha B, Kumar KNS (2009) Investigations on mechanical and two-body abrasive wear behaviour of glass/carbon fabric reinforced vinyl ester composites. *Mater Des* 30. https://doi.org/10.1016/j.matdes.2008.08.038.

Suresha B, Shiva Kumar K, Seetharamu S, Sampath Kumaran P (2010) Friction and dry sliding wear behavior of carbon and glass fabric reinforced vinyl ester composites. *Tribol Int* 43. https://doi.org/10.1016/j.triboint.2009.09.009.

Taj S, Khan S (2007) Natural fiber-reinforced polymer composites. *Proc Pakistan Acad Sci* 38:129–144.

Tokiwa Y, Calabia BP (2006) Biodegradability and biodegradation of poly(lactide). *Appl Microbiol Biotechnol.* 72:244–251

Toray (2012) Toray's strategy for carbon fiber composite materials. https://www.toray.com/global/ir/pdf/lib/lib_a136.pdf.

Tridico SR (2009) Natural animal textile fibres: Structure, characteristics and identification. In: *Identification of Textile Fibers*, Houck MM (Ed.), Amsterdam: Elsevier Ltd., pp. 27–67.

Umachitra G, Kumar MS, Sampath PS (2019) Effect of fiber length on the mechanical properties of banana fiber - Vinyl ester composites. *Mater Test* 61. https://doi.org/10.3139/120.111300.

Velasco-Parra JA, Ramón-Valencia BA, Mora-Espinosa WJ (2021) Mechanical characterization of jute fiber-based biocomposite to manufacture automotive components. *J Appl Res Technol* 19. https://doi.org/10.22201/ICAT.24486736E.2021.19.5.1220.

Waghmare PM, Bedmutha PG, Sollapur SB (2020) Investigation of effect of hybridization and layering patterns on mechanical properties of banana and kenaf fibers reinforced epoxy biocomposite. *Mater Today: Proc* 46:3220–3224.

Warid MNM, Yasim-Anuar TAT, Ariffin H, et al (2020) Static mechanical, thermal stability, and interfacial properties of superheated steam treated oil palm biomass reinforced polypropylene biocomposite. *Pertanika J Sci Technol* 28. https://doi.org/10.47836/pjst.28.S2.22.

Wawner FE (2017) Boron and silicon carbide fibers (CVD). In: *Comprehensive Composite Materials II*, Chou, T.-W. (Ed.). Oxford, UK: Elsevier Science, Ltd., pp. 85–105.

Weber E (2016) Case study: Not a normal hair case- an alpaca hair comparison. *Microsc Microanal* 22. https://doi.org/10.1017/s1431927616010965.

Wille VKD, Gentil M, Nunes GRS, et al (2021) Totora fibers as a new source for papermaking. *Biomass Convers Biorefinery*. https://doi.org/10.1007/s13399-021-01547-1.

Xue Q, Lv C, Shan M, et al (2013) Glass transition temperature of functionalized graphene-polymer composites. *Comput Mater Sci* 71. https://doi.org/10.1016/j.commatsci.2013.01.009.

Yadav SK, Schmalbach KM, Kinaci E, et al (2018) Recent advances in plant-based vinyl ester resins and reactive diluents. *Eur. Polym. J.* 98:199–215.

Yang HN, He SJ, Zhang T, et al (2018) Glass transition temperatures in pure and composite organic thin-films. *Org Electron* 60. https://doi.org/10.1016/j.orgel.2018.05.025.

Yang Y, Haurie L, Wen J, et al (2019) Effect of oxidized wood flour as functional filler on the mechanical, thermal and flame-retardant properties of polylactide biocomposites. *Ind Crops Prod* 130:301–309. https://doi.org/10.1016/j.indcrop.2018.12.090.

Yu Y, Yang W, Wang B, et al (2015) Fabrication & characterization of animal hair and human hair reinforced epoxy composite. *Polym Compos* 2:51–58.

Yusriah L, Sapuan SM, Zainudin ES, et al (2016) Thermo-physical, thermal degradation, and flexural properties of betel nut husk fiber reinforced vinyl ester composites. *Polym Compos* 37. https://doi.org/10.1002/pc.23379.

Zhang J, Chevali VS, Wang H, Wang CH (2020) Current status of carbon fibre and carbon fibre composites recycling. *Compos. Part B Eng.* 193:1–15.

Zin MH, Abdan K, Mazlan N, et al (2019) Automated spray up process for Pineapple Leaf Fibre hybrid biocomposites. *Compos Part B Eng* 177. https://doi.org/10.1016/j.compositesb.2019.107306.

5 Vinyl Ester-Based Biocomposites

Influence of Agro-Wastes on Thermal and Mechanical Properties

W. S. Chow
Universiti Sains Malaysia

CONTENTS

5.1 INTRODUCTION

Vinyl ester (VE) resin is produced by the esterification between unsaturated carboxylic acid (e.g., methacrylic acid, acrylic acid) and an epoxy resin, and dissolved in a vinyl monomer (e.g., styrene, vinyl acetate, dimethacrylates, etc.) (Cook et al., 1997). VE combines the good mechanical and thermal properties of epoxy with the ease of processability and cost-effectiveness of polyester (Correa et al., 2017). VE and its composite can be processed by compression molding of sheet molding compounds, vacuum bagging pressure method, resin transfer molding (RTM), vacuum-assisted RTM, co-injection RTM, and pultrusion (Gillio et al., 1999; McConnell, 2010; Sathishkumar et al., 2017; Fairuz et al., 2018). The uniqueness of VE includes good corrosion resistance, good heat resistance, high impact strength, superior hydrolytic stability, and good water resistance (Chen et al., 2009). The final properties of VE resins are influenced by the chemical nature of the precursors, while the presence of the unsaturated vinyl end-groups controls the crosslinking reaction and the processing ability (Amendola et al., 2002). Often, VE is used to produce large structural composite parts that can be manufactured and cured at room temperature (Wilson

DOI: 10.1201/9781003270997-5

et al., 2008). VE is widely used in building and construction, structural components in transportation and marine, storage tanks, pipelines, laminates, radiation-curable inks, and coatings.

Agro-waste is the waste from plants (e.g., plant stalks, leaves, hulls) due to farming and agriculture activities (Lim and Matu, 2015). The non-edible plant waste (e.g., mainly from starchy roots, cereal, etc.) is recorded approximately 250 million tons annually (Heredia-Guerrero et al., 2017). Examples of agro-waste include sugarcane bagasse, wheat straw, peanut shells, rice husks, cotton stalks, sorghum leaves, corn cobs, coffee husks, millet stovers, coconut shells, and so on (Abba et al., 2013). In addition, the agricultural and food processing industry produces a large amount of waste rich in fibers, such as sugar extraction residues, corn or wheat stalks (Rouilly and Rigal, 2002). In the direction of sustainable development and circular economy, efforts have been used to turn agro-waste into biofuel, biofertilizer, and fillers for biocomposites. The use of agro-waste is one of the alternatives in "greening" polymer composites.

Adding filler into VE is one of the approaches to enhance its performance and minimize the cost (Ku et al., 2012). Moreover, the incorporation of agro-waste filler into VE, as well as other polymers, can add value for both the agricultural and polymer-based manufacturing industries. Polymer biocomposites filled with agro-waste filler can widen the reusability of natural resources, reduce pollutant emissions, improve energy recovery, reduce greenhouse gas effects, and un-earth the potential of the waste generated from agricultural activities (Chow et al., 2015; Navaneethakrishnan et al., 2021). The increasing awareness on environmental protection, as well as utilization of sustainable resources, had made the agriculture residue or agro-waste a choice in materials development. The components in the agro-waste, e.g., cellulose and lignin, made it a feasible renewable resource for the biocomposite industry (Gan and Chow, 2021).

Understanding the structure–properties relationship of the polymer (e.g., VE) and filler/fibers derived from agro-waste is essential to control their processability and predict their final properties. Mechanical properties give us an idea of the stiffness, strength, and toughness of a polymer composite (Rama and Rai, 2011), while thermal properties let us know the heat resistance, thermal processing ability, thermal characteristics (e.g., glass transition temperature, melting temperature, crystallinity), and thermal decomposition/stability. In this chapter, we highlight the mechanical and thermal properties of VE reinforced with agro-waste material.

5.2 MECHANICAL PROPERTIES OF VINYL ESTER/AGRO-WASTE BIOCOMPOSITES

The mechanical properties of VE are controlled by its composition (e.g., types and amount of epoxy, carboxylic aid, vinyl monomer, initiator concentration), curing profile and parameters, degree of curing, chemorheology, reaction kinetics, and polymerization kinetics. The viscosity of VE is relatively high, and thus reactive diluents are often required to make the processing ability better. It is known that reactive diluents can also affect the mechanical properties of the VE resin (Siva et al., 1994).

The mechanical properties of the polymer composites are also affected by the testing parameters, including testing speed, strain rate, and testing temperature. Plaseied and Fatemi (2008) reported that the tensile strength and modulus of the VE resins exhibited a linear relationship with strain rate and temperature. Both of the tensile properties increase linearly with the strain rate while decreasing linearly as the test temperature (from T = –35°C to T = 100°C) increases. The extended Menges model is well fitted to the experimental value of the VE resin for a wide range of strain rates and test temperatures. Besides, the use of different curing methods could be one of the factors that influence the final properties of the VE resin and/or composites. Alfano and Pagnotta (2009) found that the microwave-assisted (power input = 1.5–6.0 kW, operating frequency = 2450 MHz) cured VE exhibited higher strength and stiffness compared to the counterpart cured naturally at room temperature. Accordingly, the comparison and benchmarking of the mechanical properties of the VE composites should be made scientifically.

The mechanical properties (e.g., flexural strength, tensile strength) of VE composites depend on the fiber content and the interfacial bonding. Often, the strength of the polymer composites is attributed to the loading and reinforcing efficiency of the filler/fiber. The improvement in mechanical properties is associated with the optimum filler content and good interfacial interaction between the matrix and reinforcement filler. On the other hand, limited wettability would reduce the stress transfer effectiveness from the matrix to the filler and consequently reduce the strength of the composite (Alhuthali and Low, 2013a). The mechanical properties of the biocomposites are governed by the types of fillers, e.g., agro-waste. In general, the mechanical properties of the polymer include stiffness, strength, elongation at break, fracture toughness, etc. It is quite reasonable that some products might need high stiffness, while others may require high toughness. According to Bharathiraja et al. (2020), comparing coconut fibers and banana bagasse, the latter exhibited higher fracture toughness when incorporated into polymer composites. This chapter mainly focuses on the effects of agro-waste on the mechanical properties of VE.

Mohamed et al. (2010a) had prepared VE/pineapple leaf fibers (PLFs) using liquid compression molding and hand lay-up technique. The mechanical properties of the VE were increased by the bleached PLFs, which was attributed to the better interfacial bonding. Moreover, Mohamed et al. (2010b) investigated the interfacial shear stress of VE/untreated PLF, VE/alkali-treated PLF, and VE/bleached PLF composites using a single-fiber pull-out test. It was found that both alkali-treated and bleached PLF can contribute to higher interfacial shear stress for the VE composites due to the effectiveness of the chemical treatment and better wettability. Raju et al. (2012) investigated the feasibility of the groundnut shell as the filler for VE and studied its mechanical properties. The groundnut shell consists of 35.7 wt.% cellulose, 18.7 wt.% hemicelluloses, 30.2 wt.% lignin, and 5.9 wt.% ash. The modulus and tensile strength of the VE were increased by the incorporation of 40 wt.% groundnut shell filler. This is due to the reinforcement of the groundnut shell filler (average particle size = 600 μm).

Nagaraja Ganesh and Murali Kannan (2015) had prepared rice straw (cut into approximately 20 mm) and added it into VE using the hand-lay-up method. The flexural strength and impact strength of the VE were increased remarkably by the incorporation of rice straw fiber (especially at the loading of 30%–40% of rice straw fiber). The enhancement in the mechanical properties is often associated with the interfacial interaction of the VE matrix and fiber. Ogah and James (2018) investigated the effect of different types of agricultural waste fibers (i.e., sugarcane bagasse fibers, oil palm fibers, coconut coir fibers, corn husk, rice husk, groundnut shell fibers) on the tensile properties and hardness of VE biocomposites. The enhancement of the tensile strength and modulus are dependent on the fibers types and fiber loading. In their study, the highest tensile strength of VE composites was achieved at the fiber loading of 10 wt.%. This could be associated with the good dispersion of the fibers in the VE matrix. However, excessive loading (i.e., 15 and 20 wt.%) of the fibers reduces the tensile strength VE, regardless of the fiber types. This is often related to limited dispersibility and agglomeration of fibers. In addition, the lowering of the wettability of the fibers and higher void content could lead to the reduction in mechanical properties. Besides, the increase of viscosity due to the reduction of the flowability of the VE resin in the high fiber-content region would make the processability difficult and consequently reduce the quality of the VE composites.

Zin et al. (2019) reported that the hybridization of PLF into a VE/glass fiber composite prepared using an automated spray-up method gave a higher tensile strength and thermal stability due to the better load transferability of the reinforcement in the VE matrix. Stalin et al. (2019) prepared tamarind seed filler (particle size = 25–60 μm) using the milling technique. The flexural and tensile properties of the VE/tamarind seed filler composites were investigated. The mechanical properties of the VE were improved significantly, especially at the filler loading of 15 wt.%. The hardness of the VE improved about 37% in the presence of the tamarind seed filler, which is related to the stiffness and rigidity of the filler. Raman et al. (2021) prepared VE/walnut shell filler biocomposites using gravity die casting and centrifugal casting. The walnut shell filler (average particle size = 300 μm) improved the tensile strength of VE, especially for the specimen prepared using the gravity die casting method, which was attributed to the uniform and homogeneous distribution of the walnut shell filler in the VE matrix (c.f. Figure 5.1).

Nagaraja Setty et al. (2021) prepared agro-waste filler from *Limonia acidissima* (LA, wood apple) shell powder using the pulverizing method. The alkali-treated LA shell powder (loading: 5–20 wt.%) was used to reinforce the VE composite. Adding 15 wt.% of the LA agro-waste filler enhanced the tensile, flexural, and impact strength of the VE due to the improved interfacial bonding. The alkali treatment removes the non-cellulose components and impurities of the LA filler and thus improves their wettability and reinforcing ability. Stalin et al. (2020) investigated the impact strength and hardness of VE composites filled with *Polyalthia longifolia* seed powder (filler loading: 5–50 wt.%). The impact strength of unfilled VE was 11.83 kJ/m^2. The impact strength of the VE increased to approximately 32 kJ/m^2 by adding 25 wt.% of *P. longifolia* seed powder (c.f. Figure 5.2). The hardness of the VE was 26.3, and it increased to 36.5 by the incorporation of 35 wt.% of the filler. The improvement of the impact strength and hardness is attributed to the homogenous distribution and good interfacial interaction.

Stalin et al. (2021) hybridized the banana fibers (extracted from the bark of the banana tree) and *Typha angustata* fibers (extracted from the *Typha angustata* grass plant stem) in the VE matrix and found that the impact strength of the VE composites

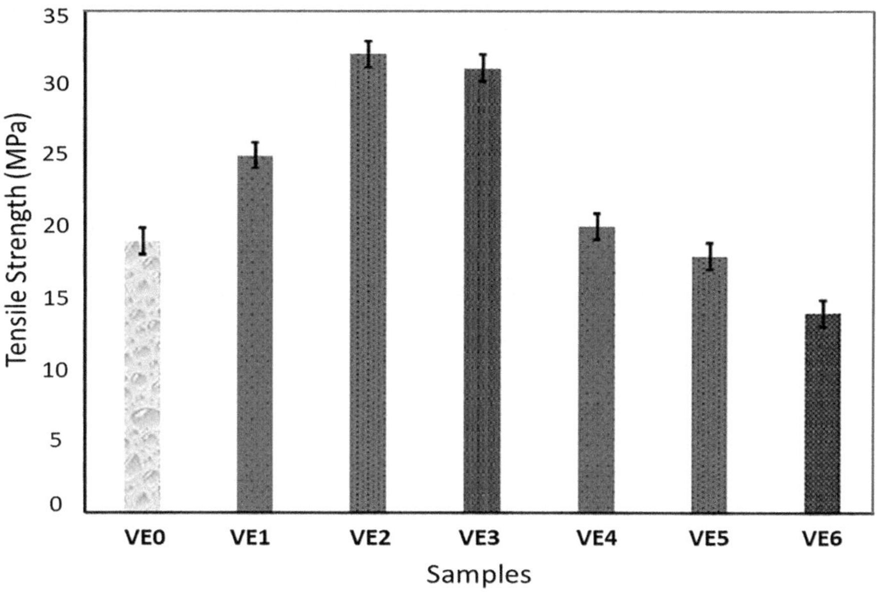

FIGURE 5.1 Tensile strength of the walnut shell powder (WSP)-reinforced VE composites (Raman et al., 2021). Note that VE0: unfilled VE; VE1: VE/WSP5%-gravity die casting; VE2: VE/WSP10%-gravity die casting; VE3: VE/WSP15%-gravity die casting; VE4: VE/WSP5%-centrifugal casting-graded; VE5: VE/WSP10%-centrifugal casting-graded; VE6: VE/WSP15%-centrifugal casting-graded.

Source: Permission obtained from Elsevier via RightsLink licence number: 5207381217985.

improved significantly due to the higher cohesive force between the fibers, as well as the higher impact absorption. Livingston et al. (2021) had utilized both untreated and alkali-treated coconut shell particles (particle size = 100–200 micron) for the preparation of hand lay-up VE biocomposite (curing temperature = room temperature; curing time = 24 h). The flexural properties (i.e., modulus and strength) of the VE were improved significantly by the addition of the alkali-treated coconut shell particles. This is due to the better wettability, better mechanical interlocking, and higher interfacial bonding between the VE and the coconut shell particles.

The durability and environmental effects on the mechanical properties of VE composites are also important aspects that are worth investigating. Manickam et al. (2015b) studied the effect of different types of water environments (i.e., groundwater, distilled water, and seawater) on the mechanical properties of VE/roselle fiber (the fiber obtained from the dried stems of the plant). The water uptake of the VE composites depends on the fiber loading and types of water. Both higher fiber loading and seawater conditions lead to higher water absorption of the VE composites and thus affect the mechanical properties. Alhuthali and Low (2015) reported that prolonged water absorption reduces the fracture toughness and elastic modulus of the VE composites. Accordingly, improving the durability and long-term properties could be one of the interesting aspects for the development of high-performance VE biocomposites.

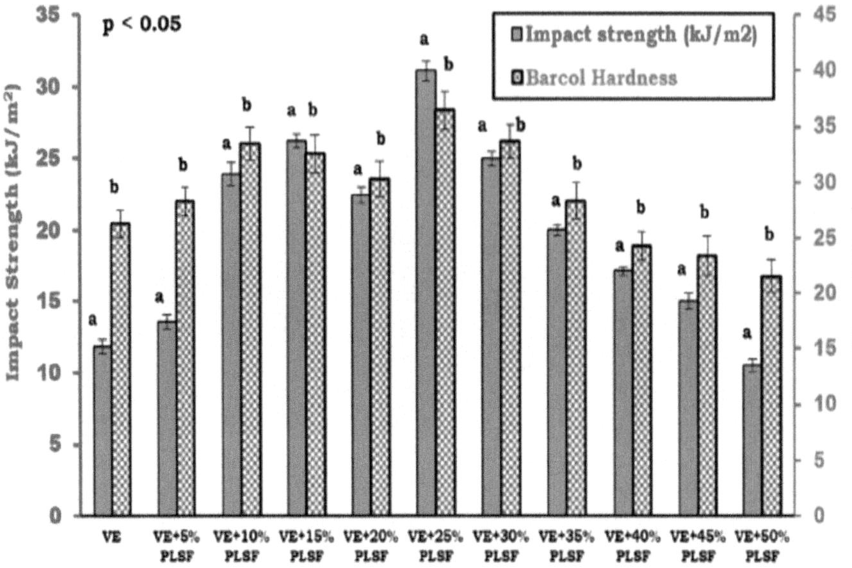

FIGURE 5.2 Effects of filler loadings on impact strengths and hardness of the various VE/*Polyalthia longifolia* seed filler (PLSF) samples (Stalin et al., 2020).

Source: Permission obtained from Elsevier via RightsLink licence number: 5207400351154.

5.3 THERMAL PROPERTIES OF VINYL ESTER/AGRO-WASTE BIOCOMPOSITES

The thermal properties of the VE biocomposites can be characterized using differential scanning calorimeter (DSC), thermogravimetric analyzer (TGA), and dynamic mechanical analyzer (DMA). The curing of VE is carried out similar to that of unsaturated polyester resin, i.e., through radical copolymerization of the unsaturated methacrylate units with the styrene-reactive diluent (Penczek et al., 2006). The commercial VE contains up to 50 wt.% styrene co-monomers. The curing of VE is induced by peroxides with or without accelerators. Different co-reactions could occur during the curing, as VE is tetrafunctional while styrene is bifunctional, and the crosslinking agent is the VE itself in the common VE/styrene formulations (Grishchuk et al., 2013). Besides, multifunctional VE can be synthesized using photopolymerization (Lee et al., 2004). We should understand some of the curing and thermal characteristics (e.g., glass transition temperature, decomposition temperature) of polymer composites so that we can use them as a reference to set up the processing parameters of the materials, as well as to design the final application.

The thermal properties of VE can be adjusted using different resin, styrene content, and types of accelerators or activators. The styrene content affects the network structure and final morphology of the VE. According to Patel et al. (1997), adding styrene to the VE resin system prior to curing reduces the curing temperature. The increase of styrene content increases the damping while reducing the glass transition temperature (T_g) of VE, associated with the higher amount of styrene reaction as well as free volume that gives higher chain mobility and network flexibility (Rodriguez et al., 2006). This information is important for us to design VE biocomposites, as we can control the basic properties of the VE.

The determination of the curing condition and parameters are important for the manufacturing of thermosetting materials. Adding filler/fiber into the VE thermoset could further influence the processing conditions. Measurement and analysis of the gel time, curing kinetics, and residual reactivity of the thermoset are useful to correlate with the quality control and optimization of the properties (Abadie et al., 2002). The glass transition temperature (T_g) of thermosetting polymers is highly dependent on its chain structure rigidity and crosslink density (Desnoes et al., 2021). The thermal properties (e.g., T_g) of VE is strongly depending on the degree of cure (Cook et al., 1997). Zhang et al. (2014) also reported that the T_g of the VE depends on the curing extent, while the crosslinking density affects the tensile strength of the VE. The post-curing also plays an essential role in the processing and properties of thermosetting composites. Thus, the mechanical property of the thermoset also depends on the post-curing temperature and time (Li et al., 2004). By understanding the thermal profile of the unfilled VE resin, it will help us better for the development of agro-waste-reinforced VE biocomposites.

Alhuthali and Low (2013b) used TGA (temperature range = 30°C–700°C; atmosphere = nitrogen) to characterize the thermal stability of VE and found that the thermal decomposition of VE occurred in a single stage starting at 260°C, while the T_{max} (the temperature at maximum weight loss) recorded at 430°C. The decomposition temperature of VE composites can be shifted to a higher temperature by the incorporation of a high thermal resistance filler. Also, it was found that the char formation ability of the filler can give higher thermal stability for the polymer composite. Pashaei et al. (2011) prepared VE composite laminates using turmeric spent as filler. Turmeric spent consists of insoluble solids after solvent extraction of curcumin. The authors dried and grounded the turmeric spent into powder form (particle size < 240 μm). The VE/turmeric spent powder/nanoclay composites were prepared using an in-situ polymerization technique. The thermal decomposition, thermal degradation kinetics, and oxidation index of the VE composites were characterized using TGA. The thermal stability and oxidation index of the VE were improved in the presence of both the turmeric spent and nanoclay fillers. In addition, the *tan δ* at T_{max} (characterized by DMA) of the VE composites was reduced by the adding of turmeric spent filler (c.f. Figure 5.3). This is associated with the restriction of the VE molecular movement in the presence of the filler. The reduction in the damping behavior is well correlated to the increase of the storage modulus of the VE composites reinforced with turmeric spent filler.

Yusriah et al. (2016) investigated the thermo-physical properties (i.e., thermal diffusivity, thermal conductivity, specific heat) of the VE/betel nut husk fiber composite using a hot disk thermal constants analyzer. The betel nut husk fiber was obtained from the betel nut fruit after exposure to the water retting process (duration = 5 days). The betel nut husk fiber mat was prepared using the cold-press method. The specific heat of the unfilled VE is 1.248 kJ/(kg K). The addition of the betel nut husk fiber increased the specific heat of the VE. In their study, the specific heat of 40 wt.% betel nut husk fiber (ripe) recorded the specific heat value of 1.834 kJ/(kg K). This indicates

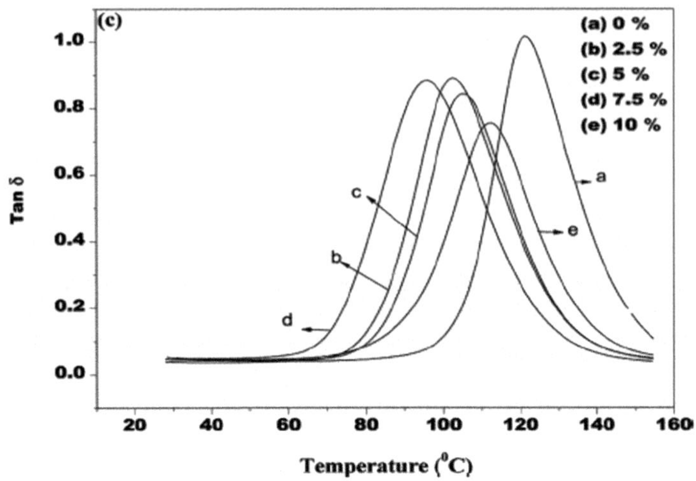

FIGURE 5.3 Plots of tan δ as a functional of temperature of VE/montmorillonite/turmeric spent filler green composites (Pashaei et al., 2011).

Source: Permission obtained from Taylor & Francis via RightsLink licence number: 5207360332621.

that more heat is needed to increase the temperature of the VE biocomposites by the incorporation of the fiber. In addition, the thermal conductivity and thermal diffusivity of the VE composites reduce with the increase of the betel nut husk fiber. The VE composites with lower thermal diffusivity indicate that the materials need a longer cooling and heating period, which is desirable for thermal insulation applications.

Nagaprasad et al. (2020) prepared biobased filler from date palm seed (the remaining agro-waste from date fruit (*Phoenix dactylifera* L.)) using the ball milling technique. The date seed filler (particle size of 30–60 μm) was mixed with VE (consisting of 1.5% dimethylaniline accelerator, 1.5% methyl ethyl ketone peroxide catalyst, and 1.5% cobalt naphthenate promoter). The VE/date seed filler (loading = 30%) biocomposites showed higher heat deflection temperature (HDT = 84°C) compared to unfilled VE (HDT = 53°C). This indicates that the date seed filler can enhance the thermal properties of the VE, and the HDT of the VE biocomposites achieved at an optimum loading of date seed filler at 30% (c.f. Figure 5.4).

Stalin et al. (2020) utilized the Mast tree (*P. longifolia*) seed as the biobased filler for VE composites. The *P. longifolia* seed filler (PLSF)-reinforced VE biocomposites were prepared using compression molding. The thermal decomposition of the VE biocomposites was characterized using a TGA (in a nitrogen atmosphere). The unfilled VE resins recorded thermal decomposition at 385°C while adding 25% PLSF shifted the decomposition temperature to 430°C. This indicates that the PLSF can enhance the thermal stability of the VE matrix.

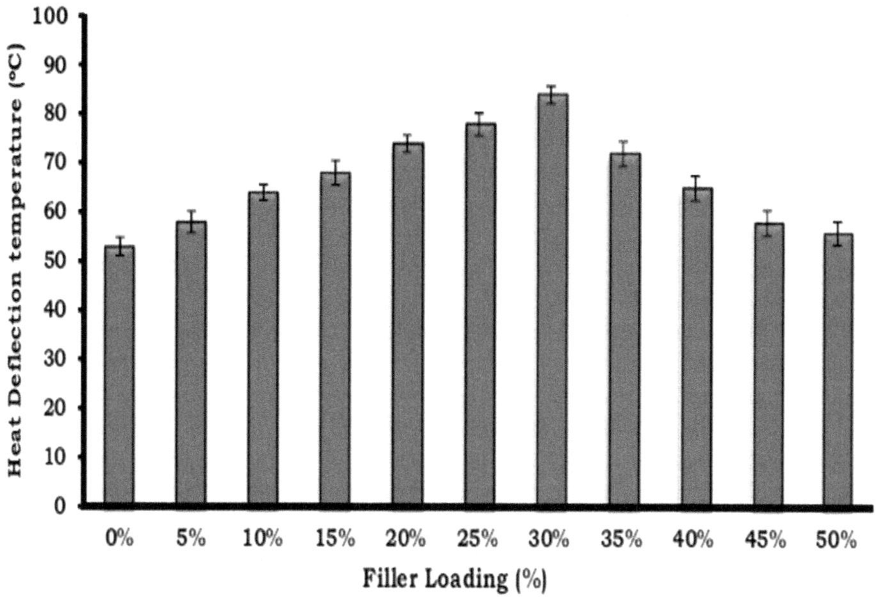

FIGURE 5.4 Effect of filler loadings on heat deflection temperature of the VE/date seed filler composites (Nagaprasad et al., 2020).

Source: Permission obtained from Elsevier via RightsLink licence number: 5207380315080.

5.4 STRATEGIES FOR THE MECHANICAL AND THERMAL PROPERTIES ENHANCEMENT

The performance of the polymer composites (e.g., mechanical and thermal properties) often depends on the interfacial adhesion between fiber/filler and matrix. The use of coupling agents, compatibilizers, and chemical treatment are some of the approaches to improve the interfacial bonding between the natural fibers or agro-waste and the polymer matrix (Nadlene et al., 2018; Senthilraja et al., 2020; Venkata Chalapathi et al., 2020). The surface chemical modification of fibers includes alkali treatment, graft copolymerization, etherification, acetylation, and treatment with isocyanates, and maleated polymer offers better fiber–matrix adhesion and thus improves the mechanical properties of the composite materials (Mohanty et al., 2001). The chemical treatment (e.g., acetylation) can enhance the interfacial adhesion between VE and bamboo fiber, as well as the moisture resistance of the VE composites in a humid environment (Chen et al., 2011).

Udaya Kumar et al. (2018) treated coconut shell powder using trichlorovinyl silane agent. The silane-treated coconut shell powder was added to VE composites. The hardness and tensile properties of the VE composites increased due to the enhancement in the interfacial bonding, which is evidenced from the fractography study through scanning electron microscopy. Ranjan and Goswami (2020) prepared VE/polyurethane (PU)/kenaf fiber interpenetrating polymer network (IPN)-based composites using the hand lay-up technique. The tensile and flexural properties of the VE/PU IPN-based composites are higher compared to the VE composites attributing to the better stress transfer during the mechanical deformation. The modulus of the VE/PU IPN-based composites was further enhanced by fiber treated with silane coupling agent (triethoxyvinylsilane) associated with the better interfacial interaction between the chemical-treated fiber and the VE/PU IPN-based composites. Accordingly, a similar concept can be applied to improve the interaction between VE and agro-waste material.

The mechanical properties of the VE composites can be tuned based on the application requirement. Adding rigid filler (organic or inorganic), core–shell-type rubber, or functionalized rubber are some of the feasible approaches to improve the toughness of the VE thermosetting materials (Gryshchuk et al., 2002). The selection of the toughening agents or impact modifiers depends on the processability and desired properties of the thermosetting materials. For example, using carboxyl-terminated liquid nitrile rubber improves the toughness significantly but reduces the stiffness and T_g of the VE-based thermoset system. However, the epoxy-functionalized hyperbranched polyesters improve the toughness of the VE slightly without reduction in the stiffness and T_g.

The hybridization and IPN concept can be used to tailor the properties of VE biocomposites. Fan et al. (1997) found that the mechanical properties of simultaneous interpenetrating networks (e.g., IPN of PU and VE) depend on the chemical binding properties of the inter-components and the molding conditions. According to Karger-Kocsis and Gryshchuk (2006), the temperature resistance of VE can be improved by the reaction of their hydroxyl groups with the polyisocyanates. According to Tang et al. (2009), the IPN exhibit higher mechanical properties attributing to the interpenetrating structure formed between the two networks (e.g., VE and PU network). Grishchuk and Karger-Kocsis (2011) prepared hybrid thermosetting materials from VE and acrylated epoxidized soybean oil using the free radical-induced crosslinking method. The VE/functionalized epoxidized soybean oil thermoset exhibited higher thermal stability compared to VE attributing to the possible IPN structure that can delay and hinder the decomposition process. Accordingly, the development of agro-waste reinforced VE IPN-structured thermosetting materials is feasible to make the VE biocomposites with a better range of mechanical properties.

The combination and hybridization of fiber/filler derived from agro-waste could enhance the selected mechanical properties of the polymer composites (Vinayagamoorthy, 2020). Hybridization of fiber/filler can be designed for desired and specific applications; for example, if corrosion resistance and durability are required, the combination of natural fiber and glass fiber can be an alternative to achieve a high-performance composite. According to Mandal et al. (2010), incorporation of 25 wt.% of bamboo fiber into VE/glass fiber composites increased its inter-laminar shear strength; however, excessive loading of bamboo fiber (>25 wt.%) reduced the mechanical properties of the VE hybrid composites. Thus, a suitable loading and combination of fiber are important to achieve the desired mechanical properties of VE biocomposites. A similar approach can be used to tailor the properties of the

VE/agro-waste biocomposites, including hybridization of agro-waste filler and synthetic fiber or hybridization of different types of agro-waste filler and fiber.

As we know, VE is produced from petroleum-based feedstock, and thus partially or full replacement of VE by biobased resin derived from renewable resources could be quite challenging. The "greening" of VE using functionalized plant/vegetable oil can be one of the feasible approaches. In the research work done by Grishchuk and Karger-Kocsis (2012) on the modification of VE resin using acrylated epoxidized soybean and linseed oil, they found that the fracture toughness of the VE can be adjusted by the amount of the functionalized vegetable oil. The development of biobased VE resin could be one of the good initiatives to partially replace the conventional VE (which consists of bisphenol A and/or styrene as reactive diluent).

Jaillet et al. (2016) synthesized epoxidized dicyclopentadiene prepolymer with biobased methacrylic monomers from soybean oil and cashew nutshell liquid. The epoxidized dicyclopentadiene prepolymer was functionalized with methacrylic groups, and it can function as solvent (for methacrylation process) and reactive diluent (for the crosslinking process), which make it a greener approach.

Shah et al. (2017) prepared green VE resin using monodisperse VE using bio-waste materials from the paper and bio-diesel industries. The green VE resin (containing 20 wt.% styrene) recorded a comparable viscosity (approximately 427 cps) with the commercial VE (containing 40 wt.% styrene; initial viscosity = 458 cps). The resin infusion times for the VE composites processing are often related to the viscosity, which is typically about 500 cps. Thus, the use of green VE resin can reduce half of the styrene-reactive diluent content. This will reduce the volatile organic compounds and their negative impact on the environment. The research work done by Desnoes et al. (2021) on the styrene-free thermoset cardanol VE resins demonstrated the feasibility of the development of more sustainable thermosetting materials.

Optimization of the processing and properties of the polymer composites (i.e., VE composites) can be realized using the design of experiments and mathematical modeling. Velumani et al. (2014) used the genetic algorithm method to optimize the mechanical properties (e.g., tensile, impact, flexural) of VE composited reinforced with coir fiber (extracted from coconut husk). It was found that the fiber length (37–45 mm) plays an important role in achieving the optimized mechanical properties of the VE composites prepared by the authors. In the case of polymer/fiber composites, a better response can be obtained using the combination of the optimization of processing parameters. A gray-based Taguchi method is used to solve the multi-objective problems (Navaneethakrishnan and Athijayamani, 2017). Manickam et al. (2015a) used the Taguchi method with gray relational analysis to optimize the flexural, impact, and tensile properties of VE composites (prepared using the hand lay-up technique). The significant effect of the process parameter on the mechanical properties' improvement was followed in the trend: fiber content > fiber length > fiber diameter. According to the author's finding, the response surface mathematical model can be used to predict the mechanical properties of the VE composites. Athijayamani et al. (2016) applied Taguchi experimental design to optimize the mechanical properties of VE/sugarcane bagasse fiber composites. It was found that the most essential processing parameters that control the strength of the VE composites are the fiber loading, the fiber length, and the alkali treatment duration. The optimum processing parameters to achieve an optimum tensile strength are fiber loading (35.6 wt.%), fiber length (15 mm), fiber diameter (0.27 mm), NaOH concentration (10%), and NaOH treatment duration (1 h). Thus, it is quite reasonable to mention that the mechanical properties (e.g., strength, modulus) are governed by the processing parameters. It is always beneficial when we can obtain optimum processing/manufacturing parameters in order to get the desired final properties of the polymer composites.

5.5 CHALLENGES AND FUTURE PERSPECTIVE

The challenges of the VE biocomposites are (1) eco-friendliness of VE/agro-waste formulation, (2) processing ability and manufacturing-ability, and (3) applicability and durability of VE biocomposites. The VE resin formulation that contains a high amount of styrene-reactive diluent for the benefit of low viscosity-based processability and moldability might face environmental pollution and health issue since the styrene is a hazardous air pollutant (Chatterjee and Gillespie, 2010). The development of plant-based or biobased VE crosslinkers and diluents could make the materials more environmentally friendly (Yadav et al., 2018). The incorporation of coupling agent and dispersing agent can improve the dispersibility of filler (micron- or nano-size) while reducing the compound viscosity so that higher filler loading is feasible and manageable to achieve VE biocomposites with higher cost-effectiveness. According to Yong and Thomas Hahn (2006), adding dispersing agent (at optimum loading) reduces the viscosity of VE nanocomposites by 50%. To obtain an appropriate viscosity is essential to control the flowability and processability of the agro-waste filler-reinforced VE biocomposites. The durability of the VE biocomposites, especially for those applications that need to meet the stringent requirements and harsh environments (e.g., moisture, UV, thermal, oxidation, weathering), is worth investigating. A balance of properties, cost-effectiveness, and environmental friendliness of VE biocomposites can be achieved by a proper design of the formulation and optimizing the processing. By understanding the structure–property–processing relationship of the materials, we are able to develop high-performance VE/agro-waste biocomposites and widen their applications.

ACKNOWLEDGMENT

The author appreciates the financial support from Universiti Sains Malaysia.

REFERENCES

Abba H.A., Nur I.Z., Salit S.M. "Review of agro waste plastic composites production." *Journal of Minerals and Materials Characterization and Engineering* 1 (2013): 271–279.

Abadie M.J.M., Mekhissi K., Burchill P.J. "Effects of processing conditions on the curing of a vinyl ester resin." *Journal of Applied Polymer Science* 84 (2002): 1146–1154.

Alfano M., Pagnotta L. "An investigation of the mechanical behaviour of vinyl ester resins cured by microwave irradiation." *Materials and Design* 30 (2009): 4537–4542.

Alhuthali A., Low I.M. "Mechanical properties of cellulose fibre reinforced vinyl-ester composites in wet conditions." *Journal of Material Science* 48 (2013a): 6331–6340.

Alhuthali A., Low I.M. "Multi-scale hybrid eco-nanocomposites: Synthesis and characterization of nano-SiC-reinforced vinyl-ester eco-composites." *Journal of Material Science* 48 (2013b): 3097–3106.

Alhuthali A.M., Low I.M. "Effect of prolonged water absorption on mechanical properties in cellulose fiber reinforced vinyl-ester composites." *Polymer Engineering and Science* 55 (2015): 2685–2697.

Amendola E., Giamberini M., Carfagna C., Ambrogi V. "Self-toughening liquid crystalline vinyl ester adhesives." *Macromolecular Symposia* 180 (2002): 153–167.

Athijayamani A., Stalin B., Sidhardhan S., Boopathi C. "Parametric analysis of mechanical properties of bagasse fiber-reinforced vinyl ester composites." *Journal of Composite Materials* 50 (2016): 481–493.

Bharathiraja G., Karunagaran N., Jayakumar V., Ganesh S. "Investigation on fracture toughness of algae filler vinyl ester composite." *Materials Today: Proceedings* 22 (2020): 1233–1235.

Chatterjee A., Gillespie, Jr. J.W. "Room temperature-curable VARTM epoxy resins: Promising alternative to vinyl ester resins." *Journal of Applied Polymer Science* 115 (2010): 665–673.

Chen H.Y., Miao M.H., Ding X. "Influence of moisture absorption on the interfacial strength of bamboo/vinyl ester composites." *Composites: Part A* 40 (2009): 2013–2019.

Chen H.Y., Miao M.H., Ding X. "Chemical treatments of bamboo to modify its moisture absorption and adhesion to vinyl ester resin in humid environment." *Journal of Composite Materials* 45 (2011): 1533–1542.

Chow W.S., Mohd Ishak Z.A. "Polyamide blend-based nanocomposites: A review." *Express Polymer Letters* 9 (2015): 211–232.

Cook W.D., Simon G.P., Burchill P.J., Lau M., Fitch T.J. "Curing kinetics and thermal properties of vinyl ester resins." *Journal of Applied Polymer Science* 64 (1997): 769–781.

Correa C.E., Betancourtb S., Vázqueze A., Gañand P. "Wear performance of vinyl ester reinforced with Musaceae fiber bundles sliding against different metallic surfaces." *Tribology International* 109 (2017): 447–459.

Desnoes E., Toubal L., Thibeault D., Bouazza A.H., Montplaisir D. "Bio-sourced vinyl ester resin reinforced with microfibrillar cellulose: Mechanical and thermal properties." *Polymer and Polymer Composites* (2021). DOI: 10.1177/09673911211002046.

Fairuz A.M., Sapuan S.M., Zainudin E.S., Jaafar C.N.A. "The effect of pulling speed on mechanical properties of pultruded kenaf fiber reinforced vinyl ester composites." *Journal of Vinyl Additive and Technology* 24 (2018): E13–E20.

Fan L.H., Hu C.P., Ying S.K. "Mechanical properties of hand-cast and reaction injection molded polyurethane and vinyl ester resin interpenetrating polymer networks." *Polymer Engineering and Science* 37 (1997): 338–345.

Gan I., Chow W.S. "Tailoring chemical, physical, and morphological properties of sugarcane bagasse cellulose nanocrystals via phosphorylation method." *Journal of Natural Fibers* 18 (2021): 1448–1459.

Gillio E.F., McKnight G.P., Gillespie Jr. J.W., Advani S.G., Bernetich K.R., Fink B.K. "Processing and properties of co-injected resin transfer molded vinyl ester and phenolic composites." *Polymer Composites* 20 (1999): 780–788.

Gryshchuk O., Jost N., Karger-Kocsis J. "Toughening of vinylester–urethane hybrid resins through functionalized polymers." *Journal of Applied Polymer Science* 84 (2002): 672–680.

Grishchuk S., Bonyár A., Elsäßer J., Wolynski A., Karger-Kocsis J., Wetzel B. "Toward reliable morphology assessment of thermosets via physical etching: Vinyl ester resin as an example." *Express Polymer Letters* 7 (2013): 407–415.

Grishchuk S., Karger-Kocsis J. "Hybrid thermosets from vinyl ester resin and acrylated epoxidized soybean oil (AESO)." *Express Polymer Letters* 5 (2011): 2–11.

Grishchuk S., Karger-Kocsis J. "Modification of vinyl ester and vinyl ester–urethane resin-based bulk molding compounds (BMC) with acrylated epoxidized soybean and linseed oils." *Journal of Materials Science* 47(2012): 3391–3399.

Heredia-Guerrero J.A., Heredia A., Domínguez E., Cingolani R., Bayer I.S., Athanassia A., Benítez J.S. "Cutin from agro-waste as a raw material for the production of bioplastics." *Journal of Experimental Botany* 68 (2017): 5401–5410.

Jaillet F., Nouailhas H., Boutevin B., Caillol S. "Synthesis of novel bio-based vinyl ester from dicyclopentadiene prepolymer, cashew nut shell liquid, and soybean oil." *European Journal of Lipid Science and Technology* 118 (2016): 1336–1349.

Karger-Kocsis J., Gryshchuk O. "Morphology and fracture properties of modified bisphenol A and novolac type vinyl ester resins." *Journal of Applied Polymer Science* 100 (2006): 4012– 4022.

Ku H., Prajapati M., Trada M. "Fracture toughness of vinyl ester composites reinforced with sawdust and postcured in microwaves." *International Journal of Microwave Science and Technology* 2012 (2012): Article ID 152726/1–152726/8.

Lee T.Y., Kaung W., Jönsson E.S., Lowery K., Guymon C.A., Hoyle C.E. "Synthesis and photopolymerization of novel multifunctional vinyl esters." *Journal of Polymer Science Part A: Polymer Chemistry* 42 (2004): 4424–4436.

Li P., Yang X.P., Yu Y.H., Yu D.S. "Cure kinetics, microheterogeneity, and mechanical properties of the high-temperature cure of vinyl ester resins." *Journal of Applied Polymer Science* 92 (2004): 1124 –1133.

Lim S.F., Matu S.U. "Utilization of agro-wastes to produce biofertilizer." *International Journal of Energy and Environmental Engineering* 6 (2015): 31–35.

Livingston T., Athijayamani A., Alavudeen A. "Evaluation of mechanical properties of coconut shell particle/vinyl ester composite based on the untreated and treated conditions." *Materials Research Express* 8 (2021): 035309/1–035309/9.

Mandal S., Alam S., Varma I.K., Maiti S.N. "Studies on bamboo/glass fiber reinforced USP and VE resin." *Journal of Reinforced Plastics and Composites* 29 (2010): 43–51.

Manickam C., Kumar J., Athijayamani A., Karthik K. "Modeling and multiresponse optimization of the mechanical properties of roselle fiber-reinforced vinyl ester composite." *Polymer-Plastics Technology and Engineering* 54 (2015a): 1694–1703.

Manickam C., Kumar J., Athijayamani A., Samuel J.E. "Effect of various water immersions on mechanical properties of roselle fiber–vinyl ester composites." *Polymer Composites* 36 (2015b): 1638–1646.

McConnell V.P. "Vinyl esters get radical in composite markets." *Reinforced Plastics* 54 (2010): 34–38.

Mohamed A.R., Sapuan S.M., Khalina A. "Selected properties of hand-laid and compression molded vinyl ester and pineapple leaf fiber (PALF)-reinforced vinyl ester composites." *International Journal of Mechanical and Materials Engineering* 5 (2010a): 68–73.

Mohamed A.R., Sapuan S.M., Shahjahan M., Khalina A. "Effects of simple abrasive combing and pretreatments on the properties of pineapple leaf fibers (PALF) and PALF-vinyl ester composite adhesion." *Polymer-Plastics Technology and Engineering* 49 (2010b): 972–978.

Mohanty A.K., Misra M., Drzal L.T. "Surface modifications of natural fibers and performance of the resulting biocomposites: An overview." *Composite Interfaces* 8 (2001): 313–343.

Nadlene R., Sapuan S.M., Jawaid M., Ishak M.R., Yusriah L. "The effects of chemical treatment on the structural and thermal, physical, and mechanical and morphological properties of roselle fiber-reinforced vinyl ester composites." *Polymer Composites* 39 (2018): 274–287.

Nagaprasad N., Stalin B., Vignesh V., Ravichandran M., Rajini N., Ismail S.O. "Effect of cellulosic filler loading on mechanical and thermal properties of date palm seed/vinyl ester composites." *International Journal of Biological Macromolecules* 147 (2020): 53–66.

Nagaraja Ganesh B., Murali Kannan R. "Mechanical behavior of agricultural residue reinforced composites." *Scientific Review* 1 (2015): 1–4.

Nagaraja Setty V.K.S., Govardhan G., Rangappa S.M., Siengchin S. "Raw and chemically treated bio-waste (*Limonia acidissima* shell powder) vinyl ester composites: Physical, mechanical, moisture absorption properties, and microstructure analysis." *Journal of Vinyl Additive and Technology* 27 (2021): 97–107.

Navaneethakrishnan S., Athijayamani A. "Taguchi method for optimization of fabrication parameters with mechanical properties in sisal fibre–vinyl ester composites," *Australian Journal of Mechanical Engineering* 15 (2017): 74–83.

Navaneethakrishnan S., Sivabharathi V., Ashokraj S. "Weibull distribution analysis of roselle and coconut-shell reinforced vinylester composites." *Australian Journal of Mechanical Engineering* 19 (2021): 457–466.

Ogah A.O., James T.U. "Mechanical behavior of agricultural waste fibers reinforced vinyl ester bio-composites." *Asian Journal of Physical and Chemical Sciences* 5 (2018): Article ID 35841/1–35841/10.

Pashaei S., Siddaramaiah, Syed A.A. "Investigation on mechanical, thermal and morphological behaviors of turmeric spent incorporated vinyl ester green composites." *Polymer-Plastics Technology and Engineering* 50 (2011): 1187–1198.

Patel J.M., Thakkar J.R., Patel R.O. "Novel vinylester resin based on hydroquinone as coating materials." *International Journal of Polymeric Materials*, 38(1997):315–320.

Penczek P., Sodhi J., Ostrysz R. "Unsaturated ester resins." *Journal of Applied Polymer Science* 101 (2006): 2627–2631.

Plaseied A., Fatemi A. "Strain rate and temperature effects on tensile properties and their representation in deformation modeling of vinyl ester polymer." *International Journal of Polymeric Materials*, 57(2008):463–479.

Raju G.U., Kumarappa S., Gaitonde V.N. "Mechanical and physical characterization of agricultural waste reinforced polymer composites." *Journal of Materials and Environmental Science* 3 (2012): 907–916.

Rama S.R., Rai S.K. "Performance analysis of waste silk fabric-reinforced vinyl ester resin laminates." *Journal of Composite Materials* 45 (2011): 2475–2480.

Raman A., Kushwah K., Goyal S., Gangil B., Ranakoti L. "Development and characterization of walnut/vinyl ester filled homogeneous and their functionally graded composites." *Materials Today: Proceedings* 47 (2021): 7121–7126.

Ranjan J.K., Goswami S. "Effect of surface treatment of natural reinforcement on thermal and mechanical properties of vinyl ester/polyurethane interpenetrating polymer network-based biocomposites." *Journal of Elastomers & Plastics* 52 (2020): 29–52.

Rodriguez E., Larrañaga M., Mondragón I., Vázquez A. "Relationship between the network morphology and properties of commercial vinyl ester resins." *Journal of Applied Polymer Science* 100 (2006): 3895–3903.

Rouilly A., Rigal L. "Agro-materials: A bibliographic review." *Journal of Macromolecular Science, Part C: Polymer Reviews* 42 (2002): 441–479.

Sathishkumar S., Suresh A.V., Nagamadhu M., Krishna M. "The effect of alkaline treatment on their properties of Jute fiber mat and its vinyl ester composites." *Materials Today: Proceedings* 4 (2017): 3371–3379.

Senthilraja R., Sarala R., Godwin Antony A., Seshadhri. "Effect of acetylation technique on mechanical behavior and durability of palm fibre vinyl-ester composites." *Materials Today: Proceedings* 21 (2020): 634–637.

Shah P.N., Dev S., Lee Y.W., Hansen C.J. "Processing and mechanical properties of bio-derived vinyl ester resin-based composites." *Journal of Applied Polymer Science* 134 (2017): 44642/1–44642/10.

Siva P., Varma I.K., Patel D.M., Sinha T.J.M. "Effect of structure on properties of vinyl ester resins." *Bulletin of Materials Science* 17 (1994): 1095–1101.

Stalin A., Mothilal S., Vignesh V., Nagarajan K.J., Karthick T. "Mechanical properties of Typha Angustata/vetiver/banana fiber mat reinforced vinyl ester hybrid composites." *Journal of Natural Fibers* (2021). DOI: 10.1080/15440478.2021.1875366.

Stalin B., Nagaprasad N., Vignesh V., Ravichandran M. "Evaluation of mechanical and thermal properties of tamarind seed filler reinforced vinyl ester composites." *Journal of Vinyl Additive and Technology* 25 (2019): E114–E128.

Stalin B., Nagaprasad N., Vignesh V., Ravichandran M., Rajini N., Ismail S.O., Mohammad F. "Evaluation of mechanical, thermal and water absorption behaviors of *Polyalthia longifolia* seed reinforced vinyl ester composites." *Carbohydrate Polymers* 248 (2020): 116748/1–116748/12.

Tang D.Y., Zhang X.H., Liu L.L., Qiang L.S. "Simultaneous and gradient IPN of polyurethane/vinyl ester resin: Morphology and mechanical properties." *Journal of Nanomaterials* 2009 (2009): Article ID 514124/1–514124/6.

Udaya Kumar P.A., Ramalingaiah, Suresha B, Rajini N., Satyanarayana K.G. "Effect of treated coir fiber/coconut shell powder and aramid fiber on mechanical properties of vinyl ester." *Polymer Composites* 39 (2018): 4542–4550.

Velumani S., Navaneetha Krishnan P., Jayabal S. "Mathematical modeling and optimization of mechanical properties of short coir fiber-reinforced vinyl ester composite using genetic algorithm method." *Mechanics of Advanced Materials and Structures* 21 (2014): 559–565.

Venkata Chalapathi K., Song J.I., Prabhakar M.N. "Impact of surface treatments and hybrid flame retardants on flammability, and thermal performance of bamboo fabric composites." *Journal of Natural Fibers* (2020). DOI: 10.1080/15440478.2020.1798849.

Vinayagamoorthy R. "Trends and challenges on the development of hybridized natural fiber composites." *Journal of Natural Fibers* 17 (2020): 1757–1774.

Wilson G.O., Moore J.S., White S.R., Sottos N.R., Andersson H.M. "Autonomic healing of epoxy vinyl esters via ring opening metathesis polymerization." *Advanced Functional Materials* 18 (2008): 44–52.

Yadav S.K., Schmalbach K.M., Kinaci E., Stanzione J.F., Palmese G.R. "Recent advances in plant-based vinyl ester resins and reactive diluents." *European Polymer Journal* 98 (2018): 199–215.

Yong V., Thomas Hahn H. "Rheology of silicon carbide/vinyl ester nanocomposites." *Journal of Applied Polymer Science* 102 (2006): 4365–4371.

Yusriah L., Sapuan S.M., Zainudin E.S., Mariatti M., Jawaid M. "Thermo-physical, thermal degradation, and flexural properties of betel nut husk fiber-reinforced vinyl ester composites." *Polymer Composites* 37 (2016): 2008–2017.

Zhang X., Bitaraf V., Wei S.Y., Guo Z.H., Colorado H.A. "Vinyl ester resin: Rheological behaviors, curing kinetics, thermomechanical, and tensile properties." *AlChe Journal* 60 (2014): 266–274.

Zin M.H., Abdan K., Mazlan N., Zainudin E.S., Liew K.E., Norizan M.N. "Automated spray up process for pineapple leaf fibre hybrid biocomposites." *Composites Part B* 177 (2019): 107306/1–107306/9.

6 Natural Fiber-Reinforced Vinyl Ester Composites

Influence of Hybridization on Mechanical and Thermal Properties

K. M. Faridul Hasan, Zsuzsanna Mária Mucsi, Czók Csilla, Péter György Horváth, Csilla Csiha, László Bejó, and Tibor Alpár
University of Sopron

CONTENTS

DOI: 10.1201/9781003270997-6

6.1 INTRODUCTION

Fibers are continuous filament materials having hair-like appearances which could function as dissemination phase. Fibers have the immense capability to perform as reinforcement material in the composite structure. Mainly, there are two categories of fibers: natural fiber (plant/protein) and man-made fiber (synthetic/regenerated) [1–8]. Natural fibers are being accepted by the composite community nowadays as prominent filler/reinforcement materials to develop sustainable polymeric composites for numerous applications. The reason behind the immense recognition of natural reinforcement materials is their environmentally friendly sustainable features, especially renewability, biodegradability, and naturally grown and available throughout the world [9–13]. On the other hand, synthetic fibers are considered as inconvenient due to their nonbiodegradability and increased environmental threats. The most popularly considered natural fibers are flax, hemp, ramie, sisal, cotton, coir, jute, and so on [5,14,15]. Interestingly, the natural fibers could be used not only as the individual reinforcement materials [16–19] but also as hybrid composite reinforcements (other types of natural fibers) to improve the thermomechanical properties of inferior one by the better one. The hybrid composite materials are comprised of different materials, therefore showing different characteristics.

Numerous polymeric materials (thermoplastic/thermosets) are used for producing the composite products. However, monomers of thermoset polymers provide lower viscosity, which is feasible for blending the monomers with reinforcing materials which could facilitate to optimize the performance characteristics of thermosets in order to be applicable for suitable application field [20]. The most used thermosetting polymers used for composites production are epoxy, polyester, MUF (melamine urea formaldehyde), PF (phenol formaldehyde), VE, MDI (methylene diphenyl diisocyanate), and so on [21]. VE can be applied in versatile routes to produce composite materials. However, VE is produced by esterification reactions between the epoxy resin and acrylic or methacrylic resin. Stalin et al. [22] investigated the reinforcement effects on vetiver/banana fiber mats on VE resin at 45° and 90° directions for several combinations and found that highest tensile strengths of 47 and 25 MPa were provided by the hybrid composites, respectively, for 90° and 45° directions, when compared to their individual reinforcements. Moreover, the obtained flexural properties were 86 and 45 MPa, respectively for 90° and 45° directions, by the same hybrid composites [22]. The morphological images also displayed a stronger bonding among the reinforcements and VE resin [22].

Natural fibers are already becoming important raw materials for the development of individual fiber or hybrid composites either as natural/natural or natural/synthetic fiber-reinforced composites [23–27]. The current chapter is focused on the natural fiber-reinforced VE polymeric composites, because, till now, not many research studies are found in this topic/research area; however, there are some reports found on different natural fibers reinforced with various other polymers like epoxy, PLA, MUF, and so on [28–34]. In this regard, there is a tremendous need for finding the effects of VE polymers on different natural fibers to make them feasible for industrial productions.

6.2 DIFFERENT NATURAL FIBERS USED FOR HYBRIDIZATION

Natural fibers have been used for materials reinforcement since 3000 years [35]. However, with the advancement of science and technology, natural fibers are also reinforced with various polymers for producing composite products. There are numerous reasons responsible for the selection of natural fibers instead of synthetic materials. The reasons are illustrated as follows:

- Economical
- Abundant worldwide availability
- Superior mechanical properties
- Superior insulation properties
- Biodegradability
- Renewability
- Environmental sustainability.

Although naturally derived fiber materials are inferior to synthetic fibers in terms of offering mechanical performance and durability, they are still drawing significant attention due to their eco-friendly features. Unfortunately, natural fibers also possess some critical challenges [36] as well:

- Uneven properties due to the growing environmental conditions
- Irregular maturity of the fibers
- Different parts of the plants/flowers/stems providing different characteristics
- Uneven properties due to growing regions of the plants
- Increased costs in some regions of the world where less plants are grown.

6.3 PHYSICAL AND CHEMICAL PROPERTIES OF NATURAL FIBERS

Cells of natural fibers are typically surrounded by rigid cell walls which have made them distinct from the cells found in the case of animal fibers. The cell walls of the plants provide superior mechanical properties in some case which ensures the structural performances. The general dimensions of plant fibers are variable; however, the length is ranged from 1 to 35 mm and diameter from 15 to 30 μm [37]. The plant fibers could be broadly classified in two categories: wood-based and nonwood-based (Figure 6.1). The wood-based fibers are also termed as short fibers, generally stands between 1 and 5 mm comprising 1 soft and hardwood materials. Conversely, nonwood fibers are termed as long fibers too, ranged between 5 and 50 mm in length. Nonwood fibers are further categorized as seed/fruits, leaf, stalk, grass, and bast fibers. The widely used nonwood fibers are cotton, sisal, abaca, flax, hemp, bagasse, kenaf, ramie, rice, wheat, and so on [38].

Chemically, plant fibers are comprised of cellulose, hemicellulose, lignin, protein, pectin, and so on [39,40]. Furthermore, plant fibers are also termed as lignocellulosic materials as governed by cellulose, lignin, and hemicellulose. Generally, cellulosic fibers contain glucose monomer units forming long chains, representing nearly

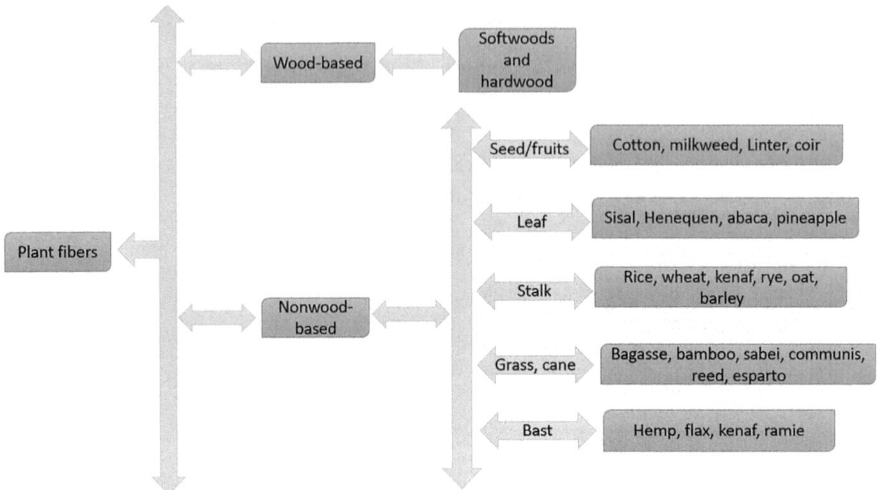

FIGURE 6.1 Natural plant-based fiber classifications.

Source: Adapted with permission from Elsevier [37]. Copyright, Elsevier 2017.

40%–45% of the plants. Each glucose unit further entails 44.4% C, 49.4% O, 6.2% H, and a slight portion of fatty acids [37]. Cellulose framework consists of repeating units ($C_6H_{11}O_5$, D-anhydro-glucose) connected to each other by β-1,4-glycosidic linkages having nearly around 10,000 degree of polymerizations [32]. Lignocellulosic fibers are also dominated by crystalline regions over amorphous regions, hence absorbing more moisture. However, interestingly, cellulose microfibrils determine the mechanical properties of plant-based natural fibers [41]. The chemical elements, –OH, –COOH, and NH_2 present in the fiber structures are responsible for water absorption by the fibers when exposed to weather/environmental conditions [42]. The physical properties of cellulosic fibers are governed by the crystalline packing factor which is directed for the hydrogen bonding in the –OH group of cellulose [43]. The unit cells of a cellulose are shown in Figure 6. 2.

6.4 SURFACE MODIFICATION OF NATURAL FIBERS

Natural fibers have inherent compatibility with the polymers due to the presence of some impurities like oil, wax, and so on. Therefore, it is imperative to remove such unwanted impurities before the fabrication, which will enhance the interaction in the composite system. Moreover, the treatments also break the atomic bond in the fiber surfaces, which will enable to functionalize them [45,46]. However, the surfaces of the natural fibers could be modified in numerous ways like chemically, biologically, and physically.

6.4.1 Physical Methods

There are several physical methods available for treating the plant-based fiber materials like ultrasound, UV light, and plasma. In case of physical method, the bundle of fibers is separated, and the structure of the fiber surfaces is modified prior to

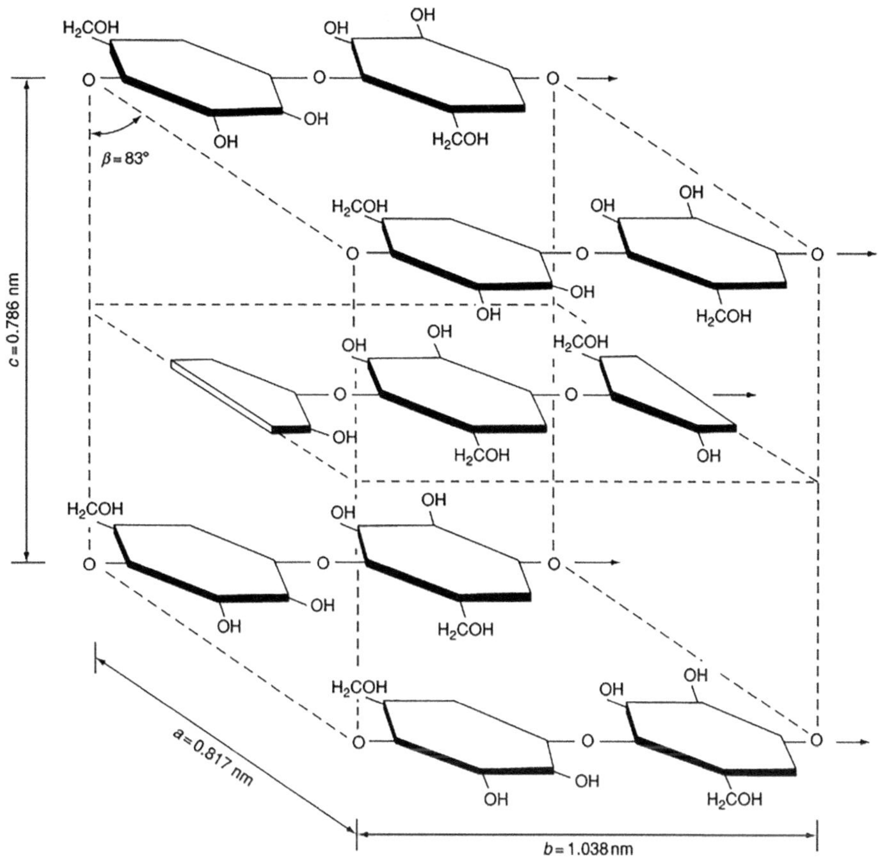

FIGURE 6.2 Unit cell of cellulose I (parallel).

Source: Reprinted with permission from Elsevier [44]. Copyright, Elsevier 2003.

composites fabrication. Among them plasma treatment is the widely used physical approach of surface modification. This method is highly effective especially for the modifications of natural fibers surfaces. However, plasma method is also divided into two more groups: thermal and nonthermal methods [47]. Plasma is ionized gas (partially), comprising radical, ionic, molecular, and excited atomic species along with the photons and free electrons [48].

6.4.2 Chemical Methods

Alternative to the physical methods is the chemical technology like silane treatment, mercerization/alkali, maleated coupling, acetylation, permanganate and peroxide, etherification, graft polymerization, benzoylation, isocyanate, and so on. The surface and mechanical properties of the fibers are improved through chemical-based modifications. A covalent bond is formed between the hydroxyl groups of cellulosic fibers and the chemical reagents used for pretreatment [49]. Crystalline structures of the natural fibers could also be improved through removing the weakly adhering polymers like lignin from the fiber surfaces [50]. Moreover, water absorption and moisture contents of fibers and associated composites could also be minimized through surface modifications, for example, using water repellents.

6.4.3 Biological Methods

Although physical and chemical methods are widely used technologies nowadays, there are still disadvantages for such approaches due to the usage of hazardous chemicals, generation of wastages and pollutants, consumption of excessive energies, and additional cost for purchasing of chemicals and machineries/equipment usages. Therefore, more eco-friendly and cost effective methods are also taken into consideration. In this regard, microorganisms like enzyme, bacteria, or fungi could facilitate to overcome such challenges through modifying the natural fiber surfaces even consuming comparatively lower energy [51,52]. Hence, such kinds of treatments are also called as "green" treatment methods. For example, enzymatic treatments could remove the hydrophilic hemicellulose, lignin, and pectin from the surfaces of plant fibers through removing the hydrophilicity from the cellulose substances [53]. The principal enzymes behind this catalytic process are oxidoreductases and hydrolases [54].

6.5 POLYMERS USED FOR COMPOSITES FABRICATION

Polymers and associated polymeric composites are gaining importance extensively in the field of automotives, furniture, aerospace, marine, telecommunications, and so on, for their outstanding thermomechanical performances, corrosion resistance, and lightweight property. The polymers are selected depending on the required performances and application of the final products. The polymers are broadly divided into two main classes as follows:

a. Thermoplastic; and
b. Thermosetting.

Among thermoplastic polymers are polylactic acid (PLA), polypropylene (PP), PE (polyethylene), polyvinyl chloride, polystyrene, low-density PE, high-density PE, and so on, whereas thermoplastic polymers are epoxy, MUF, MDI, PF, VE, and so on [30,55,56]. Both types of polymers show numerous potential to produce natural fiber-reinforced composite materials. The polymers like VE can be used mainly for laminated composites/panels from wood fibers/particles. Moreover, cements are also gaining interest for producing cementitious panels from different cellulosic fibers [57,58] to produce green structural materials.

6.6 VARIOUS COMPOSITES DEVELOPED FROM NATURAL FIBERS REINFORCEMENT

Different synthetic fiber-reinforced composites [59–62] are used by the manufacturers since long times. However, they create burdens to the environment due to inherent nonbiodegradability. Composites produced from naturally derived cellulosic fibers show new routes of green products development through minimizing greenhouse gas generations which will keep the environment safer. The natural fibers are not only providing environmental sustainability but also superior thermomechanical features to the developed products. Different thermoplastic, thermosetting, or cementitious matrices are gaining attention to reinforce such kinds of fibers [45,63–65]. Recently, besides the virgin natural fibers, there are also wastage natural fibers gaining popularity to produce sustainable composite materials [66]. Similarly, Hassan et al. conducted a research on natural fiber waste-reinforced epoxy composites in order to provide insulation performances besides the significant acoustic, thermal, and mechanical properties [66]. The perceived thermal conductivity values in case of 20% sugarcane, coconut, and cotton fiber-reinforced epoxy composites were 0.342 ± 0.02, 0.303 ± 0.01, and 0.373 ± 0.01 W/(m.K), respectively [66]. Additionally, the similar percentage of fibers provided the impact strengths by 4.43 ± 0.18, 4.83 ± 0.22, and 9.73 ± 0.38 kJ/m^2, respectively, and flexural properties by 29.7 ± 1.03, 39.9 ± 1.5, and 81.7 ± 4.03 MPa, respectively [66]. A recent study by Ibrahim et al. [67] reported about the development of flax fiber and date palm-reinforced starch hybrid composites, where they mixed different weight percentages (20–80) of fibers to understand the reinforced characteristics [67]. The composites were preheated at 160°C for 30 min and 5 MPa pressure by hot pressing machine, whereas the tensile strength and modulus values were increased from 3.8 and 0.4 MPa, respectively, to 32.7 and 2.8 MPa, respectively [67]. Similarly, plenty of researches are on-going worldwide from different point of aspects to develop hybrid composites through reinforcing natural fibers/cellulosic materials with versatile matrix [68–70].

6.7 TECHNOLOGY BEHIND VINYL ESTER POLYMER

Vinyl polymers are cheaper and easy to process. VEs are organic chemical ingredient containing ester group (R-COO-R′) with the functional group R′(vinylic) [71]. The formulation of VE involves the condensation of methacrylic acid with epoxy and styrene. However, the obtained condensation products function as crosslinking agent to improve the fibers adhesion. The structure of VE polymers with some examples are displayed in Figure 6.3. Although it was said that VE performs better with synthetic glass fiber reinforcements [20], nowadays, natural fibers are also gaining popularity for the fabrication with VE resin [35,72]. However, matrix can be tuned to produce suitable categories of composites. For example, by using VE, the stiffness, resistance to chemicals, dimensional stability, and most importantly the mechanical performances can be maximized [73]. Additionally, VE is also comparatively cheaper than epoxy resin [73]. Moreover, VE is also easy to handle at ambient temperature, and hence easy to control the curation rate and reactivity than epoxy resin too. Interestingly, VE is also compatible with epoxy resin in terms of mechanical properties, especially hydrolytic stability [35]. Conversely, there are also challenges to use VE resin as it is brittle in nature [35]. In this regard, natural fibers are mixed with VE resins during shape casting in composites fabrication [35]. Therefore, VE resins are also sometimes considered as the polyaddition products of monounsaturated monomers like polyvinyl benzoate and polyvinyl acetate. Initially, VE resins were used and reported for dental applications in the United States as the adhesive made bondings with acrylic dental prostheses, although such a resin was too reactive

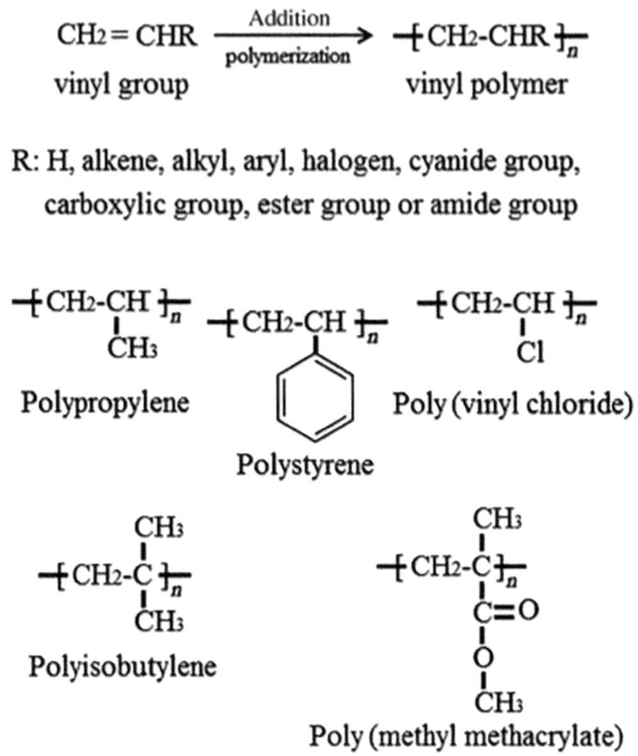

FIGURE 6.3 Vinyl polymers' chemical structure examples.

Source: Adapted with permission from Elsevier [75]. Copyright, Elsevier 2018. Properties of vinyl ester (according to Vasavibala Resins Pvt., Ltd., India) [76].

during this time [71]. However, nowadays, VE resins are also gaining popularity for producing highly performed fiber-reinforced composites. The technology behind this performance is the presence of C=C bond in the molecular chain, which is feasible during crosslinking to cure the polymeric composites. A schematic linkage is shown in equation (6.1) for the linkage of VE polymeric resin in order to link two unsaturated monomer elements (chain polymerizations) [74].

$$n[CH_2 = CHX \rightarrow -[-CH_2 - CHX -]_2 \quad (6.1)$$

Interestingly, VE has recycling capability, hence facilitating pollution-free environmental features. Additionally, VE-based products could be recycled for multiple times even after the decades of life span. However, there are three different characteristics of vinyl polymers [75]:

- Thermosets
- Vinyl plastics, and
- Rubber.

- Appearance: Between pale yellow and pale brown (liquid liquid)
- Viscosity at room temperature, cP (Pro Spindle 62, rpm 60 + Brookfield Viscometer LV DV II: 300–400)
- Specific gravity at room temperature: 1.04–1.06
- Volatile content in %: 07–11
- Shelf life at room temperature, months: 3
- Exothermal peak (°C, under insulated conditions for 100 g mixture): 160–170
- Value of acid, mg of KOH/g: 7–11

6.8 NATURAL FIBERS REINFORCEMENT WITH VINYL ESTER POLYMERS

Natural fibers will degrade biologically; at least they would not remain undegraded for thousands of years like other synthetic fibers like glass, carbon, and so on [18,77]. Comparatively, synthetic fibers have higher density (2.54 g/cm^3 in case of carbon and 1.8–2.1 g/cm^3 for glass fibers) compared to natural fibers (1.25–1.5 g/cm^3) [78,79]. Therefore, composites produced from cellulosic fiber-reinforced polymers are lighter than that of synthetic fibers. Ammar et el. developed sugar palm natural fiber-reinforced VE composites through hand lay-up technology and found that unidirectional fiber arrangements provided superior mechanical properties (2501 MPa tensile modulus, 93.08 flexural strength, 3328 MPa flexural modulus, and 33.66 kJ/m^2 impact strength) [36].

Generally, natural fibers are hydrophilic in nature, whereas matrix is hydrophobic [81]. However, the hydrophilic characteristics of natural fibers negatively affect the mechanical performances of the composites, whereas significant chemical modifications/pretreatments are necessary for improving the fiber–matrix interactions in order to minimize the risk of strength loss [82]. Likewise, acetylation treatment of palm fiber facilitated to improve the durability and mechanical properties of VE polymeric composites (Figure 6.4) [80]. The presence of moisture strongly needs to be taken into consideration for producing biocomposites. Otherwise, water could be diffused in polymeric composites structure in the following three ways [81,83]:

- The micro-gaps present in the inter-polymer chain.
- Capillary transport gaps in the interface between the fiber to matrix structure maybe for the imperfect wetting and impregnations.
- Micro-cracks occurred in the matrix system due to the swelling of cellulosic fibers.

Generally, the performance characteristics of the composites depend on the properties of polymer used, adhesion capability between the fiber and polymer, and the interfacial bonding.

In addition, nanoparticles can also take significant role for enhancing the thermomechanical performances of VE resin-reinforced composites. The improved performances could lead to improvement in a wide variety of properties like resistance against flammability, barrier properties, thermal stability, mechanical

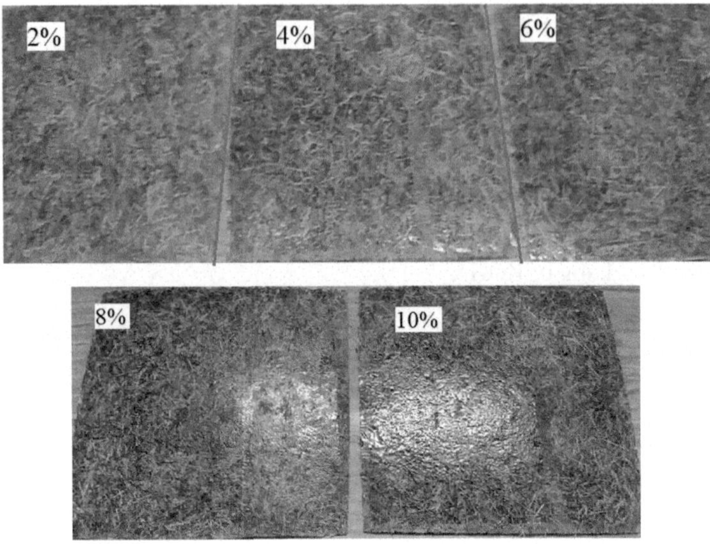

FIGURE 6.4 Palm fiber-reinforced VE polymeric composite plates.

Source: Adapted with permission from Elsevier [80]. Copyright, Elsevier 2020.

performances, dimensional stability, and so on. The optimum nanoparticles like ZnO, TiO_2, silica, montmorillonite, silver, nanoclay, and carbon nanotube concentrations have the inherent capability to enhance the performance of polymers significantly [84,85]. In a recent study reported by Alhuthali et al. [79] on recycled cellulosic fiber-reinforced VE composites and nanocomposites (nanoclay), it is seen that nanocomposites displayed higher strengths compared to control composites (where no nanoparticles used). The same study also showed that, after a certain level of nanosilica loading (more than 3 wt.%), the flexural and impact properties start to decline (Figure 6.5).

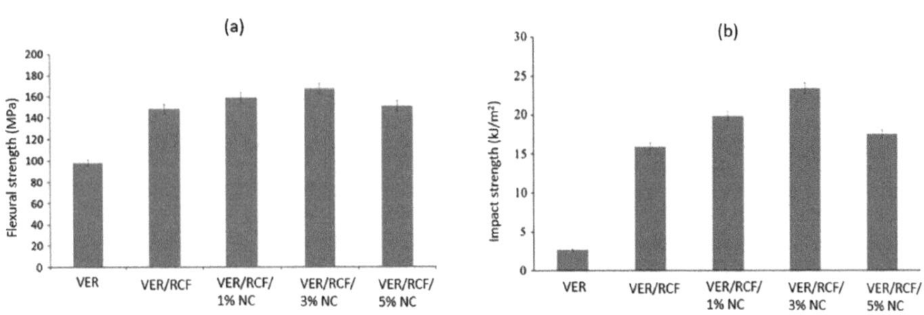

FIGURE 6.5 Mechanical properties of recycled cellulosic fiber-reinforced VE polymeric composites and nanocomposites (nanoclay): (a) flexural strength and (b) impact strength. VER: Vinyl ester resin; RCF: recycled cellulosic fiber sheets; NC: nanocomposites.

Source: Adapted with permission from Elsevier [79]. Copyright, Elsevier 2012.

6.9 HYBRID COMPOSITES DEVELOPED FROM VINYL ESTER POLYMERS

A hybrid composite is comprised of at least two different elements/materials in the system. When various fibers (at least two or more) are incorporated in the polymeric composite system, it is termed as hybrid composite [33]. Matrix could be single- or multi-phase. Recently, hybrid composites are utilized in order to tailor and improve the mechanical properties and long-term durability of the products through minimizing some negative drawbacks of single-fiber-reinforced composites [86]. However, the performances of the hybrid composites are dependent on the reinforcement fiber materials and the constituent polymers used. In a recent study, it was reported that the hybrid composites developed from sugar palm/roselle fiber reinforced with VE polymeric resin provided superior mechanical performances compared to the neat VE resin [87], where the optimum tensile performance was shown by 50% roselle/50% sugar palm. The same study also revealed that nonuniform distributions of fiber could create significant gaps between the fibers and VE resin, which may result in weaker interfacial bonding in the composite system as well as decline in the flexural properties. Not only natural/natural fiber but also natural/synthetic fiber-reinforced VE resin composites are also researched by the scientists, whereas VE also functions as the excellent resin in composite system [88]. However, through investigating the fiber to matrix compatibility, fracture characteristics, natural variability, structural property behavior, and so on perfect combination of fiber proportions could be identified and selected.

6.10 MECHANICAL AND THERMAL PROPERTIES OF DEVELOPED HYBRID COMPOSITES

Natural fibers can be successfully reinforced with the thermosetting polymers like VE resin which will provide significant thermomechanical performances [22]. Nasimudeen et al. developed some hybrid composite plates through reinforcing banana, jute, and kenaf fiber with VE polymeric resin and found that the produced materials provided significant mechanical performances, whereas jute/kenaf/banana/kenaf/banana/jute hybrid laminates provided a maximum tensile strength off 34.12 MPa and stiffness of 1.67 MPa [89]. Moreover, the produced plates also showed significant resistance against flexural loading [89]. Besides, natural fiber-reinforced VE polymeric composites also show improved flame retardancy, i.e., the decomposition temperature becomes higher and the char yield too [90]. The TGA (thermogravimetric analysis) and DTG (Derivative TG) studies generally show the thermal stability of the materials like weight loss, residual char yield, and the decomposition temperatures related to certain polymeric materials like cellulose/hemicellulose, or lignin present in the tested products [91]. A major weight loss was noticed in the case of cotton fibers at 345°C due to α-cellulose degradation with 19% char yield, whereas the principal weight loss for VE resins was found at 355°C and for their reinforced composites at 349°C, demonstrating an improvement after the fabrication into composites between the two (reinforcement and filler), which was reported by a recent study [90]. The addition of nanofiller in the VE resin-reinforced hybrid composites from natural fibers could also show substantial thermomechanical and physical properties which may be increased with the increase in nanoparticles loading. Similarly, results were also found and reported by HPS et al., in case of coconut/kenaf fiber-reinforced VE resin-reinforced nanocomposites with nanoparticles obtained from coconut hells [92]. The tensile strength was found as 47.66 MPa in case of composites developed without nanofillers loading, whereas the values increased up to 58.69 MPa at 3% nanofiller incorporations [92]. However, the mechanical properties started to fall upon increased loading after 3% [92]. The similar improved stability against the heat is also noticed in terms of TGA analysis for the reported products [92].

6.11 APPLICATION OF VINYL ESTER POLYMERS

Natural fibers have drawn tremendous attention due to their low-cost features, environment-friendliness, better thermomechanical performances, and so on. Therefore, composite manufacturers utilize natural fiber-reinforced polymeric composites. Additionally, a composite product is greatly influenced by the matrix. However, VE resin has profound history in the field of polymer chemistry and industrial manufacturing units. The natural fiber-reinforced hybrid composites could be used for structural framing and sub-flooring, roof system (load bearing), windows, doors, furnitures, and so on. Natural fibers have extensive insulation characteristics; hence, composites made of cellulosic fibers are used as cabin lining, door panels, and cushioned seats. Interestingly, cotton-based textile recycled materials could also be a big sources of such insulation panel developments through reinforcing with VE resin. Moreover, thermosetting polymer like VE also has extreme potential in the construction and building materials sector like false ceiling, partitioning, door frames, surface frame/panels, and so on. Furthermore, natural fiber-reinforced VE composites could also be used for packaging, fishing baskets, geotextiles, dartboards, and so on [74,93]. Some of the applications of natural fiber-reinforced composites are tabulated in Table 6.1.

TABLE 6.1
Natural Fiber-Reinforced Polymeric Composites Application [51,94–96]

Parts of Automotive	Manufacturer	Model of the Automotive
Headliner panel, seat backs, door panels, door trim, noise insulation panel	BMW	BMW I series and 3, 5, 7 series
Seat-back lining, door cladding, back cushions, seat bottoms	Fiat	Brava, punto, Alpha Romeo 146, Marea
Boot lining, seat backs, back and side door panels, spare tire lining	Audi	Roadster, A2, A4, A6, A8, Coupe
Door paneling (interior)	Citroen	C5
Windshield, door panel, pillar cover panel, business table, dashboard	Chrysler	S-class, A, C, and E models
Wheel box, sun visor, roof cover, internal engine cover, bumper, interior insulation	Mercede-Benz	Mercedes A and Truck
Rear storage shelf and insulation	Rover	2000
Rear parcel shelf	Renault	Clio and Twingo
Seat back and parcel shelf	Peugeot	406

6.12 CONCLUSION

This chapter includes the discussions on various natural fiber-reinforced VE polymeric hybrid composites and their thermomechanical performances. Different technologies like hand lay-up, hot pressing, and so on are implemented for producing hybrid composites, whereas VE was used as matrix material. The developed products are typically tested for tensile, flexural, impact, and morphological properties, using FTIR (Fourier transform Infrared spectroscopy), XRD (X-ray Diffraction), TGA (Thermogravimetric analysis)/DTG (Derivative thermogravimetry), DSC (Differential scanning calorimetry), XPS (X-ray photoelectron spectroscopy), DMA (Dynamic mechanical analysis), and so on. The composites developed by different research studies showed display not only significant mechanical performances but also considerable thermal stability. VE resin contains slightly better thermal stability than natural fibers; therefore, their combined reinforcement in the composite structures' positive attribution toward improved resistance against thermal degradation. The morphological studies most often may exhibit fiber distributions in the composite system, and any flaws that occur during the fabrication may negatively influence the mechanical performances. However, extensive research is still needed to look for more novel routes to produce hybrid composites through combining various natural fibers with VE polymers, in order to make them more economical to the industries. In the future, VE resin will replace some other thermosetting polymers like as epoxy, polyester, MUF, and so on, in the manufacturing industries.

REFERENCES

1. Mohammed, L., et al., A review on natural fiber reinforced polymer composite and its applications. *International Journal of Polymer Science*, 2015, 243947.
2. Hasan, K.M.F., P.G. Horváth, and T. Alpár, Silk protein and its nanocomposites, in *Biopolymeric Nanomaterials: Fundamentals and Applications*. 2021, Elsevier: Amsterdam, Netherlands. pp. 309–323.
3. Hasan, K.F., et al., Nanosilver coating on hemp/cotton blended woven fabrics mediated from mammoth pine bark with improved coloration and mechanical properties. *The Journal of the Textile Institute* 2021, **113**(12): pp. 2641–2650.
4. Hasan, K.F., et al., Hemp/glass woven fabric reinforced laminated nanocomposites via in-situ synthesized silver nanoparticles from *Tilia cordata* leaf extract. *Composite Interfaces*, 2021.
5. Hasan, K.F., et al., Coloration of flax woven fabric using *Taxus baccata* heartwood extract mediated nanosilver. *Colouration Technology*, 2021. **138**(2): pp. 1–11.
6. Hasan, K.F., et al., Coloration of woven glass fabric using biosynthesized silver nanoparticles from *Fraxinus excelsior* tree flower. *Inorganic Chemistry Communications*, 2021. **126**: p. 108477.
7. Hasan, K., et al., Colorful and facile in situ nanosilver coating on sisal/cotton interwoven fabrics mediated from European larch heartwood. *Scientific Reports*, 2021. **11**(1): pp. 1–13.
8. Bhatia, J.K., B.S. Kaith, and S. Kalia, Recent developments in surface modification of natural fibers for their use in biocomposites, in *Biodegradable Green Composites*, Susheel Kaliam, Editor 2016, John Wiley & Sons: Hoboken, NJ. pp. 80–117.
9. Ahmad, F., H.S. Choi, and M.K. Park, A review: natural fiber composites selection in view of mechanical, light weight, and economic properties. *Macromolecular Materials and Engineering*, 2015. **300**(1): pp. 10–24.

10. Al-Oqla, F.M. and S. Sapuan, Natural fiber reinforced polymer composites in industrial applications: feasibility of date palm fibers for sustainable automotive industry. *Journal of Cleaner Production*, 2014. **66**: pp. 347–354.
11. Alam, M., M. Maniruzzaman, and M. Morshed, *Application and Advances in Microprocessing of Natural Fiber (Jute)–Based Composites*. 2014. pp. 243–260.
12. Alpár, T., G. Markó, and L. Koroknai, Natural fiber reinforced PLA composites: Effect of shape of fiber elements on properties of composites, in *Handbook of Composites from Renewable Materials, Design and Manufacturing*. 2017, Wiley and Sons: Hoboken, NJ. pp. 287–309.
13. Balea, A., A. Blanco, and C. Negro, Nanocelluloses: natural-based materials for fiber-reinforced cement composites. A critical review. *Polymers*, 2019. **11**(3): Article number: 518.
14. Mahmud, S., et al., In situ synthesis of green AgNPs on ramie fabric with functional and catalytic properties. *Emerging Material Research*, 2019. **8**(4): pp. 623–633.
15. Hasan, K.F., et al., Effects of sisal/cotton interwoven fabric and jute fibers loading on polylactide reinforced biocomposites. *Fibers and Polymers*, 2021. **23**: pp. 3581–3595.
16. Faruk, O., et al., Biocomposites reinforced with natural fibers: 2000–2010. *Progress in Polymer Science*, 2012. **37**(11): pp. 1552–1596.
17. Gallos, A., et al., Lignocellulosic fibers: a critical review of the extrusion process for enhancement of the properties of natural fiber composites. *RSC Advances*, 2017. **7**(55): pp. 34638–34654.
18. Holbery, J. and D. Houston, Natural-fiber-reinforced polymer composites in automotive applications. *JOM*, 2006. **58**(11): pp. 80–86.
19. Kalia, S., et al., Natural fibers, bio-and nanocomposites. *International Journal of Polymer Science*, 2011, 735932.
20. Mullins, M., D. Liu, and H.-J. Sue, Mechanical properties of thermosets, in *Thermosets*, G. Qipeng, Editor. 2018, Elsevier: Radarweg, Amsterdam, the Netherlands. pp. 35–68.
21. Hasan, K.F., et al., Rice straw and energy reed fibers reinforced phenol formaldehyde resin polymeric biocomposites. *Cellulose*, 2021. **28**: pp. 7859–7875.
22. Stalin, A., et al., Mechanical properties of hybrid vetiver/banana fiber mat reinforced vinyl ester composites. *Journal of Industrial Textiles*, 2020: p. 1528083720938161.
23. Hasan, K., et al., Novel insulation panels development from multilayered coir short and long fiber reinforced phenol formaldehyde polymeric biocomposites. *Journal of Polymer Research*, 2021. **28**(12): pp. 1–16.
24. Hasan, K.F., et al., A state-of-the-art review on coir fiber-reinforced biocomposites. *RSC Advances*, 2021. **11**(18): pp. 10548–10571.
25. Hasan, K.M.F., P.G. Horváth, and T. Alpár. Effects of alkaline treatments on coconut fiber reinforced biocomposites in *9th Interdisciplinary Doctoral Conference*. 2020. Pecs, Hungary: Doctoral Student Association of the University of Pécs.
26. Hasan, K.F., et al., Thermomechanical characteristics of flax-woven-fabric-reinforced poly (lactic acid) and polypropylene biocomposites. *Green Materials*, 2021. **40**: pp. 1–10.
27. Hasan, K., P.G. Horváth, and T. Alpár, Potential fabric-reinforced composites: a comprehensive review. *Journal of Materials Science*, 2021. **56**: pp. 14381–14415.
28. Hasan, K.F., et al., Industrial Flame Retardants for Polyurethanes, in *Materials and Chemistry of Flame-Retardant Polyurethanes Volume 1: A Fundamental Approach*. 2021, ACS Publications: Washington, DC, USA. pp. 239–264.
29. Hasan, K.M.F., et al., Energy reed fiber reinforced thermosetting polymeric biocomposite, in *Springwind Conference*. 2021: Sopron, Hungary.
30. Hasan, K.M.F., G.H. Péter, and L.A. Tibor, Design and fabrication technology in biocomposite manufacturing, in *Toward the Value-Added Biocomposites: Technology, Innovation and Opportunity*. 2021, CRC Press: Boca Raton, FL.

31. Amiandamhen, S., M. Meincken, and L. Tyhoda, Natural fibre modification and its influence on fibre-matrix interfacial properties in biocomposite materials. *Fibers and Polymers*, 2020. **21**: pp. 677–689.
32. Bismarck, A., S. Mishra, and T. Lampke, Plant fibers as reinforcement for green composites, in *Natural Fibers, Biopolymers, and Biocomposites*. 2005, CRC Press: New York. pp. 52–128.
33. Mahmud, S., et al., Comprehensive review on plant fiber-reinforced polymeric biocomposites. *Journal of Materials Science*, 2021. **56**: pp. 7231–7264.
34. Adeniyi, A.G., et al., A review of coir fiber reinforced polymer composites. *Composites Part B: Engineering*, 2019. **176**: p. 107305.
35. Razali, N., et al., Mechanical and thermal properties of Roselle fibre reinforced vinyl ester composites. *BioResources*, 2016. **11**(4): pp. 9325–9339.
36. Ammar, I.M., et al., Mechanical properties of environment-friendly sugar palm fibre reinforced vinyl ester composites at different fibre arrangements. *Environment Asia*, 2019. **12**(1):pp. 25–35.
37. Petroudy, S.D., Physical and mechanical properties of natural fibers, in *Advanced High Strength Natural Fibre Composites in Construction*. 2017, Elsevier: Duxford. pp. 59–83.
38. Tibor, L.A., G.H. Péter, and K.M.F. Hasan, Introduction to biomass and biocomposites, in *Toward the Value-Added Biocomposites: Technology, Innovation and Opportunity*. 2021, CRC Press: Boca Raton, FL
39. Hasan, K., et al. Effects of cement on lignocellulosic fibres. in *9th Hardwood Proceedings*. 2020. Sopron, Hungary: University of Sopron Press.
40. Hasan, K.M.F., P.G. Horváth, and T. Alpár, Lignocellulosic fiber cement compatibility: a state of the art review. *Journal of Natural Fibers*, 2021: pp. 1–26.
41. Petroudy, S.R.D., et al., The effect of xylan on the fibrillation efficiency of DED bleached soda bagasse pulp and on nanopaper characteristics. *Cellulose*, 2015. **22**(1): pp. 385 395.
42. Hasan, K.M.F., H. Péter György, and A. Tibor, Thermomechanical behavior of methylene diphenyl diisocyanate-bonded flax/glass woven fabric reinforced laminated composites. *ACS Omega*, 2020. **6**(9): pp. 6124–6133.
43. Sjostrom, E., *Wood Chemistry: Fundamentals and Applications*. 1993, New York: Gulf Professional Publishing.
44. Holtzapple, M.T., Cellulose, in *Encyclopedia of Food Sciences and Nutrition*, C. Benjamin, Editor. 2003, Academic Press, Elsevier: Maryland, USA. pp. 998–1007.
45. Kalia, S., B. Kaith, and I. Kaur, Pretreatments of natural fibers and their application as reinforcing material in polymer composites—a review. *Polymer Engineering & Science*, 2009. **49**(7): pp. 1253–1272.
46. Mukhopadhyay, S. and R. Fangueiro, Physical modification of natural fibers and thermoplastic films for composites—a review. *Journal of Thermoplastic Composite Materials*, 2009. **22**(2): pp. 135–162.
47. Zille, A., F.R. Oliveira, and A.P. Souto, Plasma treatment in textile industry. *Plasma Processes and Polymers*, 2015. **12**(2): pp. 98–131.
48. Sparavigna, A., *Plasma treatment advantages for textiles*. arXiv preprint arXiv:0801.3727, 2008.
49. Khoathane, M.C., E.R. Sadiku, and C.S. Agwuncha, Surface modification of natural fiber composites and their potential applications, in *Surface Modification of Biopolymers*. 2015, Hoboken, NJ: John Wiley & Sons, Inc., pp. 370–400.
50. Li, X., L.G. Tabil, and S. Panigrahi, Chemical treatments of natural fiber for use in natural fiber-reinforced composites: a review. *Journal of Polymers and the Environment*, 2007. **15**(1): pp. 25–33.
51. Ferreira, D.P., J. Cruz, and R. Fangueiro, Surface modification of natural fibers in polymer composites, in *Green Composites for Automotive Applications*. 2019, Elsevier: Duxford, United Kingdom. pp. 3–41.

52. Kalia, S., et al., Surface modification of plant fibers using environment friendly methods for their application in polymer composites, textile industry and antimicrobial activities: a review. *Journal of Environmental Chemical Engineering*, 2013. **1**(3): pp. 97–112.
53. George, M., P.G. Mussone, and D.C. Bressler, Surface and thermal characterization of natural fibres treated with enzymes. *Industrial Crops and Products*, 2014. **53**: pp. 365–373.
54. Araujo, R., M. Casal, and A. Cavaco-Paulo, Application of enzymes for textile fibres processing. *Biocatalysis and Biotransformation*, 2008. **26**(5): pp. 332–349.
55. Georgopoulos, S.T., et al., Thermoplastic polymers reinforced with fibrous agricultural residues. *Polymer Degradation and Stability*, 2005. **90**(2): pp. 303–312.
56. Pascault, J.-P., et al., *Thermosetting Polymers,* in *Handbook of Polymer Synthesis, Characterization, and Processing.* 2002: John Wiley & Son: New York, USA. pp. 519-533.
57. Akhavan, A., J. Catchmark, and F. Rajabipour, Ductility enhancement of autoclaved cellulose fiber reinforced cement boards manufactured using a laboratory method simulating the Hatschek process. *Construction and Building Materials*, 2017. **135**: pp. 251–259.
58. Gwon, S., Y.C. Choi, and M. Shin, Effect of plant cellulose microfibers on hydration of cement composites. *Construction and Building Materials*, 2021. **267**: p. 121734.
59. Hejazi, S.M., et al., A simple review of soil reinforcement by using natural and synthetic fibers. *Construction and Building materials*, 2012. **30**: pp. 100–116.
60. Mohanavel, V., et al., Influence of stacking sequence and fiber content on the mechanical properties of natural and synthetic fibers reinforced penta-layered hybrid composites. *Journal of Natural Fibers*, 2021: pp. 1–13.
61. Pakravan, H.R. and T. Ozbakkaloglu, Synthetic fibers for cementitious composites: a critical and in-depth review of recent advances. *Construction and Building Materials*, 2019. **207**: pp. 491–518.
62. Unterweger, C., O. Brüggemann, and C. Fürst, Synthetic fibers and thermoplastic short-fiber-reinforced polymers: properties and characterization. *Polymer Composites*, 2014. **35**(2): pp. 227–236.
63. Shubhra, Q.T., et al., Characterization of plant and animal based natural fibers reinforced polypropylene composites and their comparative study. *Fibers and Polymers*, 2010. **11**(5): pp. 725–731.
64. Chattopadhyay, S.K., et al., Biodegradability studies on natural fibers reinforced polypropylene composites. *Journal of Applied Polymer Science*, 2011. **121**(4): p. 2226–2232.
65. Hasan, K.F., P.G. Horváth, and T. Alpár, Development of lignocellulosic fiber reinforced cement composite panels using semi-dry technology. *Cellulose*, 2021. **28**: pp. 3631–3645.
66. Hassan, T., et al., Acoustic, mechanical and thermal properties of green composites reinforced with natural fibers waste. *Polymers*, 2020. **12**(3): p. 654.
67. Ibrahim, H., et al., Characteristics of starch-based biodegradable composites reinforced with date palm and flax fibers. *Carbohydrate Polymers*, 2014. **101**: pp. 11–19.
68. Karaduman, Y., L. Onal, and A. Rawal, Effect of stacking sequence on mechanical properties of hybrid flax/jute fibers reinforced thermoplastic composites. *Polymer Composites*, 2015. **36**(12): pp. 2167–2173.
69. Bledzki, A., H.P. Fink, and K. Specht, Unidirectional hemp and flax EP-and PP-composites: influence of defined fiber treatments. *Journal of Applied Polymer Science*, 2004. **93**(5): pp. 2150–2156.
70. Hasan, K.F., et al., Semi-dry technology-mediated coir fiber and Scots pine particle-reinforced sustainable cementitious composite panels. *Construction and Building Materials*, 2021. **305**: p. 124816.
71. Kandelbauer, A., et al., Unsaturated polyesters and vinyl esters, in *Handbook of Thermoset Plastics*. 2014, Elsevier: San Diego, CA. pp. 111–172.

72. Ghosh, R., et al., Effect of fibre volume fraction on the tensile strength of Banana fibre reinforced vinyl ester resin composites. *International Journal of Advanced Engineering Science*, 2011. **4**(1): pp. 89–91.
73. Aprilia, N.S., et al., Exploring material properties of vinyl ester biocomposites filled carbonized jatropha seed shell. *BioResources*, 2014. **9**(3): pp. 4888–4898.
74. Mohamed, S., et al., Introduction to natural fiber reinforced vinyl ester and vinyl polymer composites, in *Natural Fibre Reinforced Vinyl Ester and Vinyl Polymer Composites*. 2018, Elsevier: Duxford, United Kingdom. pp. 1–25.
75. Hao, L., et al., Natural fiber reinforced vinyl polymer composites, in *Natural Fibre Reinforced Vinyl Ester and Vinyl Polymer Composites*. 2018, Woodhead Publishing, Elsevier: Duxford. pp. 27–70.
76. Johnson, R.D.J., V. Arumugaprabu, and T.J. Ko, Mechanical property, wear characteristics, machining and moisture absorption studies on vinyl ester composites–a review. *Silicon*, 2019. **11**(5): pp. 2455–2470.
77. Anuar, H. and A. Zuraida, Improvement in mechanical properties of reinforced thermoplastic elastomer composite with kenaf bast fibre. *Composites Part B: Engineering*, 2011. **42**(3): pp. 462–465.
78. Sgriccia, N., M. Hawley, and M. Misra, Characterization of natural fiber surfaces and natural fiber composites. *Composites Part A: Applied Science and Manufacturing*, 2008. **39**(10): pp. 1632–1637.
79. Alhuthali, A., I.M. Low, and C. Dong, Characterisation of the water absorption, mechanical and thermal properties of recycled cellulose fibre reinforced vinyl-ester eco-nanocomposites. *Composites Part B: Engineering*, 2012. **43**(7): pp. 2772–2781.
80. Senthilraja, R., R. Sarala, and A.G. Antony, Effect of acetylation technique on mechanical behavior and durability of palm fibre vinyl-ester composites. *Materials Today: Proceedings*, 2020. **21**: pp. 634–637.
81. Dhakal, H.N., Z.A. Zhang, and M.O. Richardson, Effect of water absorption on the mechanical properties of hemp fibre reinforced unsaturated polyester composites. *Composites Science and Technology*, 2007. **67**(7–8): pp. 1674–1683.
82. Herrera-Franco, P. and A. Valadez-Gonzalez, A study of the mechanical properties of short natural-fiber reinforced composites. *Composites Part B: Engineering*, 2005. **36**(8): pp. 597–608.
83. Espert, A., F. Vilaplana, and S. Karlsson, Comparison of water absorption in natural cellulosic fibres from wood and one year crops in polypropylene composites and its influence on their mechanical properties. *Composites Part A: Applied science and Manufacturing*, 2004. **35**(11): pp. 1267–1276.
84. Pavlidou, S. and C. Papaspyrides, A review on polymer–layered silicate nanocomposites. *Progress in Polymer Science*, 2008. **33**(12): pp. 1119–1198.
85. Hasan, K.F., P.G. Horváth, and T. Alpár, Nanotechnology for waste wood recycling, in *Nanotechnology in Paper and Wood Engineering*, Rajeev Bhat, Ashok Kumar, Tuan Anh Nguyen, and Swati Sharma, Editors. 2021, Woodhead Publishing: Duxford. pp. 61–80.
86. Paturel, A. and H.N. Dhakal, Influence of water absorption on the low velocity falling weight impact damage behaviour of flax/glass reinforced vinyl ester hybrid composites. *Molecules*, 2020. **25**(2): p. 278.
87. Razali, N., S. Sapuan, and N. Razali, Mechanical properties and morphological analysis of Roselle/sugar palm fiber reinforced vinyl ester hybrid composites, in *Natural Fibre Reinforced Vinyl Ester and Vinyl Polymer Composites*. 2018, Elsevier. pp. 169–180.
88. Kumar, C.N., M. Prabhakar, and J.-I. Song, Effect of interface in hybrid reinforcement of flax/glass on mechanical properties of vinyl ester composites. *Polymer Testing*, 2019. **73**: pp. 404–411.
89. Nasimudeen, N.A., et al., Mechanical, absorption and swelling properties of vinyl ester based natural fibre hybrid composites. *Applied Science and Engineering Progress*, 2021. **14**(4): pp. 680–688.

90. Shahedifar, V. and A.M. Rezadoust, Thermal and mechanical behavior of cotton/vinyl ester composites: effects of some flame retardants and fiber treatment. *Journal of Reinforced Plastics and Composites*, 2013. **32**(10): pp. 681–688.
91. Hasan, K.F., et al., Thermo-mechanical properties of pretreated coir fiber and fibrous chips reinforced multilayered composites. *Scientific Reports*, 2021. **11**(1): pp. 1–13.
92. HPS, A.K., et al., Incorporation of coconut shell based nanoparticles in kenaf/coconut fibres reinforced vinyl ester composites. *Materials Research Express*, 2017. **4**(3): p. 035020.
93. Mohanty, A., et al., Natural fibers, biopolymers and biocomposite: An introduction, in *Natural Fibers, Biopolymers, and Biocomposites*, A.K. Mohanty, M. Misra, and L.T. Drzal, Editors. 2005, CRC Press: Boca Raton, FL. pp. 1–31.
94. Dunne, R., et al., A review of natural fibres, their sustainability and automotive applications. *Journal of Reinforced Plastics and Composites*, 2016. **35**(13): pp. 1041–1050.
95. Baltazar-Y-Jimenez, A. and M. Sain, Natural fibres for automotive applications, in *Handbook of Natural Fibres*. 2012, Woodhead Publishing Limite: Cambridge, UK. pp. 219–253.
96. Cristaldi, G., et al., Composites based on natural fibre fabrics. *Woven Fabric Engineering*, 2010. **17**: pp. 317–342.

7 Natural Fiber-Reinforced Vinyl Ester Composites

Influence of CNT Nanofillers on Thermal and Mechanical Properties

Soubhik De, Tanaya Sahoo, B. N. V. S. Ganesh Gupta K, and Rajesh Kumar Prusty
National Institute of Technology

CONTENTS

7.1 INTRODUCTION TO NATURAL FIBER-REINFORCED POLYMER (NFP) COMPOSITES

Natural fibers have played an important role in constituting fundamental materials for textile, automation, construction, and bio-applications [1]. NFs can range from fibrous plant matter to biomass, agro-mass or any other photosynthetic matter. NFs such as flax, jute, kenaf, hemp and sisal, wherein the cellulose is the primary

DOI: 10.1201/9781003270997-7

constituent, offer significant weight reduction, cost, sustainability, and recyclability benefits. Though NFs have been incorporated into several engineering, scientific, and commercial domains, primary technical considerations have to be addressed to garner wide-scale acceptance.

7.1.1 Natural Fiber Types

The most basic grouping of NFs is the botanical type: (1) bast fibers like jute, flax, cannabis, ramie, and kenaf; (2) leaf fibers like banana, sisal, agave, and pineapple; (3) seed fibers such as coir, cotton, and kapok; (4) grass and reeds such as wheat, maize, sugarcane (bagasse), and rice; and (5) all other types are common wood fibers and roots. Wood, by far, is the most bountiful fiber obtained from trees with an annual world production of 1.75×10^9 tons per year from over 10,000 species. In comparison, cotton is produced 18.5×10^6 tons per year, while kenaf, flax, and hemp 9.7×10^5, 8.3×10^5, and 2.1×10^5 tons per year, respectively. Agri-based fibers on the mode of their cultivation can fit into one or more types. For instance, coconut or palm can have both stem and shell fibers. Different varieties of NFs are shown in Figure 7.1.

7.1.2 Properties

NFP composites give the manufacturing and automotive sectors impetus, where durability is a primary concern. Depending on the composition of constituents of fiber-building components like cellulose, water content, or lignin, the fiber's morphological, physical, and mechanical characteristics vary. Moreover, the aspect ratio of individual fibers and experimental conditions highly influence the strength of the fibers [2,3]. While NFs tend to have lower tensile strength than synthetic fibers, they are significantly rigid with comparable stiffness and strength [4–6]. For instance, Alavudeen et al. [7] studied the mechanical properties of hybrid NFP composites like kenaf fiber (KF) and banana/KF hybrid. The mechanical strength of woven banana/KF hybrid composites was found to have increased because of the hybridization effect of kenaf with banana fibers. Impact strength, tensile and flexural properties of the woven hybrid composite of banana/KFs were also superior to those of the individual fibers. Moreover, surface treatments like the sodium lauryl

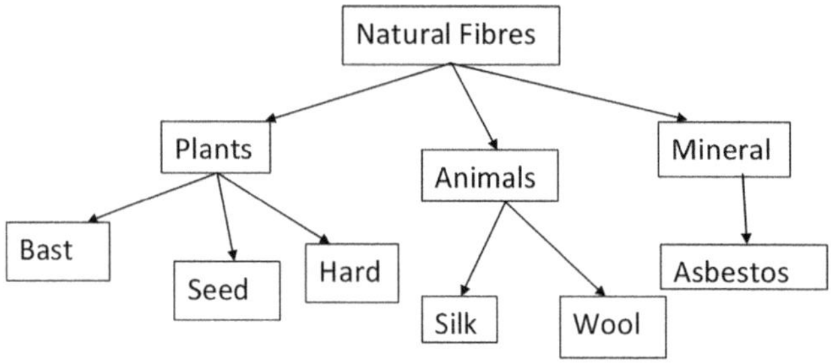

FIGURE 7.1 NF types.

sulfate treatment appeared to provide an additional factor in enhancing mechanical strength through improved interfacial bonding. The advantages of NFs to be used on such a wide scale are their abundance, inexpensive and renewable nature. They are lightweight materials with high toughness, low density, and biodegradable properties, contributing to their environmental application. They are potential alternatives for traditionally used reinforcements with composites for applications that require high strength to weight ratio and further weight reduction. Certain factors affect the mechanical performance of an NFP composite in structural applications and are listed below:

- Fiber selection: The difference in the strength of fibers is highly contingent on the origin of the fibers. Plant-based NFs constitute cellulose, whereas animal-based fibers are composed of proteins. Microfibrils and higher cellulose content that get aligned in the direction of the fiber give higher structural requirements that enhance their performance [8]. Moreover, one has to be selective about property preferences depending on the application of such fibers. For instance, through mechanical characterization silk was found to have a very high strength, but lower stiffness [9].
- Matrix selection: Both thermoset and thermoplastic polymers are used as matrix. The ability of the matrix for load transfer and withstanding adverse environmental and temperature conditions significantly affects their mechanical performance [10,11]. Among thermosets, epoxy resin, polyester, vinyl ester (VE) and phenol formaldehyde resins are commonly used [12].
- Fiber dispersion: Agglomeration is a major challenge in processing fibers, especially with hydrophilic fibers and hydrophobic matrices. The optimum dispersion of fibers is crucial to augment their strength and most other mechanical properties of NFP composites [13]. Increasing weight percentage and length of fibers tend to agglomerate more. Fiber dispersion is vital for interfacial adhesion and preventing delamination failures. Figure 7.2 displays the fractured surface of NF-reinforced VE composite, and matrix fracture caused the overall failure of the composite due to inefficient bonding at the interface [14].

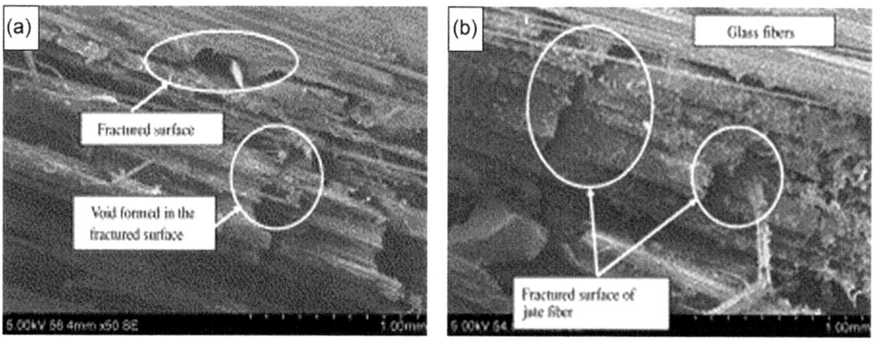

FIGURE 7.2 (a and b) SEM images of fracture surface of NF-reinforced composites [14].

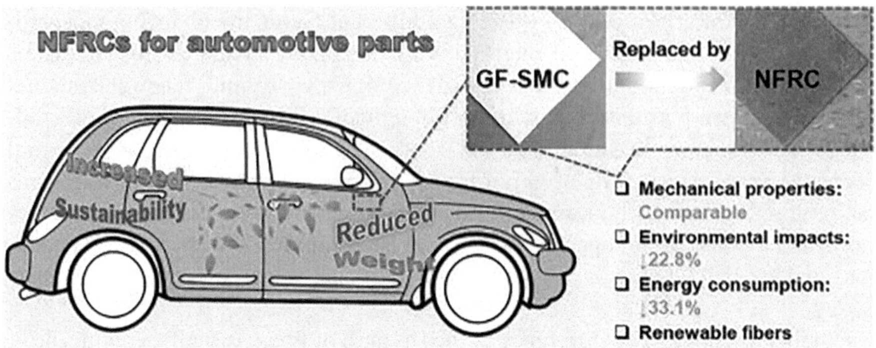

FIGURE 7.3 Advantage of using NFP composites in automotive parts [15].

7.1.3 Natural Fibers and Their Application in Industries

The utilization of inexpensive and lightweight NFs offers an enormous scope in replacing a large segment of mineral fillers and glass fibers in numerous exterior as well as interior manufactured parts. Especially the automotive industry, over the past decade, NFP composites with thermoplastic and thermoset matrices have garnered the European car manufacturers' and suppliers' attention for headliners, panels, seat backs, dashboards, and interiors (Figure 7.3). Hemp, kenaf, flax, jute, and sisal have found their utility in providing automobile parts reinforcement not only for the reasons stated above but also because of their lesser reliance on resources of oil, and recyclability has added value benefit for being eco-friendly.

NFP composites have been thriving because of their robust recyclable and environment-friendly applications compared to synthetic polymer composites. Moreover, with nanofiller infusion, the incompatibility between the matrix and NF is primarily taken care of while improving the surface properties of the polymers. Since the addition of CNT greatly enhances the polymers' tensile properties and interfacial adhesion, it finds its application in automotive and manufacturing industries, because of its high fatigue resistance, superior strength, high fracture toughness to weight ratio, and improved corrosion resistance to extreme environments [16]. Owing to their outstanding stiffness and strength combined with low density and high energy absorption, CNT-reinforced polymers are invaluable in ballistic and armory applications [17].

7.1.4 Limitations

NFs can be highly hydrophilic constituents, absorbing moisture and being incompatible with composite matrices. Poor interfacial adhesion and poor wettability are significant challenges that lead to ineffective performance of NFP composites [18]. The interphase which distinguishes chemically distinct materials may also be subjected to differential thermal expansions, especially in curing processes where microcracking is likely to develop in high-modulus resin systems. The matrix part of a composite is crucial for transferring the load to the stiffer fiber part through shear stresses at their interface. An efficient bond between the fiber and the polymeric matrix at these interfaces is required to carry out the process otherwise the composite becomes vulnerable to environmental exploits resulting in underperformance in mechanical and environmental applications and a reduced life span. This is especially undesirable in manufacturing since longevity of the product is a basis of its progressing economy. To overcome poor adhesion challenges, NFP composites are incorporated with nanofillers.

7.2 IMPROVE INTERFACE/INTERPHASE OF NFPs THROUGH ADDITION OF NANOFILLERS

Nanofillers have been considered good addendums in composites to significantly accentuate the mechanical performance of the materials when they are needed to tailor their performance in different environmental mediums. Numerous attempts are being made to improve the interphase and especially the compatibility of the hydrophilic fibers with the hydrophobic matrix by various strategic surface treatments [20]. The addition of nanofillers has been found to be especially effective considering the enormous interfacial area they create, giving rise to exceptional material properties. Wetting, adsorption, electrostatic interaction, chemical and physical bonding are various controlling mechanisms of fiber/matrix bonding that influence the overall interface properties. Moreover, the bulk behavior of the composite is not just a function of its constituents but is also influenced to a great extent by the existing interface/interphase arising from their bonding nature [21]. The hardness, for example, was found to increase in a study that showed graphene-based nanofiller addition by weight percent to bamboo fibers (Figure 7.4). Corresponding increase in tensile was observed to be 30% for functionalized graphene as shown in Figure 7.5.

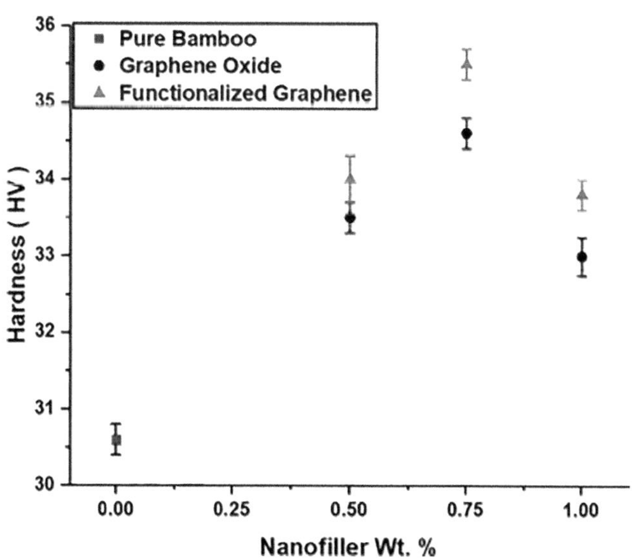

FIGURE 7.4 Shows nanofillers (graphene oxide and functionalized graphene) in bamboo up to 0.75 wt.%; the ascent of the hardness value to the proportionality limit is increased and then decreased [19].

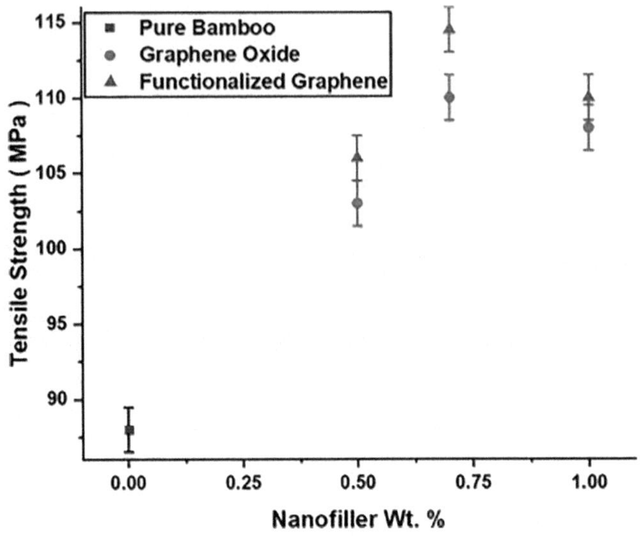

FIGURE 7.5 Shows nanofiller (graphene oxide & functionalized graphene) in bamboo up to 0.75 wt.%, the ascent of the tensile strength value to the proportionality limit is increased and then decreased [19].

The interphase discussion regarding the performance of the polymer is also essential because the region is considered to influence bulk properties. Moreover, since the interphase plays a vital role in the load-bearing abilities of the NFP composite, the dominant failure mode in fiber-reinforced polymers is attributed to the interphase where debonding between fiber and matrix occurs. Nanofillers enable roughening of the fiber surface, reducing stress-transfer length, thus redistributing stress via stress field homogenization through diminishing stress concentration [22,23]. Various types of nanofillers have been tailored with NFP to cater to a wider range of utilities. Their ability to modify the composites' miscibility and morphology depends on their interaction with polymer components, and their structure and composition in the polymer determine their functionality to a great extent [24,25].

7.2.1 Carbon Nanotubes (CNTs): Types, Structure, and Functionalization

CNTs, since their discovery by Iijima in 1991, have revolutionized material science technology as a typical nanofiller in composites for structural and high-end applications. They have given an impetus to mechanical properties like hardness, toughness, interlaminar shear strength (ILSS), and even electrical and thermal conductivity in fiber-reinforced composites and are being utilized for various high-end applications [26,27]. CNTs have shown promising results, mainly because of their inherent structures that result in strong bonding phenomena and conductivity. Such dramatic improvements in the strengths of CNTs-incorporated composites have also been attributed to the directional load-bearing abilities of such nanofillers [28]. Compared to inorganic fibers like glass, aramid, or carbon fibers, carbon nanofibers show better yield strength as their diameter decreases in nanoscale, following the Hall–Petch

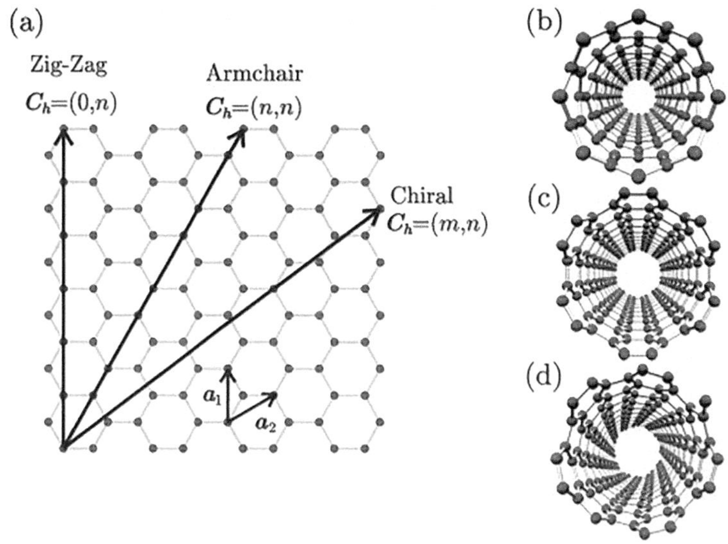

FIGURE 7.6 (a) Geometrical configurations of CNTs; (b) zigzag; (c) armchair; (d) chiral [29].

relationship. Moreover, the nanoscale fillers have a higher contact area to interact with the matrix.

CNTs are made up of rolled-up graphene sheets in concentric cylinders. The macromolecules have their atoms oriented in a hexagonal lattice and are capped at their ends by one-half of a fullerene-like moiety. Figure 7.6 shows the different orientations and modes of folding that classify them into various types, namely armchair, zigzag, or chiral. The electrical conductivity of the nanotubes depends on the nature of folding.

Structurally, CNTs can be classified as multi-walled nanotubes (MWCNTs) and single-walled nanotubes (SWCNTs). SWCNTs are one atom-layer thick, whereas MWCNTs have several concentric layers of carbon atoms. As much as CNTs are widely recognized for their invaluable load-bearing capabilities, their feeble dispersibility due to strong intermolecular bonding poses a challenge in exploring their full potential in polymeric composites. This hinders their processing in the industrial sphere and gives rise to the agglomeration of the CNTs. Over the years, various techniques have been employed to tune CNTs for optimum processing and render them soluble [30]. This includes modifying the surface of CNTs to improve their solubility in solvents and stability. In this regard, the functionalization of CNTs has been conducted extensively. In the presence of strong acids, functionalization occurs by the oxidation of CNTs leading to unrolling of the end caps of CNTs to associate functional groups like ketone, hydroxyl, carboxyl, carbonyl, amine, or ester onto the surface of the CNTs. Functionalized CNTs are then obtained as the final product and it can be seen in Figure 7.7. This functionalization helps to an extent the application in various industries, i.e., biosensors, supercapacitors, hydrogen storage, catalyst, aircraft structure, waste water treatment, and polymer composites due to their superior performance.

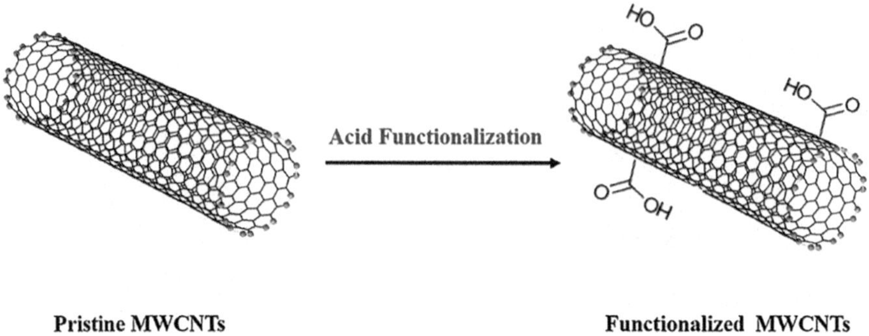

FIGURE 7.7 Acid-functionalized structure of CNTs [31].

In the case of covalent functional groups, free radical grafting promisingly substitutes alkyl groups, anilines, and diazonium salts as larger functionalizing molecules facilitate CNT dispersion in solvents even at low degrees of functionalization [32]. Figure 7.8 shows scanning electron microscope (SEM) images, and Figure 7.9 shows the energy dispersive spectroscopy (EDS) of functionalized CNTs [33]. EDS result shows no strong detection of metal impurities following acid functionalization. However, chemical functionalization that results in sp^2 to sp^3 hybridization may decrease the length-to-diameter ratio modifying the electronic structure of the CNTs leading to bandgap widening, and this widening decreases the electrical and thermal conductivity of the CNTs.

7.3 CNTs AND FUNCTIONALIZED CNTs INFLUENCE ON THE MECHANICAL AND THERMAL PROPERTIES OF NATURAL FIBER (NF)/VE COMPOSITES

The nature of mechanical properties of NFP polymeric composite depends upon the source of fiber, the volume fraction of fiber/polymer matrix, and interfacial adhesion between fiber and polymer. Ghosh et al. [34] have reported that banana fiber-reinforced VE composite containing 35% volume fraction showed 38.65% and 65% increment in tensile strength and tensile modulus, respectively, compared to neat VE. The increment in properties at 35% volume fraction of the fiber ensures good interfacial bonding between fiber and matrix at the interface. Shen et al. [35] have studied the effect of CNT addition and the role of its content in wt.% in a polymeric matrix on the flexural and interlaminar properties of NFPs. They found that CNT at the interface increased flexural strength and modulus, respectively, by 30% without compromising the strain at failure. Figure 7.10 depicts the SEM imaging of short jute fibers (JFs) with incorporated CNTs, denoted as JF-CNT, that appear to have substantially lesser structural defects than pristine JFs. After alkali functionalization the cementing layer between JFs was found to be hydrolyzed. The fractured surfaces are further shown in Figure 7.11 after tensile testing [33].

The addition of CNTs in a superior molding resin has given an impetus to advancements in surface modification, production efficiencies, and inherently

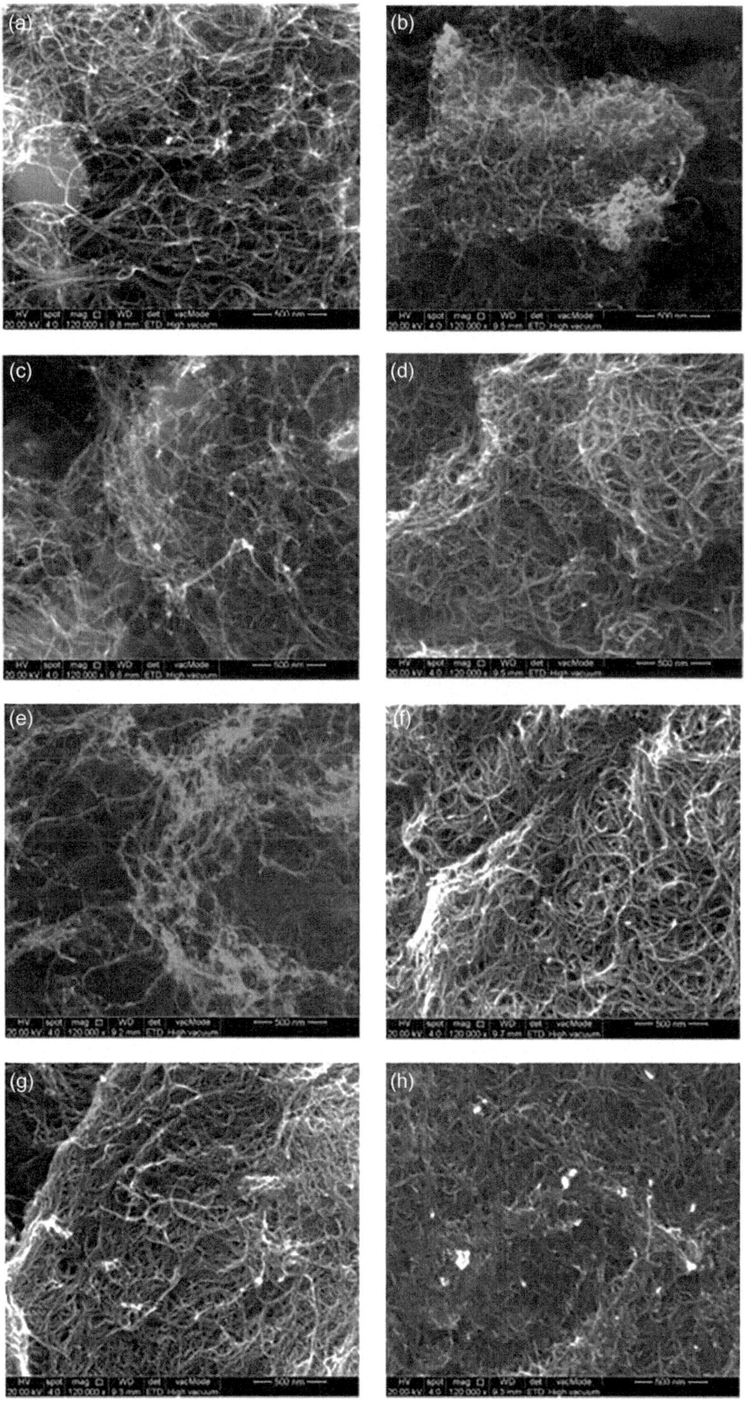

FIGURE 7.8 SEM images of CNTs samples with scale bar of 500 nm: (a) as-synthesized CNTs; (b–h) functionalized CNTs via different acid treatment methods [33].

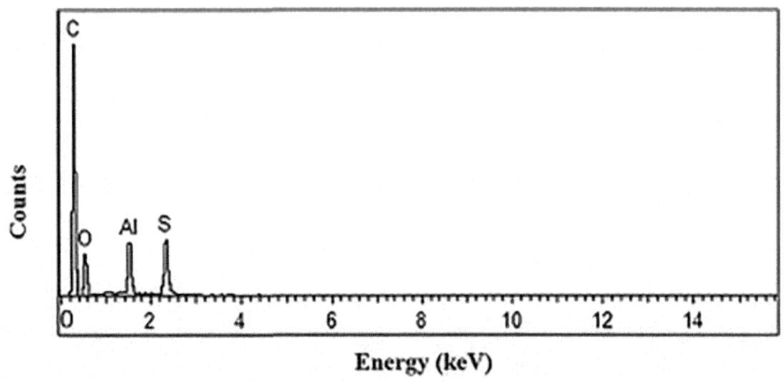

FIGURE 7.9 EDS spectrum of (Sample E) f-CNTs [33].

FIGURE 7.10 (a) SEM images showing the surface morphology of (a) pristine JF (b) and (c) JF-CNT [33].

excellent interfacial features. The use of CNTs which inherently possess a substantially high aspect ratio is an essential factor in determining the toughness of the polymeric system. Shen et al. [35] illustrate that the 0.6 wt.% of CNT content showed 38% improvement in ILSS, which defines efficient stress transfer through the interface. The tremendous surface area of CNT was changed over to an enormous CNT/VE interfacial region with more grounded interfacial holding between the polymeric chain of VE and CNT, promoting efficient stress-transfer decisive pressure move between delicate polymer and firm CNTs (Figure 7.11). However, their homogeneous

FIGURE 7.11 SEM images of the fracture surfaces of NFP composites after tensile testing (a, c, e, and g) showing JF and (b, d, f, and h) CNT-JF and (i) Transmission Electron Microscope (TEM)image of JF-CNT composite [33].

dispersion facilitates stress distributions that reduce stress concentrations and improve the composites' mechanical performance.

Due to the aggregation and solvent chemistry of CNTs, nanomaterials can have their size and shape regulated. With agglomeration and restrictive dispersion took care of by functionalization, NFP composites have bettered at exhibiting superior structural performance. Secondly, the type of bonding of CNTs with the composite constituents and the mechanical load transfer from the surrounding to the CNTs filler greatly influence the strength of the interface. CNTs, in particular, have yielded an increased impact energy absorption value. Their tubular shape especially helps in limiting crack propagation by sustaining tensile strength. Constructions of simulated molecular behavior have shown CNT incorporation an increased impact energy absorption value, particularly in CNTs which contributes toward increased toughening of the resins. Microscopic examinations under SEM have shown that CNTs are responsible for forming networked structures in the matrix that impart an enhanced flexural characteristic to the polymer bends. However, the increased weight percentage of CNTs leads to substantial agglomeration that causes poor dispersibility of the nanofillers. The lumping of CNTs leads to stress concentration areas which prevent stress transfer from the matrix to the nanofillers. Hence, rendering the nanofillers active while dispersion and employing functionalization to counter poor adhesion directly provide a mechanical advantage to the polymers.

The amount of functionalized CNT incorporation also determines the overall performance of the composites. The nature of functionalized groups influences the extent of improvement of microstructural properties of the overall polymer. For instance, thermal analysis of a lower content of CNT functional groups, which form a CNT-coating compared to the higher content of hydroxyl functionalities in JF-reinforced composites, showed a hydrophobic character, the inherent hydrophilicity of the NFs in the composite. Although functionalized CNTs in natural fiber-reinforced VE composites have pointed toward such potential high-end applications, much research has to be done to establish proper modes of processing to tailor the reinforcements. As functionalization is introduced with CNTs, the wettability of the otherwise non-polar NFs improves.

The presence of van der Waals force and hydrogen bond between functionalized CNTs and flax fiber helped CNT-COOH to cover the outer layer of the yarns. Fourier-transform infrared (FT-IR) spectra characterization has shown the presence of OH-stretching of yarns coated with functionalized CNTs (Figure 7.12a). The results show better dispersion abilities of functionalized CNTs and hence better structural reinforcements in the composites. Figure 7.12 shows the FT-IR spectrum treated KF, an excellent NF because of a very high cellulose content.

Russo et al. [37] have studied a number of surface treatment techniques with hydroxyl and carboxyl groups that have resulted in high-performance NFP composites that, through micrographic and thermal analysis, proved to have an enhanced interfacial strength probably because of the inter-entanglement of functional groups used. To not compromise the structural advantage of CNTs, lesser destructive techniques like plasma functionalization have become popular considering their cost-effectiveness in bio-applications. In one such study, the mechanical and thermal behavior of functionalized CNTs in JFs were reported to have enhanced. This was attributed to the formation of a continuous networked CNTs structure onto the polymeric surface, creating a protective layer, reducing the flammability of the nanocomposites.

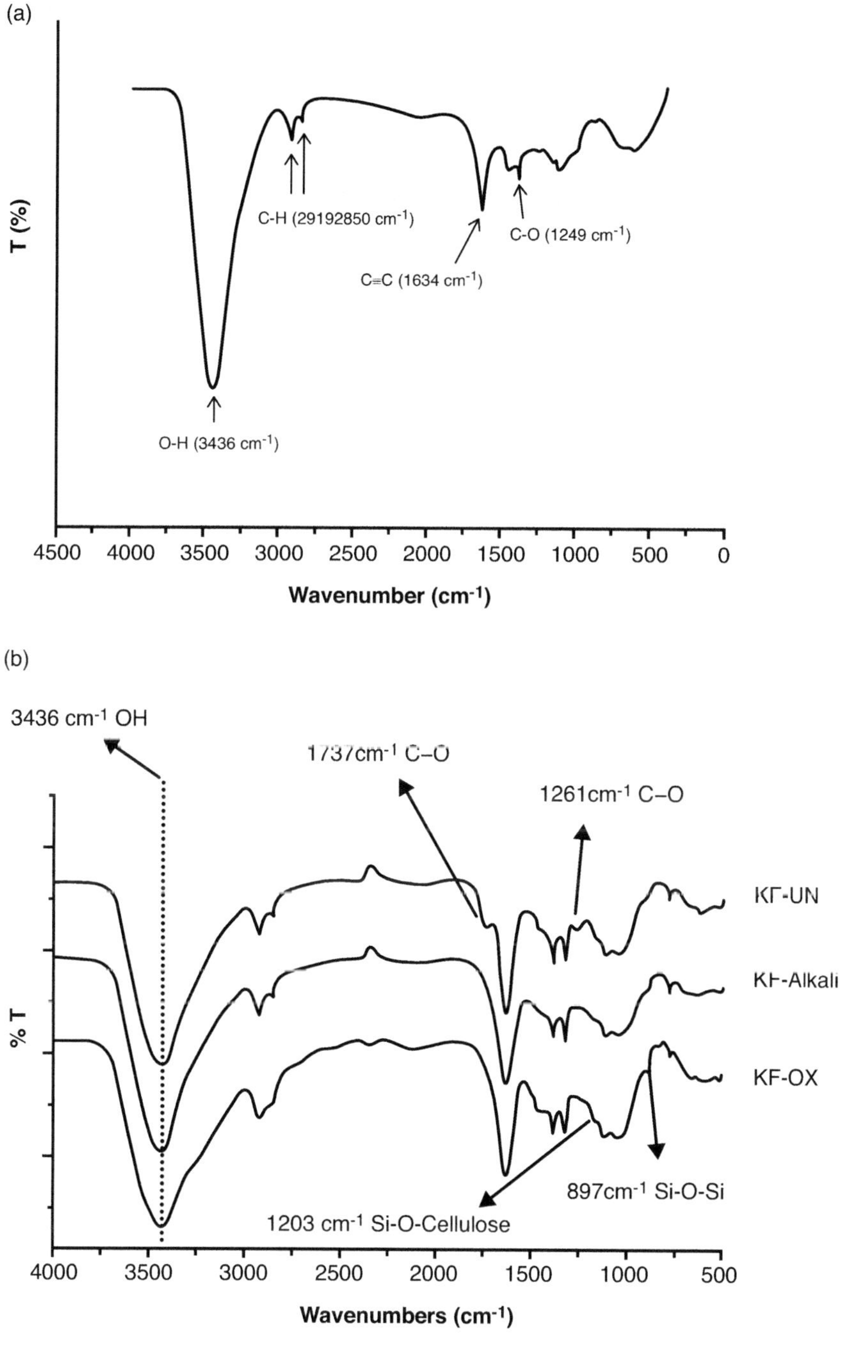

FIGURE 7.12 FT-IR spectrum of (a) surface-modified MWCNTs and (b) raw KF (KF-UN), purified KF (KF-Alkali), and KF modified with OX-silane (KF-OX) [36].

Mazlan et al. have studied the thermal properties of pineapple leaf (PALF), KF, VE resin, and PALF/KF/VE hybrid composites [38]. They have observed that the loss in weight is slower at first due to the vaporization of water molecules in the composites since NFs are known for their notoriety of moisture retention. They also reported that at higher temperatures (250°C–370°C), drastic weight loss occurred due to the decomposition of hemicellulose. Bassyouni et al. have studied thermo-mechanical properties of JF-VE reinforced with CNTs. The results revealed that the addition of CNTs improved the glass transition temperature by 112.4°C [39]. They also reported that addition of CNT in jute fiber/VE composite showed improvement in storage modulus response.

7.4 CONCLUSIONS

This chapter presents an overview of NFP composites, properties, advantages, and limitations. Also, it emphasizes the mechanical and thermal properties of CNT-embedded NFP composites.

- NFs provide the manufacturing and automotive sector impetus, where durability is a primary concern. The advantages of NFs to be used on such a wide scale are their abundance, inexpensive and renewable nature.
- An efficient bond between the fiber and the polymeric matrix at these interfaces is required to carry out the process. Otherwise, the composite becomes vulnerable to environmental exploits resulting in underperformance in mechanical and environmental applications and a reduced life span.
- CNTs have been considered advantageous addendums in composites to significantly accentuate the mechanical performance of the materials when they are needed to tailor their performance in different environmental mediums.
- The addition of CNT greatly enhances the interfacial adhesion strength of the NFP composite, and it finds its application in automotive and manufacturing industries, because of its high fatigue resistance, high fracture toughness, and superior strength to weight ratio.

REFERENCES

1. Natesan K, Kumaresan K, Sathish S, Gokulkumar S, Prabhu L, Vigneshkumar N. An overview: Natural fiber reinforced hybrid composites, chemical treatments and application areas. *Mater Today Proc* 2020;27:2828–34.
2. Hao X, Zhou H, Mu B, Chen L, Guo Q, Yi X, et al. Effects of fiber geometry and orientation distribution on the anisotropy of mechanical properties, creep behavior, and thermal expansion of natural fiber/HDPE composites. *Compos Part B Eng* 2020;185:107778.
3. Symington MC, Banks WM, West OD, Pethrick R. Tensile testing of cellulose based natural fibers for structural composite applications. *J Compos Mater* 2009;43:1083–108.
4. Wongsa A, Kunthawatwong R, Naenudon S, Sata V, Chindaprasirt P. Natural fiber reinforced high calcium fly ash geopolymer mortar. *Constr Build Mater* 2020;241:118143.
5. da Luz FS, da Garcia Filho FC, Del-Rio MTG, Nascimento LFC, Pinheiro WA, Monteiro SN. Graphene-incorporated natural fiber polymer composites: A first overview. *Polymers* 2020;12:1601.

6. Li H, Sain MM. High stiffness natural fiber-reinforced hybrid polypropylene composites. *Polym-Plast Technol Eng* 2003;42:853–62.
7. Alavudeen A, Rajini N, Karthikeyan S, Thiruchitrambalam M, Venkateshwaren N. Mechanical properties of banana/kenaf fiber-reinforced hybrid polyester composites: Effect of woven fabric and random orientation. *Mater Des 1980–2015* 2015;66:246–57. https://doi.org/10.1016/j.matdes.2014.10.067.
8. Mohanty A, Wibowo A, Misra M, Drzal L. Effect of process engineering on the performance of natural fiber reinforced cellulose acetate biocomposites. *Compos Part Appl Sci Manuf* 2004;35:363–70.
9. Shah DU, Porter D, Vollrath F. Can silk become an effective reinforcing fibre? A property comparison with flax and glass reinforced composites. *Compos Sci Technol* 2014;101:173–83.
10. Holbery J, Houston D. Natural-fiber-reinforced polymer composites in automotive applications. *JOM* 2006;58:80–6.
11. Bakri MKB, Jayamani E, Hamdan S, Rahman M, Kakar A. Potential of Borneo Acacia wood in fully biodegradable bio-composites' commercial production and application. *Polym Bull* 2018;75:5333–54.
12. Raja T, Anand P, Karthik M, Sundaraj M. Evaluation of mechanical properties of natural fibre reinforced composites: A review. *Int J Mech Eng Technol* 2017;8:915–24.
13. Kureemun U, Ravandi M, Tran L, Teo W, Tay T, Lee H. Effects of hybridization and hybrid fibre dispersion on the mechanical properties of woven flax-carbon epoxy at low carbon fibre volume fractions. *Compos Part B Eng* 2018;134:28–38.
14. Ramesh M, Palanikumar K, Reddy KH. Mechanical property evaluation of sisal–jute–glass fiber reinforced polyester composites. *Compos Part B Eng* 2013;48:1–9. https://doi.org/10.1016/j.compositesb.2012.12.004.
15. Wu Y, Xia C, Cai L, Garcia AC, Shi SQ. Development of natural fiber-reinforced composite with comparable mechanical properties and reduced energy consumption and environmental impacts for replacing automotive glass-fiber sheet molding compound. *J Clean Prod* 2018;184:92–100. https://doi.org/10.1016/j.jclepro.2018.02.257.
16. Rajak DK, Pagar DD, Kumar R, Pruncu CI. Recent progress of reinforcement materials: A comprehensive overview of composite materials. *J Mater Res Technol* 2019;8:6354–74.
17. Benzait Z, Trabzon L. A review of recent research on materials used in polymer–matrix composites for body armor application. *J Compos Mater* 2018;52:3241–63.
18. Kalia S, Kaith B, Kaur I. Pretreatments of natural fibers and their application as reinforcing material in polymer composites—a review. *Polym Eng Sci* 2009;49:1253–72.
19. Khatua SK, Sahoo PK, Kumari K, Srivatsava M, Sahu D, Dalai N. Behavioural study of graphene oxide/functionalized graphene on bamboo fiber reinforced composite. *Mater Today Proc* 2021;47:3633–6. https://doi.org/10.1016/j.matpr.2021.01.100.
20. Pommet M, Juntaro J, Heng JY, Mantalaris A, Lee AF, Wilson K, et al. Surface modification of natural fibers using bacteria: Depositing bacterial cellulose onto natural fibers to create hierarchical fiber reinforced nanocomposites. *Biomacromolecules* 2008;9:1643–51.
21. Haghdan S, Smith GD. Natural fiber reinforced polyester composites: A literature review. *J Reinf Plast Compos* 2015;34:1179–90.
22. Xiong X, Shen SZ, Hua L, Liu JZ, Li X, Wan X, et al. Finite element models of natural fibers and their composites: A review. *J Reinf Plast Compos* 2018;37:617–35.
23. Rowell RM, Sanadi AR, Caulfield DF, Jacobson RE. Utilization of natural fibers in plastic composites: Problems and opportunities. *Lignocellul-Plast Compos* 1997;13:23–51.
24. Jawaid M, Bouhfid R. *Nanoclay Reinforced Polymer Composites: Natural Fibre/Nanoclay Hybrid Composites*. Singapore: Springer; 2016.

25. Sen B, Fulmali AO, K BNVSGG, Prusty RK, Ray BC. A study of the effect of carbon nanotube/nanoclay binary nanoparticle reinforcement on glass fibre/epoxy composites. *Mater Today Proc* 2020;26:2026–31. https://doi.org/10.1016/j.matpr.2020.02.440.
26. Nurazzi N, Sabaruddin F, Harussani M, Kamarudin S, Rayung M, Asyraf M, et al. Mechanical performance and applications of CNTs reinforced polymer composites—A review. *Nanomaterials* 2021;11:2186.
27. Ferreira FV, Pinheiro IF, de Souza SF, Mei LH, Lona LM. Polymer composites reinforced with natural fibers and nanocellulose in the automotive industry: A short review. *J Compos Sci* 2019;3:51.
28. Sapiai N, Jumahat A, Mahmud J. Flexural and tensile properties of kenaf/glass fibres hybrid composites filled with carbon nanotubes. *J Teknol* 2015;76:115–120.
29. Basheer BV, George JJ, Siengchin S, Parameswaranpillai J. Polymer grafted carbon nanotubes—Synthesis, properties, and applications: A review. *Nano-Struct Nano-Objects* 2020;22:100429. https://doi.org/10.1016/j.nanoso.2020.100429.
30. Barhoum A, Bechelany M, Makhlouf ASH. *Handbook of Nanofibers.* Wollerau: Springer; 2019.
31. Jun LY, Mubarak NM, Yon LS, Bing CH, Khalid M, Abdullah EC. Comparative study of acid functionalization of carbon nanotube via ultrasonic and reflux mechanism. *J Environ Chem Eng* 2018;6:5889–96. https://doi.org/10.1016/j.jece.2018.09.008.
32. Islam M, Rahman MJ, Mieno T. Safely functionalized carbon nanotube–coated jute fibers for advanced technology. *Adv Compos Hybrid Mater* 2020;3:285–93.
33. Tzounis L, Debnath S, Rooj S, Fischer D, Mäder E, Das A, et al. High performance natural rubber composites with a hierarchical reinforcement structure of carbon nanotube modified natural fibers. *Mater Des* 2014;58:1–11. https://doi.org/10.1016/j.matdes.2014.01.071.
34. Ghosh R, Reena G, Krishna A, Raju BL. Effect of fibre volume fraction on the tensile strength of Banana fibre reinforced vinyl ester resin composites. *Int J Adv Eng Sci Technol* 2011;4:89–91.
35. Shen X, Jia J, Chen C, Li Y, Kim J-K. Enhancement of mechanical properties of natural fiber composites via carbon nanotube addition. *J Mater Sci* 2014;49:3225–33.
36. Chen P-Y, Lian H-Y, Shih Y-F, Chen-Wei S-M, Jeng R-J. Preparation, characterization and crystallization kinetics of Kenaf fiber/multi-walled carbon nanotube/polylactic acid (PLA) green composites. *Mater Chem Phys* 2017;196:249–55. https://doi.org/10.1016/j.matchemphys.2017.05.006.
37. Russo P, Vitiello L, Sbardella F, Santos JI, Tirillò J, Bracciale MP, et al. Effect of carbon nanostructures and fatty acid treatment on the mechanical and thermal performances of flax/polypropylene composites. *Polymers* 2020;12:438.
38. Mazlan A, Sultan M, Md Shah A, Safri S. Thermal properties of pineapple leaf/kenaf fibre reinforced vinyl ester hybrid composites. *IOP Conf Ser Mater Sci Eng* 2019;670:012030. https://doi.org/10.1088/1757-899X/670/1/012030.
39. Bassyouni M, Abdel-Hamid S, Abdel-Aziz MH, Zoromba MS. *Characterization of Vinyl Ester/Jute Fiber Bio-Composites in the Presence of Multi-Walled Carbon Nanotubes.* vol. 730, Wollerau: Trans Tech Publ; 2017, pp. 221–5.

8 Vinyl Ester-Based Biocomposites

Influence of Nanoclay on Thermal and Mechanical Properties

Santhosh N.
MVJ College of Engineering

Anand G.
Achariya College of Engineering Technology

CONTENTS

8.1 INTRODUCTION

Biocomposites comprises one or more constituents or phases of natural origin, i.e., either matrix or reinforcement phase. Further, when it comes to reinforcement, this can include plant fibres, viz., flax, kenaf, jute, hemp, cotton, and flax or fibres made from recycled wood. By-products of used paper and food crops may potentially be utilized as reinforcements. According to Fowler et al. (2006), agricultural by-products have garnered more attention in recent years. Alternative fillers are cheap and abundant. Agricultural by-products, depending on the polymer, can be used in thermoplastic and thermosetting polymer matrices. Composite material manufacturing is heavily dependent on the material's intended purpose. In recent years, the growing usage of agricultural filler reinforcements in thermoplastics and thermosetting resins, as well as the adoption of these composites in a diverse application, has stimulated research in the field of vinyl ester composites. Natural fibres of various

DOI: 10.1201/9781003270997-8

sorts are commonly employed in biocomposites. The polymer material is added to improve texture and finish and further improve the capacity or reduce production costs (Ibrahim et al., 2012).

The evolution of biocomposites has gained significant momentum due to the need for newer approaches for synthesizing biocomposites. There are numerous methods for making biocomposites; however, including agricultural fibres into the composite is the most cost-effective approach. The use of jute fibres, banana fibres, pineapple leaf fibres, etc., is ever-increasing, owing to enhanced performance capabilities of the biocomposites. The matrix, on the other hand, plays a crucial part in improving the composites' performance capabilities; hence, optimal polymer selection is an important aspect of biocomposite synthesis. Among various types of polymers, vinyl ester thermosetting plastics are often used in the development of advanced composites due to their excellent properties, chemical resistance, rigidity, dimensional stability, and higher strength than polyester resin, Also, it is reported that it costs less than epoxy (Suresha and Shiva Kumar, 2009) and is easy to handle at room temperature. It has the same mechanical properties as epoxy resin, especially hydrolysis stability. In addition, the cure rate and response can be better controlled when the biocomposites are synthesized using vinyl esters. Since vinyl ester resin is fragile, one way to improve its performance while lowering its cost is to reinforce it using fillers (Ku et al., 2012). Because the structural items are cast to shape, the best choice for strengthening them is to combine the vinyl ester resin with particle fillers. Among various types of fillers, nanoclay is one among them which can improve the characteristics of the composites.

There are several types of fillers from renewable as well as nonrenewable resources. Biomass, which is carbonaceous in origin and rich in organic matter, is one of the fillers generated from renewable sources. Because lignocellulosic fibres have a high fixed carbon content, they can yield biocarbon (carbon black) after pyrolysis or carbonization (Abdul Khalil et al., 2007). Nowadays, experts have produced technological improvements in product materials. Use of carbon black as a nanofiller is also an important aspect of research that has paved way for enhanced mechanical characteristics among the biocomposites. Carbon black and the procurement of biomass as a filler for composite materials such as bamboo (Abdul Khalil et al., 2010; Onyeagoro, 2012; Acharya and Samantarai, 2012), coconut shell, oil palm empty bundle (Abdul Khalil et al., 2010), and rice husks have all been studied extensively (Acharya and Samantarai, 2012). These fillers are commonly used in composite materials to reduce costs, increase workability and mechanical properties, and improve flame resistance and electrical properties. Filler concentration, interaction with the matrix, filler size and shape, and filler dispersion all influence composite characteristics (Abdul Khalil et al., 2013). Various plant parts containing high carbon content, including seeds, seed cakes, shells, and seed coatings, can be utilized as a filler for Jatropha. In their study, carbon black filler derived from Jatropha seed coat was shown to be suitable. Vinyl ester composite composites require it as a filler. Researchers focus on the characteristics of carbon black-loaded vinyl ester composites with various load factors. The Jatropha seed coat version was compared to the unfilled vinyl ester version. The effect of carbon black filler on the mechanical and thermal properties of composite materials, as well as their shape, was examined. Furthermore, the properties of the carbon black made from biomass that was employed as a filler in the composites were investigated.

Furthermore, it is critical to comprehend the significance of thermoplastic materials in composite material production. Polymers have supplanted traditional materials in a variety of applications over the last few decades. This came about because of the advantages that polymers offer over traditional substances. The ease of processing, increased production, and lower cost are all advantages of polymers (Saheb and Jog, 1999). To lower the high modulus needs in maximal applications, many polymers have been produced with the help of fillers and/or fibres (Omrani et al., 2016). A polymer matrix composite is made out of a polymer matrix with fibres inserted in it, as well as glass, aramid, and carbon. Vinyl monomers, which are small molecules having carbon–carbon double bonds, are also utilized in the production of vinyl polymers and composites (Mallakpour and Zadehnazari, 2013). They make up the majority of polymer composition.

Vinyl polymers are classified into three types: thermoplastic, thermosetting (vinyl ester), and elastomer (Saba et al., 2014). Thermoplastic polymers are the most often utilized matrices for biofibres nowadays. The most popular thermoplastics used for this purpose are polypropylene, polyethylene, and polyvinyl chloride. In the meantime, thermosetting matrices such as phenolic, epoxy, and polyester resins are widely used (Puglia et al., 2005).

Fibre Reinforced Polymer (FRP) have exceedingly superior mechanical properties as compared to traditional materials. Therefore, several trials were made to make use of polymers in unique industrial applications. Moreover, numerous kinds of reinforcements together with fibres are incorporated into polymers to improve their performance capabilities. Lightweight, biodegradable, high stiffness, corrosion resistance, and low coefficients of friction make FRP matrix composites very appealing. These are crucial mechanical and tribological features found in a wide range of equipment, from automobiles to spacecraft. Hence, there is a need for synthesizing biodegradable polymer composites and the vinyl ester-based biocomposites are one such combination of the composites for greater performance capabilities.

In this context, the current analysis aims to assess the performance capabilities of polyester-based biocomposites as well as characterize the effect of fillers, particularly nanoclay, on thermal and mechanical parameters.

8.2 NATURAL FIBRES FOR VINYL ESTER-BASED BIOCOMPOSITES

As nonrenewable resources are becoming scarcer, there is a widespread understanding of the reinforcements and matrix. As a result, unique natural fibres that can improve the properties of composites must be investigated. Natural fibres can be used in three different ways to make biocomposites: textiles, paper, and cloth (Habibi et al., 2008). In select applications involving composite materials used in the automotive, construction, and packaging sectors, natural fibres can finally replace glass fibres as a reinforcing fabric. Natural fibres are categorized according to their origin, such as lignocellulosic materials, animal natural reinforcements, and mineral sources. Lignocellulosic fibres are also known as cellulose and are made up of both wood and non-wood or plant fibres. Compounds including cellulose, hemicellulose, lignin, and pectin make up plant fibres (Vaisanen et al., 2016). The relative content of elements can be used to approximate the features of these fibres. Seed, leaf, bast or stem, fruit,

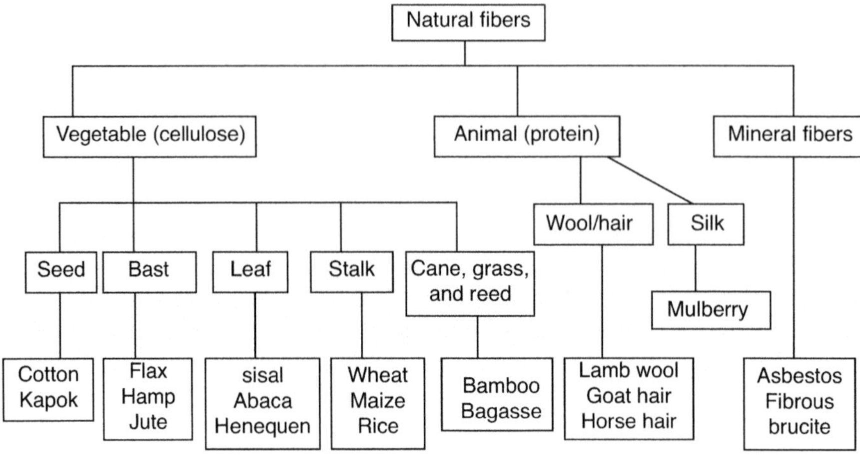

FIGURE 8.1 Classification of natural fibres (Jusoh et al., 2016).

and stalk fibres are all types of non-wood lignocellulosic fibres. Bast is the source of the majority of commercial fibres (e.g., hemp, flax, kenaf, and jute). These fibres come from the phloem that surrounds the stem and is present in plants, and they aid in the stability of high-strength fibres. Fibres from leaves (such as sisal) are, on the other hand, one of the most common low-strength reinforcements. Figure 8.1 shows a few examples of natural fibres.

There are numerous materials reinforcements that are used for synthesis of vinyl ester composites and numerous researchers throughout the world have accomplished their research purview in the area of vinylesters – natural fibre composites and the hybrid vinyl ester composites.

Plants and vegetation can provide natural cellulosic fibres. The most frequent natural reinforcements are seed (e.g., cotton and kapok), stem or bast (e.g., flax, jute, hemp, kenaf, and sugarcane), and leaf (e.g., pineapple and banana) (Namvar et al., 2014). Plants grown primarily for fibre (e.g., cotton, flax, hemp, and kenaf) or for other reasons (e.g., coconut (the strands are frequently referred to as "coir"), sugarcane, banana, and pineapple) can provide these reinforcements. However, due to limited accessibility, extraction issues, execution-related features, and limited growing areas, a few fibres are no longer widely used. Furthermore, some plants produce several types of fibre. Natural fibres such as jute, flax, hemp, and kenaf have all been frequently used. Fruit and stem fibres are found in agave, coconut, and oil palm, whereas stem and hull fibres are found in cereal grains. Furthermore, bast fibres generated from the interior bark or phloem of dicotyledonous vegetation contribute to the textural rigidity and stiffness of the plant stem. These fibres are taken from a thin bark and emerge as fibre bundles or strands while the length of the stem is parallel. Bast strands typically range in length from 1 to 100 cm and have a diameter of 1 mm.

Hemp is a plant fibre originating from the *Cannabis* genus. Hemp is also thought to be the world's earliest cultivated fibre plant. Within five months, the plants reach a height of 5 m and a stem diameter of 425 mm. Hemp is the source of two types of natural fibres: bast and core hurds from hardwood sources.

Hemp stem has 22%–42% bast fibres and 58%–78% core hurds by weight. The hurds include 42 wt.% cellulose, 19 wt.% hemicellulose, and 22 wt.% lignin (Shahzad, 2012). Aside from that, bast fibres have higher levels of polyose (57%) and hemicellulose (9%), as well as 5%–9% lignin.

The blooming plant ramie (*Boehmeria nivea*) is primarily found in Asia. It is a long-lived plant that matures in 50 days in a warm and humid region. Ramie fibres can be discovered on the stem's bark layer, especially beneath a thin coating of bark. 72% cellulose, 14% hemicellulose, 1.3% lignin, and 4.5% pectin make up this fibre (Nandi et al., 2015). The sticky quality of the ramie plant's bark prevents fibres from being entirely separated from the bark. Pounded, scraped, heated, washed, and chemical actions are some of the separation procedures. Furthermore, ramie stalks can be removed with the use of a revolutionary decorticating apparatus created in Japan (Das et al., 2010).

Flax is a member of the Linaceae family and is utilized in natural fibre composites. The flax plant can reach a height of 177 cm and a diameter of 1.6 cm. Flax is produced for the purpose of extracting natural fibres from composites and is harvested after about a hundred days or when the plant's bottom turns yellowish. Linseed oil is produced from the seeds of the plant. The flax plant stem is made up of several layers of fibre that are bound together and can be difficult to separate. The bast fibres and the inner bark are separated. Flax fibre is mostly composed of cellulose, hemicellulose, wax, lignin, and pectin. Cellulose makes up up to 70% of the total number of components in flax fibre. Flax can now be utilized as a composite reinforcement. Flax also has 20% hemicellulose content, 10% pectin, and 10% lignin content (Yan et al., 2014). With values ranging from glass to aramid fibres, it boosts tensile strength and modulus while reducing elongation. The drawn out strands are used in sewing, yarn and weaving, and geo-fabric. Flax fibres are traditionally used to make household fabrics and clothing. Flax fibres are also used in nontextile markets such as packaging, asbestos replacement reinforcements, panel boards, plastics and concrete, automotive lining materials, and insulation. Flax is now being studied for use in a range of industries, including automotive, aerospace, and biomedical (Maity et al., 2014). Flax is becoming more popular for a range of applications. Flax is easier to grow than other herbal fibres since it doesn't require special soil conditions or herbicides, and it uses a lot less water.

Hibiscus cannabinus L., popularly known as kenaf, is a cellulosic plant that has both commercial and environmental benefits. It's a tropical textile crop related to both cotton and jute. The stem is made up of an outer bark, bast strands, and a vast focal variety of middle or stick fibres. Kenaf stalks have the same cross-section as hemp stalks. By retting, kenaf bast strands are released from beneath the stem's thin bark layer. Kenaf has extraordinary mechanical properties and, because it develops in 150 days, it may broaden suddenly. This plant has 34% fibre content and 61% core fibre content that is consistent with stalk weight. Kenaf fibre, on the other hand, has a cellulose content of 65.1% and 21.1% lignin and pectin. In humid climates, the plant can reach a height of 2.9 m and a base diameter of 34 cm. The stem of the kenaf plant is made up of two parts: exterior fibrous bark and internal woody centre. Kenaf is being employed in a wide range of novel applications, such as paper, construction products, absorbents, and animal feeds. A form of addition also uses bast fibre strands for increased characteristics. This fibre is utilized in automotive dashboards, carpet padding, corrugated media as a fibreglass and other artificial fibre substitute, textiles, and injection-moulded and extruded plastic fibres (Raman Bharath et al., 2015). Kenaf bast fibre strands are also utilized commercially in a number of environmentally friendly goods, such as fibre mats, to minimize soil erosion caused by water and wind.

8.3 VINYL ESTERS FOR BIOCOMPOSITES

Vinyl polymers are the largest group of the polymers used in composites. Vinyl monomers, which are small molecules with carbon–carbon double bonds, are used to make vinyl polymers. The process of connecting unsaturated monomers to generate chain polymers is known as vinyl polymerization (Hibi et al., 2016). Vinyl is the most versatile material due to its important recycling characteristic, which is required to achieve zero pollution. Vinyl items are great because they can be recycled several times throughout the course of their useful life. Vinyl esters are recycled in large quantities thanks to polymer makers and processors. Post-industrial recycling converts the majority of synthetic vinyl compounds into finished goods. The rise in recycling packaging is a major factor in the production of vinyl polymers, which supports the growth of post-consumer vinyl recycling. Vinyl esters aid in the development of automated plastic separation and vinyl recycling, which is critical to the success of plastic recycling. Vinyl is separated through the catalytic action of chlorine in the chemical makeup of vinyl. Natural or plant fibre-reinforced vinly ester composites can be made using hand lay-up and spray-up techniques. After waxing and coating the mould with gel coat, the laminate is cured in a heated oven. The continuous fibre mat and fabric are manually positioned within the matrix phase, with catalysed resin sprayed on each layer in the hand lay-up process. Following that, post-processing is done to make a robust and tight laminate. The catalysed resin is sprayed into the mould first, and then the chopped fibre is embedded. To make a laminated composite, the latter connects the laminates together. Resin Transfer Moulding (RTM) is a technique for producing high-strength composites (Ho et al., 2012). This method's low strength enables for the most efficient composite designs. The matrix phase and chopped fibre are then spread over both half of the mould, followed by gel lining. Following the mould's foundation preparation, the resin is applied using either injection or vacuum techniques. This method's curing temperature is determined by the resin system. Compression moulding is a low-cost moulding method with a short cycle time. Sheet moulding chemicals can also be used to achieve the moulding procedure (Sheet Moulding Compound (SMC)). By using this sheet, the fibre is sandwiched between the plies of resin paste in the centre.

8.4 INFLUENCE OF NANOCLAY

In the recent times, there is a drastic improvement in the study interests of polymer nanocomposites with nanoclay fillers (Adekunle, 2015); particularly the vinyl ester thermosetting resins that are desired as matrix alongside the nanoclay fillers are making up for most of the research avenues in polymer-based nanocomposites. The vinyl ester, due to its bisphenol diepoxide content, famous for its excellent chemical, corrosion resistance, tensile and flexural properties in addition to excessive elongation, is excessively used in real-time applications ranging from automotive components to aerospace parts. Therefore, vinyl esters are broadly used for sheet moulding, pultrusion, filament winding, and resin transfer processes to fabricate FRP pipes, tanks, automotive parts, components of huge ships and construction materials subjected to excessive static and dynamic loads, machines and electric elements, vessels, spacecrafts, and windsurf, wherein the fabric selection is an important aspect (Aisyah and Tajuddin, 2014). Researchers have carried out extensive investigations to enhance the tensile, thermal, and electrical characteristics (Ahmad et al., 2012; Anwer and Bhuiyan, 2012; Ashori and Nourbakhsh, 2010; Carada et al., 2016; Carvalho et al., 2010; Cecen et al., 2009; Debnath et al., 2013; Gupta and Kumar 2012; Gupta et al., 2015; Han and Choi, 2010; Hassan and Wagner, 2016; Husseinsyah and Mostapha, 2011; Jamil et al., 2006; Kabir et al., 2011; Kumar et al., 2011; Li et al., 2007; Melo and Santos, 2009; Mengeloglu and Karakus, 2008). Nanoparticles together with nanoclay, nano $CaCO_3$, CNT, CuO, FeO, and Al_2O_3 have been embedded in resin to enhance mechanical characteristics (Moghaddam and Mortazavi, 2016; Mohammed et al., 2015; Obayi et al., 2008; Ojha et al., 2012). These nanoparticles scattered throughout the matrix generate composites with better characteristics, and hence play an important role in limiting polymer chain mobility under stress (Osita et al., 2016; Pandey, 2007; Paridah et al., 2011). As a result, the mechanical and thermal conductivity of clay nanocomposites is projected to dramatically improve. As a result, it is clear that a processing procedure that results in complete exfoliation and appropriate dispersion of clay debris is necessary for the development of strength in these nanocomposites (Ray et al., 2004). Among the nanoparticles, special interest has been paid to clay within the subject of nanocomposites. Clays (layered silicates) are determined to be one of the most appropriate nano reinforcements for polymers, due to their excessive intercalation chemistry, excessive factor ratio, ease of availability, and occasional value (Guo et al., 2008). Rosas et al., 2011, have worked on the influence of three wt.% treated Na+ montmorillonites nanoclay-strengthened epoxy nanocomposite on elasticity and tensile strength and have reported that there is a drastic improvement of 88% and 21%, respectively (Smith and Yeomans, 2009). Stevulova et al. (2014) studied the elastic modulus and strength of vinyl ester matrix nanocomposites with the addition of 1, 2, and 3 wt.% of 40 nm nanoparticles. Sepet, H., Tarakçıoglu, N., 2014, discovered that adding 4 wt.% nanoclay to vinyl ester composites boosted their elasticity by 23%. The modulus of nanoclay/vinyl ester nanocomposites will grow monotonically with increasing clay concentration, according to Tanushree, 2016. However, the change in modulus is dependent on the clay contents (up to 10 wt.% nanoclay). Sepet and Tarakçıoglu, 2014, pronounced that the mechanical performance, wear characteristics, and thermal characteristics of polymeric substances improve with the inclusion of layered silicates in the composites. The intention of this review is to identify the impact of the addition of montmorillonites nanoclay debris on the conduct of vinyl ester nanocomposites and their mechanical and thermal characteristics.

Sepet and Tarakçıoglu (2014) thoroughly investigated the effects of nanoclay on mechanical and thermal properties. They have studied the stress–strain behaviour of the nanocomposites reinforced with nanoclay. The strain to failure lowers with the addition of nanoclay to the vinyl ester, as can be seen in the Figure 8.2. Furthermore, the nanoclay-filled composites have 1.55 times the tensile strength of the clean vinyl ester composites. The significant improvement in tensile strength is due to consistent

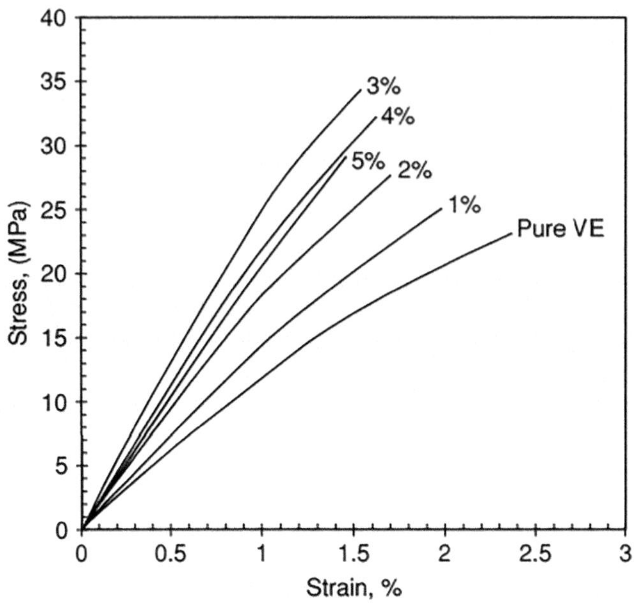

FIGURE 8.2 Stress–strain curve for different wt.% of nanoclay (Sepet and Tarakçıoglu, 2014).

clay dispersion and exfoliation, which increases bonding and tensile strength characteristics (Stevulova et al., 2014, and Webber et al., 2002). However, due to the presence of clay, the elastic modulus falls with clay content. Aggregation and void forms can be blamed for the decrease in elastic modulus with increasing clay content (Westman et al., 2010). Figure 8.2 shows the stress–strain curve for various nanoclay loading percentages, while Figures 8.3 and 8.4 illustrate the tensile strength and young's modulus for various nanoclay loading percentages, respectively.

Sepet and Tarakçıoglu, 2014, conducted extensive research on how the impact strength of composites falls substantially as the weight loading of nanoclay increases.

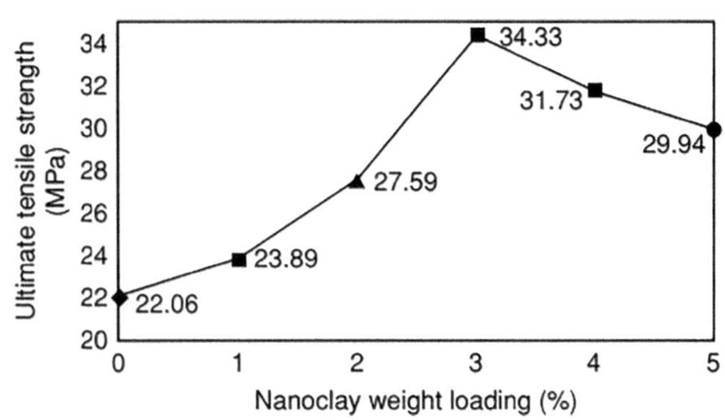

FIGURE 8.3 Ultimate tensile strength for different wt.% of nanoclay (Sepet and Tarakçıoglu, 2014).

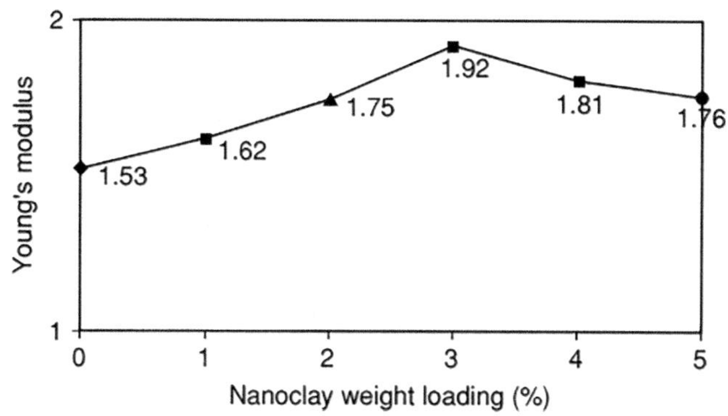

FIGURE 8.4 Young's modulus for different wt.% of nanoclay (Sepet and Tarakçıoglu, 2014).

This is due to the fact that the matrix gets more brittle as the clay component in composites increases. Micro-coring and agglomeration, which result in the creation of micro-voids, are primarily responsible for the decrease in impact strength characteristics (Liu and Wu, 2001). Figure 8.5 depicts the impact strength fluctuation for various nanoclay loading weight percents.

Differential scanning calorimetry curves for different weight percents of nanoclay are used to investigate the thermal behaviour of vinyl ester-based biocomposites. Figure 8.6 shows the glass transition temperature of vinyl ester biocomposites (T_g). From the graph, it is evident that the T_g for pure vinyl ester is 99.8°C, while the T_g value increases up to 101.5°C for 3 wt.% of nanoclay, beyond which the T_g value drastically decreases owing to localized accumulation of the filler material in the matrix phase. The T_g value of pure vinyl ester is 99.7°C. For the nanoclay-reinforced vinyl ester nanocomposites, at low concentration of nanoclay addition, T_g increased up to 101.3°C with 3 wt.% addition of nanoclay. With a 5 wt.% addition of nanoclay, the T_g value dropped to 95.8°C, which was attributed to aggregation of nanoclay content at some localized places and void formation at other parts.

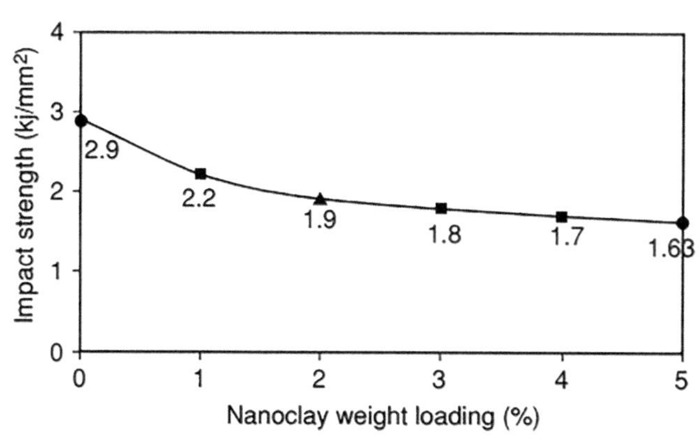

FIGURE 8.5 Impact strength for different wt.% of nanoclay (Sepet and Tarakçıoglu, 2014).

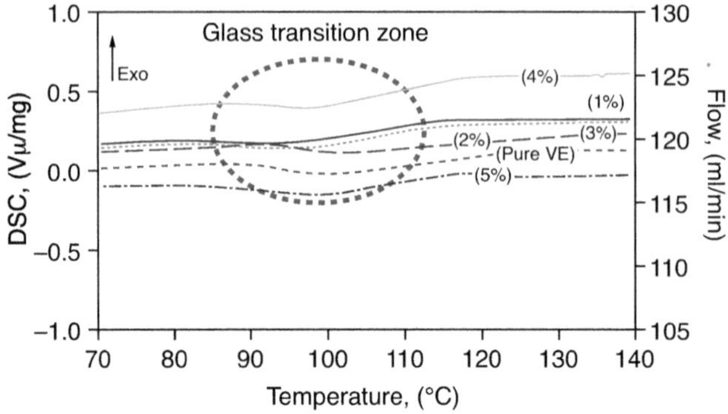

FIGURE 8.6 Glass transition zone for different wt.% of nanoclay (Sepet and Tarakçıoglu, 2014).

Nanoclay-filled roselle fibres-reinforced vinyl ester biocomposites were studied thermally by Nadlene et al. (2016). The nanoclay comprises 18 wt.% octacylamine and 3 wt.% aminopropyl-triethoxy-silane. The findings reveal that silane coupling agents in the matrix phase improve the composites' thermal properties, resulting in a higher glass transition temperature (T_g). This affects the thermal stability of roselle fibre-reinforced vinyl ester composites.

8.5 APPLICATION OF VINYL ESTER BIOCOMPOSITES

Natural fibre reinforced vinyl ester polymer composites with nanoclay fillers have been widely used in a range of technological applications because of their low cost. Automotive interiors, aircraft components, and many more applications use vinyl ester polymer composites with nanoclay fillers. Natural fibres in a viny ester matrix are a cost-effective and environmentally friendly alternative to non-degradable composite disposal (Józefiak, and Michalczyk, 2020).

Plant-based fibres such as lignocellulosic fibres were chosen for usage in polymer composites because of their unique properties. Plant-based fibres include abaca, kenaf, and jute, to name a few. Cost, availability, nonabrasiveness, low density, acceptable mechanical characteristics, and decomposability are all advantages of plant-based fibres (Nagaprasad et al., 2020).

While these fibres aren't as stout as artificial fibres (such as carbon fibre), the stiff conduct of these fibres outperforms the stiffness of E-glass fibres in terms of mechanical properties. Natural fibre stiffness, nonabrasive conduct of nanoclay, and superior properties of vinyl ester matrix all lead to the adoption of vinyl ester composites for multidomain applications.

Natural fibres that are utilized in vinyl ester biocomposites must be biodegradable, and they can perform as an excellent insulator in thermal and acoustic applications (Al-Oqla and Sapuan, 2014). Because of its insulating characteristics, plant fibre-reinforced composites are used in a range of indoor automobile components such as cushioned seats, door panels, and cabin lining. Recycled cotton fibres from fabric reinforcements are commonly used as insulation for such indoor elements. Insulation materials, in addition to cotton fibres, may be used with coconut fibres.

Coconut fibres combined with vinyl ester matrix are used to make cushioned chairs in particular. Humidity absorption also contributes to the comfortability of those seats. These plant-based fibres' potential for vinyl ester composites is a key consideration for real-time applications in automotive and aerospace components. Plant-based fibres, notably flax fibre mat impregnated in epoxy resin, are used in the manufacture of door panels for automotive applications. Additionally, the usage of these fibres reduces the load on composite components by 20 times, improving mechanical qualities. Internal trim components made of cotton or other analogous fibres bonded with thermoset polyphenolic resins are another example. Despite the fact that thermosets have a better thermal balance than thermoplastics, thermosets' decomposability issues, as a result of recycling requirements within the company, lead to a greater demand for thermoplastics from manufacturers.

The advantages of natural fibres reinforced with vinyl esters, such as lightweight, improved thermal and acoustic insulation, and rigidity, have been proven to be advantageous in the building industry. As a result, natural fibres are frequently used in construction components such as doorframes, partitions, fake ceilings, and floor panels. Natural fibres, for example, are incorporated to sandwich composite panels to reduce weight and provide thermos-acoustic insulation. Furthermore, using sisal (a type of plant-based fibre) in a hybrid composite with wollastonite and polyester increased the composite's ductility and acoustic insulation. To give strength to roof tiles and chequered floor tiles, this combination is utilized as a moulding compound. The use of jute fibre for lower strength and flexural rigidity is another example of natural fibre-reinforced vinyl ester composites. The electrical insulation and corrosion resistance capabilities of the woven jute fibre mat enhanced phenolic resin composite are well known, with no symptoms of deformation such as warping, discoloration, or swelling. Additionally, composites made of jute and coconut fibres are a less expensive alternative to hardwood for windows and doors. Due to the 46% lignin content in coconut fibre, it can provide exact tensile strength and has more corrosion resistance in moist environments than teak wood. Additionally, coconut fibre-reinforced cement composite sheet may be used as a roofing alternative to asbestos sheet.

Natural fibre reinforced biocomposites have applications not just in the automotive and aerospace industries, but also in tourism, fashion, agriculture, and handicrafts. Packaging materials and fishing baskets are typically made in rural locations. This sort of fibre is extensively used in the tourism industry to manufacture natural fibre-reinforced polyester composites for bags, suitcases, and other products. In the fashion business, natural fibre-reinforced vinyl ester composites are used to make belts and shoes. Bamboo fibre-based vinyl ester composites are also employed in the apparel sector for a variety of applications. Sheets, towels, and bags are also made using bamboo fibres-reinforced vinyl esters. Plant-based fibres are commonly used for geotextiles in the agricultural business. Geotextiles are woven fabrics that are used to protect soil from erosion and weeds. Jute fibres are used in many eco-friendly geotextiles since they are biodegradable and can be biodegradable. Jute fibres-reinforced vinyl ester-based composites are very famous, particularly with nanoclay for applications in dashboards, carpets, demonstrated by means of its use in incredibly diverse forms of merchandise and components like dartboards, panel boards, etc.

8.6 CONCLUSIONS

To increase the adherence of natural fibres to polymer matrices, chemical, additive, or physical treatments can be applied. The addition of nanoclay to the polymer matrix triggers a secondary response, resulting in a complete collation that improves the strength of natural fibre-reinforced vinyl ester composites by adding a strong cross-connected connection. In addition, reactive monomer reactions brought about by nanoclay additives and chemical additives such as glycidyl methacrylate, anhydride, maleic anhydride, stearic acid, isocyanate, silane, and methyl methacrylate are also techniques of enhancing thermal and mechanical characteristics of vinyl ester composites. Pracella et al. (2010) studied the thermal, rheological, mechanical, and morphological properties of vinyl ester composites supplemented with cellulose fibres and hemp, as well as nanoclay fillers, after functionalization and reactive processing. They discovered that adding nanoclay to a composite can change its thermal balance, making it a possible chemical cure for improving composite qualities. This is due to the presence of nanoclay as an adjuvant, which has increased hemp fibre dispersion and interface bonding. These modifications passed off because of the crystallization technique of hemp fibre nucleation in the vinyl ester matrix. Furthermore, it enhances the mechanical properties. Tensile properties are the most important factor to consider when determining the composites' strength. It relies upon the shape and composition of the plant fibre in the composite, which is encouraged through numerous factors viz., the composition, crystallography, composition, and luminal porosity. Various researches had been performed to analyse the tensile characteristics and suitability of its use within the industry. The presence of nanoclay content in the matrix phase, for example, accelerates the bonding of kenaf fibres in the matrix phase. Kenaf fibre is well known for its easy-to-extract qualities and mechanical features. Using kenaf fibres in the vinyl ester matrix can increase its mechanical and thermal properties due to the strong bonding created by the nanoclay inclusion in the matrix phase. As a result, the impact of nanoclay on the thermal and mechanical properties of vinyl ester-based biocomposites is discussed in this work.

REFERENCES

Abdul Khalil, H. P. S., Firoozian, P., Bakare, I. O., Akil, H. M., Noor, A. M., 2010. Exploring biomass based carbon black as filler in epoxy composites: flexural and thermal properties. *Mater. Des.* 31, 3419–3425.

Abdul Khalil, H. P. S., Noriman, N. Z., Ahmad, M. N., Ratnam, M. M., Nik Fuaad, N. A., 2007. Polyester composites filled carbon black and activated carbon from bamboo (*Gigantochloa scortechinii*): physical and mechanical properties. *J. Reinf. Plast. Compos.* 26 (3), 305–320.

Abdul Khalil, H. P. S., Sri Aprilia, N. A., Bhat, A. H., Jawaid, M., Paridah, M. T., Rudi, D., 2013. A Jatropha biomass as renewable materials for biocomposites and its applications. *Renew. Sustain. Energy. Rev.* 22, 667–685.

Acharya, S. K., Samantarai, S. P., 2012. Investigation in to tribo potential of biomass based carbon black filler in epoxy composite. *Int. J. Sci. & Eng. Res.* 3 (6), 1–4.

Adekunle, K. F., 2015. Surface treatments of natural fibres—a review: part 1. *Open J. Polym. Chem.* 5 (03), 41–46.

Aisyah, G. S., Tajuddin, R. M., 2014. Trends in natural fibre production and its future. Conference: 5th Brunei International Conference on Engineering and Technology (BICET 2014) DOI: 10.1049/cp.2014.1080, 1–8.

Al-Oqla, F. M., Sapuan, S. M., 2014. Natural fiber reinforced polymer composites in industrial applications: feasibility of date palm fibers for sustainable automotive industry. *J. Clean. Prod.* 66, 347–354.

Ahmad, M., Gharayebi, Y., Salit, M. S., Hussein, M. Z., Ebrahimiasl, S., Dehzangi, A., 2012. Preparation, characterization and thermal degradation of polyimide (4-APS/BTDA)/SiO_2 composite films. *Int. J. Mol. Sci.* 13, 4860–4872.

Anwer, M. M., Bhuiyan, A. H., 2012. Influence of low temperature plasma treatment on the surface. *Opt. DC Electr. Propert. Jute.* 1, 16–22.

Ashori, A., Nourbakhsh, A., 2010. Bio-based composites from waste agricultural residues. *Waste Manage.* 30 (4), 680–684.

Carada, P. T. D., Fujii, T., Okubo, K., 2016. Effects of heat treatment on the mechanical properties of kenaf fiber. In *AIP Conference Proceedings*, 1736, pp. 020029.

Carvalho, K. C. C., Mulinari, D. R., Voorwald, H. J. C., Cioffi, M. O. H., 2010. Chemical modification effect on the mechanical properties of HIPS/coconut fiber composites. *Bio Resources* 5 (2), 1143–1155.

Cecen, V., Tavman, I. H., Kok, M., Aydogdu, Y., 2009. Epoxy and polyester-based composites reinforced with glass, carbon and aramid fabrics: measurement of heat capacity and thermal conductivity of composites by differential scanning calorimetry. *Polym. Compos.* 30 (9), 1299–1311.

Das, P. K., Nag, D., Debnath, S., Nayak, L. K., 2010. Machinery for extraction and traditional spinning of plant fibres. *Ind. J. Tradit. Knowl.* 9 (2), 386–393.

Debnath, S., Nguong, C. W., Lee, S. N. B., 2013. A review on natural fibre reinforced polymer composites. *World Acad. Sci. Eng. Technol.* 73, 1123–1130.

Fowler, P. A., Hughes, J. M., Elias, R. M., 2006. Review, biocomposites: technology, environmental credentials and market forces. *J. Sci. Food. Agric.* 86, 1781–1789.

Gupta, A., Kumar, A., 2012. Chemical properties of natural fiber composites and mechanisms of chemical modifications. *Asian J. Chem.* 24 (4), 1831.

Gupta, M. K., Srivastava, R. K., Bisaria, H., 2015. Potential of jute fibre reinforced polymer composites: a review. *Int. J. Fiber Textile Res.* 5, 30–38.

Habibi, Y., El-Zawawy, W. K., Ibrahim, M. M., Dufresne, A., 2008. Processing and characterization of reinforced polyethylene composites made with lignocellulosic fibers from Egyptian agro-industrial residues. *Compos. Sci. Technol.* 68 (7), 1877–1885.

Han, S. O., Choi, H. Y., 2010. Morphology and surface properties of natural fiber treated with electron beam. *Microscopy Sci. Technol. Applicat. Educat.* 3, 1880–1887.

Hassan, M. M., Wagner, M. H., 2016. Surface modification of natural fibers for reinforced polymer composites: acritical review. *Rev. Adhes. Adhes.* 4 (1), 1–46.

Hibi, Y., Ouchi, M., Sawamoto, M., 2016. A strategy for sequence control in vinyl polymers via iterative controlled radical cyclization. *Nat. Commun.* 7, 11064.

Ho, M. P., Wang, H., Lee, J. H., Ho, C. K., Lau, K. T., Leng, J., et al., 2012. Critical factors on manufacturing processes of natural fibre composites. *Compos. Part B Eng.* 43 (8), 3549–3562.

Husseinsyah, S., Mostapha, M., 2011. The effect of filler content on properties of coconut shell filled polyester composites. *Malaysian Polym. J.* 6 (1), 87–97.

Ibrahim, M. S., Sapuan, S. M., Faieza, A. A., 2012. Mechanical and thermal properties of composites from unsaturated polyester filled with oil palm ash. *J. Mech. Eng. Sci (JMES)* 2, 133–147.

Jamil, M. S., Ahmad, I., Abdullah, I. 2006. Effects of rice husk filler on the mechanical and thermal properties of liquid natural rubber compatibilized high-density polyethylene/natural rubber blends. *J. Polym. Res.* 13, 315–321.

Józefiak, K., Michalczyk, R. 2020. Prediction of structural performance of vinyl ester polymer concrete using FEM elasto-plastic model. *Materials* 13, 4034.

Jusoh, A. F., Rejab, M. R. M., Siregar, J. P., Bachtiar, D., 2016. Natural fiber reinforced composites: a review on potential for corrugated core of sandwich structures. In *MATEC Web of Conferences*, 74, pp. 00033.

Kabir, M. M., Wang, H., Aravinthan, T., Cardona, F., Lau, K. T., 2011. Effects of natural fibre surface on composite properties: a review. In *Proceedings of the 1st International Postgraduate Conference on Engineering, Designing and Developing the Built Environment for Sustainable Wellbeing (eddBE 2011)*, University of Southern Queensland, pp. 94–99.

Ku, H., Prajapati, M., Trada, M., 2012. Fracture toughness of vinyl ester composites reinforced with sawdust and postcured in microwaves. *Int. J. Microw. Sci. Technol.* 8, 1–4.

Kumar, R., Obrai, S., Sharma, A., 2011. Chemical modifications of natural fiber for composite material. *Der Chemica Sinica* 2 (4), 219–228.

Li, X., Tabil, L. G., Panigrahi, S., 2007. Chemical treatments of natural fiber for use in natural fiber-reinforced composites: a review. *J. Polym. Environ.* 15 (1), 25–33.

Liu X., Wu Q., 2001. PP/clay nanocomposites prepared by grafting-melt intercalation. *Polymer* 42, 100–139.

Maity, S., Gon, D. P., Paul, P., 2014. A review of flax nonwovens: manufacturing, properties, and applications. *J. Nat. Fibers* 11 (4), 365–390.

Mallakpour, S., Zadehnazari, A., 2013. Thermoplastic vinyl polymers: from macro to nanostructure. *Polym. Plast. Technol. Eng.* 52 (14), 1423–1466.

Melo, J. D. D., Santos, E. A. D., 2009. Mechanical and microstructural evaluation of polymer matrix composites filled with recycled industrial waste. *J. Reinf. Plast. Compos.* 28 (20), 2459–2471.

Mengeloglu, F., Karakus, K., 2008. Thermal degradation, mechanical properties and morphology of wheat straw flour filled recycled thermoplastic composites. *Sensors* 8, 500–519.

Moghaddam, M. K., Mortazavi, S. M., 2016. Physical and chemical properties of natural fibers extracted from Typha Australis Leaves. *J. Nat. Fibers* 13 (3), 353–361.

Mohammed, L., Ansari, M. N., Pua, G., Jawaid, M., Islam, M. S., 2015. A review on natural fiber reinforced polymer composite and its applications. *Int. J. Polym. Sci.* 2015, 1–15.

Nadlene, R., Sapuan, S. M., Jawaid, M., Ishak, M. R., Yusriah, L., 2016. The effects of chemical treatment on the structural and thermal, physical, and mechanical and morphological properties of roselle fiber-reinforced vinyl ester composites. *Polym. Compos.* 13, 1–14.

Nagaprasad, N., Stalin, B., Vignesh, V., Ravichandran, M., Rajini, N., Ismail, S.O., 2020. Effect of cellulosic filler loading on mechanical and thermal properties of date palm seed/vinyl ester composites. *Int. J. Biol. Macromol.* 147, 53–66.

Namvar, F., Jawaid, M., Tanir, P. M., Mohamad, R., Azizi, S., Khodavandi, A., et al., 2014. Potential use of plant fibres and their composites for biomedical applications. *Bio Resources* 9 (3), 5688–5706.

Nandi, A. K., Banerjee, U., Biswas, D., 2015. Improvement in physical and aesthetic properties of jute fabrics by blending ramie fibre in suitable proportions. *Int. J. Textile Sci.* 4 (4), 73–77.

Obayi, C. S., Odukwe, A. O., Obikwelu, D. O. N., 2008. Some tensile properties of unsaturated polyester resin with varying volume fraction of carbon black nanoparticles. *Niger. J. Technol.* 27 (1), 20–27.

Ojha, S., Raghavendra, G., Acharya, S. K., 2012. Fabrication and study of mechanical properties of coconut raw and carbon black reinforced epoxy composite. *Int. J. Syst. Algorithms Appl.* 2 (12), 68–71.

Omrani, E., Menezes, P. L., Rohatgi, P. K., 2016. State of the art on tribological behavior of polymer matrix composites reinforced with natural fibers in the green materials world. *Eng. Sci. Technol. Int. J.* 19 (2), 717–736.

Onyeagoro, G. N., 2012. Cure characteristics and physico-mechanical properties of carbonized bamboo fibre filled natural rubber vulcanizates. *Int. J. Mod. Eng. Res. (IJMER)* 2 (6), 4683–4690.

Osita, O., Ignatius, O., Henry, U., 2016. Study on the mechanical properties of palm kernel fibre reinforced epoxy and poly-vinyl alcohol (PVA) composite material. *Int. J. Eng. Technol.* 7, 68–77.

Pandey, S. N., 2007. Ramie fibre: part I. Chemical composition and chemical properties. A critical review of recent developments. *Text. Prog.* 39 (1), 1–66.

Paridah, M. T., Basher, A. B., SaifulAzry, S., Ahmed, Z., 2011. Retting process of some bast plant fibres and its effect on fibre quality: a review. *Bio Resources* 6 (4), 5260–5281.

Pracella, M., Haque, M. M. U., Alvarez, V., 2010. Functionalization, compatibilization and properties of polyolefin composites with natural fibers. *Polymers* 2 (4), 554–574.

Puglia, D., Biagiotti, J., Kenny, J. M., 2005. A review on natural fibre-based composites—Part II: application of natural reinforcements in composite materials for automotive industry. *J. Nat. Fibers* 1 (3), 23–65.

Raman Bharath, V. R., Vijaya Ramnath, B., Manoharan, N., 2015. Kenaf fibre reinforced composites: a review. *ARPN J. Eng. Appl. Sci.* 10 (13), 5483–5485.

Ray, D., Sarkar, B. K., Basak, R. K., Rana, A. K., 2004. Thermal behavior of vinyl ester resin matrix composites reinforced with alkali-treated jute fibers. *J. Appl. Polym. Sci.* 94, 123–129.

Rosas, R. M., Orrit-Prat, J., Ramis-Juan, X., Marin-Genesca, M., Rahhali, A., 2011. Study on dielectric, thermal, and mechanical properties of the ethylene vinyl acetate reinforced with ground tire rubber. *J. Reinf. Plast. Compos.* 30 (7), 581–592.

Saba, N., Tahir, P. M., Jawaid, M., 2014. A review on potentiality of nano filler/natural fiber filled polymer hybrid composites. *Polymers* 6 (8), 2247–2273.

Saheb, D. N., Jog, J. P., 1999. Natural fiber polymer composites: a review. *Adv. Polym. Technol.* 18 (4), 351–363.

Sepet, H., Tarakçıoglu, N., 2014. Effect of nanoclay addition on mechanical and thermal behavior of vinyl ester based nanocomposites obtained by casting. *World J. Eng.* 11 (1), 1–8.

Shahzad, A., 2012. Hemp fiber and its composites - a review. *J. Compos. Mater.* 46 (8), 973–986.

Smith, P. A., Yeomans, J. A., 2009. Benefits of fiber and particulate reinforcement. *Mater. Sci Eng.* 2, 133–154.

Stevulova, N., Cigasova, J., Estokova, A., Terpakova, E., Geffert, A., Kacik, F., et al., 2014. Properties characterization of chemically modified hemp hurds. *Materials* 7 (12), 8131–8150.

Suresha, B., Shiva Kumar, K. N., 2009. Investigations on mechanical and two body abrasive wear behaviour of glass/carbon fabric reinforced vinyl ester composites. *Mater. Des.* 30, 2056–2060.

Tanushree, C. B., 2016. Characterization and mechanical properties of bast fibre. *Int. J. Home Sci.* 2 (2), 291–295.

Vaisanen, T., Haapala, A., Lappalainen, R., Tomppo, L., 2016. Utilization of agricultural and forest industry waste and residues in natural fiber-polymer composites: a review. *Waste Manage.* 54, 62–73.

Webber III, C. L., Bledsoe, V. K., Bledsoe, R. E., 2002. Kenaf harvesting and processing. *Trends New Crops New Uses* 9, 340–347.

Westman, M. P., Fifield, L. S., Simmons, K. L., Laddha, S., Kafentzis, T. A., 2010. Natural Fiber Composites: a Review (No. PNNL-19220). Pacific Northwest National Laboratory (PNNL), Richland, WA.

Yan, L., Chouw, N., Jayaraman, K., 2014. Flax fibre and its composites-a review. *Compos. Part B Eng.* 56, 296–317.

9 Natural Fibre-Reinforced Vinyl Ester Composites

Influence of Silica Nanoparticles on Thermal and Mechanical Properties

Shwetharani R. and Yatish K. V.
Centre for Nano and Material Sciences,
Jain (Deemed-to-be University)

Jyothi M. S.
AMC Engineering College

Lavanya C.
Centre for Nano and Material Sciences,
Jain (Deemed-to-be University)

Sabarish Radoor
King Mongkut's University of Technology North Bangkok

R. Geetha Balakrishna
Centre for Nano and Material Sciences,
Jain (Deemed-to-be University)

CONTENTS

DOI: 10.1201/9781003270997-9

9.1 INTRODUCTION

Natural fibres are biopolymers consisting of continuous hair-like filament material that are sourced from either plant or animal. The major component in plant-based natural fibres is cellulose and protein is the major constituent in animal-based fibres [1]. Fibres can be entangled into sheets to generate products like papers. Natural fibres are widely used in the polymeric composites preparation in order to decrease the environmental impact from synthetic/man-made fibre-reinforced composites. Natural fibre-reinforced composites exhibit unique properties like renewable, biodegradable and eco-friendly [2]. Composites are a blend of different materials and preserve different characteristics of each component. These components contribute collectively to enhance the mechanical properties and robustness. In the past, polymers have substituted various conventional materials or metals in several applications due to their unique properties like easy processing, improved productivity and lowered cost [1]. The polymer properties differ with the use of fibres or fillers to reach the high strength or high modulus. A natural fibre-reinforced composite is a material made up of polymer matrix with strong fibres. Vinyl polymers make large family of polymers, which includes VEs (thermosetting). The phase interaction that takes place at the interfaces of polymer matrix contribute to enhanced properties in these composites, such as electrical, mechanical, optical and thermal properties. Currently, NPs are gaining remarkable interest for improving the functional and mechanical performances (antibacterial, anti-corrosion, flame retardancy) in composites. The uniform distribution of nanomaterials enhances the thermal, mechanical and molecular movability. NPs such as TiO_2, SiO_2, rGO, carbon nanotubes and ZnO are compatible with other ingredients such as biofibres and matrix polymers, thus forming composites [3]. The composite material reinforced with natural fibres, polymer and nanomaterials are greatly popular due to their strength, lightweight, stiffness, damping property, fire-resistant, biodegradability and antibacterial properties. Furthermore, NPs are also receiving consideration owing to their potential to increase the functionality and mechanical performances in composite [4]. The present book chapter reviews the effect of silica NPs on improvement of thermal and mechanical properties in natural fibre-reinforced VE composites.

9.2 PREPARATION TECHNIQUES FOR VINYL ESTER SILICA NANOPARTICLES (VESiNPs) COMPOSITES

Vinyl ester (VE) resins are thermosetting polymers obtained by reacting α, β-unsaturated carboxylic acid, for instance, acrylic/methacrylic acid with an epoxy resin [5,6]. It is greatly considered as an effectual anti-corrosion coating material in various industrial applications, and it also offers considerable resistance against the hydrothermal aging [6,7]. Methyl ethyl ketone peroxide (MEKP), Trigonox and benzoyl peroxide are the widely exploited catalysts for VE resins [8]. Cobalt naphthalene/cobalt octoate of 0.2–0.4 wt.% is often used as a promoter to VE resin [9].

These resins possess greater resistance against acids and bases. The unsaturated ends of VE resins make them highly reactive, so that they can be used with styrene in order to process like unsaturated polyester. In consequence, they will bring forth the coarse strength for the moulding process [5]. VE resins' good mechanical properties (mainly elongation and tensile), good thermal resistivity, resistance against high heat deflection temperature and micro-cracks formation are exceptional. These resins pervade glass fibres as well as other reinforcements effectively [7,8].

The incorporation of nanofillers (especially silica-based NPs) could be highly beneficial in accelerating the coating performance, durability and resistance against the hydrothermal aging [7,10]. These fillers help in reducing void fraction and also prevent the crack formation throughout resin composite by lessening its shrinkage [10,11]. The addition of silica NPs to VE resins has attracted more attention because of their greater stability at higher temperatures and in strong acidic and basic environments, and they also exhibit a strong adhesion between VE resin and silica particles [11].

Thus, based on weight percent of nanocomposite reinforcement, required amount of NPs is added to VE resin. NPs in the resin are dispersed thoroughly with the help of high-shear mixer until uniform dispersion. High shear force is used in order to break down the silica aggregations. Then the coupling agent/adhesion promoter such as BYK-C8000 of certain amount is added during the mixing process. Later, the mixed samples are subjected to sonication with the help of probe ultrasonic device. However, in order to prevent the VE resin temperature from increasing, the sonication is carried out using a water bath. The trapped bubbles are removed using degassing process. In the second stage, cobalt octoate, dimethyl aniline, and MEKP are added (in weight percent with respect to VE resins) and mixed thoroughly. The obtained mixture is then transferred to a steel/silicon/aluminium mould and is left as such to cure for several hours at ambient conditions. Post-curing process are carried out at 60°C for 3–4 h [7,9,12]. However, in some cases the mixture is transferred to a spray gun rather than a mould. The spray gun with a nozzle diameter of 1 nm connected to an air compressor is used to spray on an aluminium sheet keeping the spraying distance as 20 cm. The coatings are left for curing using the same procedure as mentioned above [6]. The mechanical properties of VE thermoset polymers are greatly enhanced upon incorporation of up to 5 wt.% silica NPs. However, beyond 5 wt.% concentration, the properties decreased due to highly viscous solution [12]. The most commonly used silica nanoparticles are SiO_2 [10,13], nanoclay (0–2.5 wt.%) [7], phyllosilicates (stacks of tetrahedral Si_2O_5 layers), montmorillonite, hectorite, mica as well as numerous smectite-type nanoclay have been studied [12], octadecylamine-modified nanoclay (up to 5 wt.%) [12], silica aerogels (5–15 wt.%) [6], n-SiC [4,14–15], micro- and nano-SiC particles [16]and PANi-stabilized silica NPs [11], etc.

9.3 PROPERTIES OF VINYL ESTER SILICA NANOPARTICLES (VESiNPs) COMPOSITES

9.3.1 Properties of Natural Fibre-Reinforced Vinyl Ester

The main advantages of natural fibre are low density, biodegradability, low cost, high toughness, good thermal properties and low energy content. There are many factors that influence the composites performance, including the fibre type and its content, and type of matrix interaction with materials [17]. Fibres act as carriers of load in the matrix, which leads to uniform stress distribution [18]. Some of the studies suggested that minimal fibre content may be better in composites, which may result in better tensile strength due to good interfacial bonding between the fibre and matrix [19]. The composite is highly influenced by its matrix. The use of VE polymer may improve a composite's stiffness, chemical resistance, dimensional stability, and strength [20]. In addition, the cost of VE resin is lower than epoxy resins. The physical properties of VE resin are shown in Table 9.1. The mechanical properties of both VE resins and epoxy resins are similar in hydrolytic stability. However, since the VE resin is brittle, to improve its performance (mechanical and thermal properties) and lower the cost of the resin, it is necessary to reinforce the fillers. Therefore, several studies have reported using natural fibres as fillers to improve the performance of VE resins. Table 9.2 shows the properties of some selected natural fibres.

Several studies have reported the use of natural fibre as filler in VE to improve the properties. Nadlene et al. [21] studied the mechanical and thermal properties of roselle (*Hibiscus sabdariffa* L.) fibre-reinforced VE (RFVE), which is prepared using the hand lay-up method. RFVE composites are prepared by 5, 10, 20 and 30 wt.% of roselle fibres. Mechanical properties such as impact and tensile tests are performed

TABLE 9.1
Physical Properties of VE Resin

Properties	Vinyl Ester Resin
Appearance	Pale yellow liquid
Acid value	10 ± 2 (mg KOH/g)
Viscosity	400–600 cps @ 25°C
Specific gravity	1.06–1.08 @ 25°C
Volatile content	35% ± 5%
Gel point	30–35 min @ 82°C

TABLE 9.2
Properties of Selected Natural Fibres [3]

Fibre	Density (g/cm^3)	Tensile Strength (MPa)	Elongation (%)	Young's Modulus (GPa)	Decomposition Temperature (°C)
Kenaf	0.6–1.5	223–1191	1.6–4.3	11–60	229
Cotton	1.5–1.6	287–597	3–10	5.5–12.6	232
Hemp	1.48	550–900	1.6	70	215
Bamboo	1.2–1.5	500–575	1.9–3.2	27–40	214
Sisal	1.33–1.5	400–700	2–14	9–38	205–220
Jute	1.3–1.46	393–800	1.5–1.8	10–30	215
Sugarcane bagasse	1.1–1.6	170–350	6.3–7.9	5.1–6.2	232

to analyse the properties of fibre surface treatment. The tensile strength increases gradually to maximum, and suddenly decreased, denoting that fracture occurred in the material. The 20% fibre loading is the optimum value, which increases the tensile strength of RFVE approximately by 136% (41.50 MPa) compared to neat VE. The optimal tensile strength and modulus from 20% fibre loading are due to good interfacial bonding between fibre/matrix and identical dispersal of the fibre in matrix [22]. In addition, the impact property exhibits material ability to attract and dissipate energy under fibre loading. The impact of the roselle fibre-VE composite properties decreases with increase in the fibre loading because of incompetence transfer of stress from the matrix to the fibre when the control load is applied. Thermogravimetric analysis (TGA) is used to determine the stability and thermal behaviour of composites by evaluating the mass change when exposed to high temperature as a function of time [23]. Figure 9.1 shows the TGA of RFVE subjected to fibre loading. When compared to other composites, the 10% fibre loading in VE showed the maximum degradation with temperature. The thermal stability of composites decreased at higher fibre loading. This may be due to the percentage of moisture being increased due to an increase in fibre loading, which is the usual characteristic of natural fibre [24]. Thus, the thermal stability of RFVE increased with the addition of roselle fibre.

Sugar palm fibre (SPF) composites are prepared with different fibre loading (10, 20, 30 and 40 wt.%) to study the mechanical and thermal behaviour of SPF-reinforced vinyl ester (SPF/VE) composites [25]. The effect of SPF fibre loading on tensile modules and strength, and flexural strength and modules indicates that addition of SPF fibre to VE reduces the tensile and flexural strength of composites and for impact strength, 30 wt% SPF shows the great improvement in the composite (5.4 kJ/m^2). Further, the addition of SPF to VE reduces the thermal stability of the composites

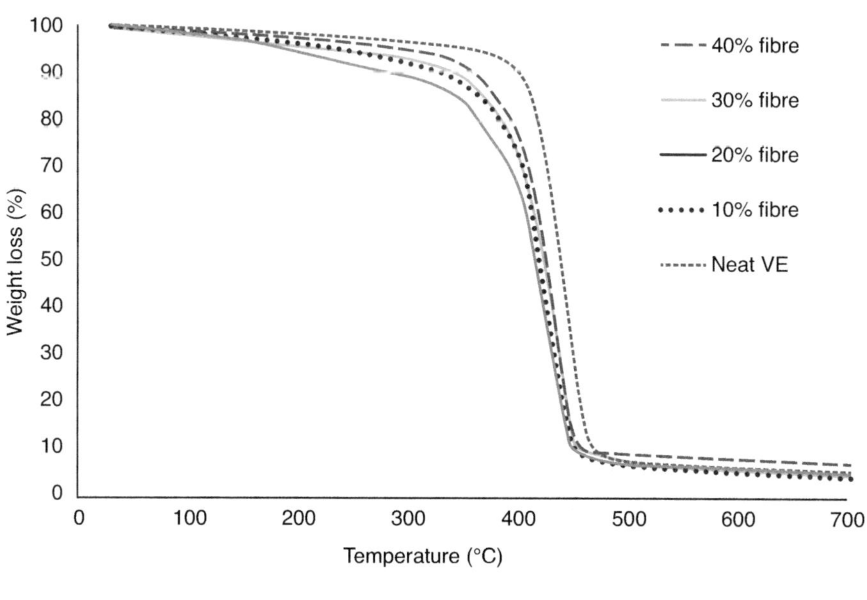

FIGURE 9.1 TGA of RFVE subjected to fibre loading [6].

and shows the thermal degradation with 10 wt.% (at 270.83°C), though 40 wt.% was at 196.67°C. In addition, Huzaifah et al. [26] investigated the effect of soil burial on neat VE and SPF/VE (10 wt.%). The specimens are buried in the soil for 0, 200, 400, 600, 800 and 1000 h to evaluate the thermal and mechanical properties. The water uptake increased by 0.92% of SPF/VE compared to VE (0.42%) after 200-hour soil burial. The mechanical properties (tensile, impact and flexural strength) decreases and thermal stability increases after 200 h of soil burial when compared to neat VE. This may be due to wettability and poor bonding caused by moisture absorption.

Similarly, 10 wt.% betel nut husk (BNH) fibre in VE increases the mechanical (178.15% tensile strength and 36.77% modulus strength), physical and thermophysical properties of BNH/VE composites [27]. Mahato et al. [28] studied the influence of alkali-treated sisal fibre in VE composites. Sisal fibre was treated with 2% sodium hydroxide solution for 2, 3 and 8 h at 35°C. The improvement in the flexural strength and flexural modulus has been observed with fibre loading from 10 to 30 wt.%. The tensile and flexural strength of alkali-treated sisal fibre-reinforced VE composites (25 wt.%) were higher compared to untreated fibre, and this might be due to inherently increased matrix–fibre bonding.

9.3.2 Silica Nanoparticles

Silica is one of most complex and prolific groups of materials, existing both as a mineral constituent and as a synthesized product. Development in nanotechnology has led to the synthesis of nanoscale silica, SiO_2, which has been extensively employed as filler in engineering composites. The silica particles obtained from natural sources are not useful for advanced industrial and scientific applications because they contain metal impurities [29]. Over the last decade, silica NPs gained more importance due to their versatile and distinctive physiochemical properties. Silica NPs are categorized as mesoporous and nanoporous, with the size of both NPs being regulated by modifying the surfactant content during NPs synthesis. Silica NPs are used in different applications, due to their hydrophilic nature, easy and low-cost preparation, large specific surface area, good biocompatibility, and controlled particle size [30].

The properties of silica NPs are usually size-dependent; if the size of silica NPs is less than 5 nm, higher Si atoms (more than half) are present on the surface. Therefore, one or more silanol groups (≡Si–OH) should be present on the surface. Consequently, the extent of chemical modification of silica, such as metal ion insertion and organofunctional group grafting, is largely dependent on the silanol group concentration in per gram of silica. The total silanol groups per unit area of silica gives the information about dispersal of silanol group on the surface of silica. The decrease in the particle size increases the silanol group concentration, which is interconnected to the specific surface area. However, the decrease in particle size with decrease in silanol number indicates that these NPs are chemically reactive and can be used in catalyst applications [31].

Silicon NPs compact quickly, and the sintering temperature is found to be much lower than that of conventional microsized powders. The silicon with a particle size of 1.6 μm required 1600°C to achieve transparency, but the transparency of ~20 nm silica fume is achieved at a temperature of 1200°C [22]. Compared to the conventional particles, the sintering property of silica NPs is credited to the large surface area which provides higher contact of particles. The optical properties of silica NPs depend on the size of the particles; especially below 10 nm, the optical property of silica NPs increased by incorporation with functionalized groups [32]. The silica NPs (fillers) has a higher boundary surface, which influences the ability to produce silica-polymer nanocomposites with enhanced properties. The benefits of silica NPs include efficient reinforcement with exceptional heat stability, mechanical strength, thermal expansion, reduced shrinkage, improved abrasion resistance, and increased electric and optical properties [33]. For silica NPs, a particle size of below 100 nm allows good optical transparency. Due to the high surface to volume ratio of silica NPs, they offer increases in the various properties at lower loadings.

9.3.3 Vinyl Ester Silica Nanoparticles (VESiNPs) Composites

The composites of polymer nanocomposites are prepared from the dispersion of NPs to polymer matrix. The properties such as mechanical strength, superior toughness, excellent thermal and electrical properties, and more importantly moisture stability are increased by the dispersion of NPs to the polymer matrix [34]. Several researchers focused on the study of the effect of silica NPs in synthetic fibre-reinforced polymer composites to improve their mechanical and thermal properties. Indeed, synthetic fibre-based polymer composites did perform better, but their environmental friendliness is now being questioned [35]. Therefore, the replacement of synthetic fibre with natural fibre in designing polymer composites has gained more importance.

Baniani et al. [8] studied the influence of nanosilica (0.3, 0.75, 1 wt.%) on mechanical properties of nanosilica-vinyl ester nanocompsites. Increasing the silica content improves the elastic modulus and decreases the tensile strength (Table 9.3). The tensile strength depends on the stress transfer between matrix and fillers of nanocomposites. However, dispersion of more SiO_2 particles results in the reduction of strength, and this might be due to the weak interfacial bonding. Adding nanosilica (0.3 wt.%) reinforcement decreases the fracture energy while improving the fracture toughness.

The mechanical and fracture behaviour of vinyl ester/nano-silicon carbide nanocomposites with varying n-SiC concentrations of 1, 3, and 5 wt.% are investigated by Alhuthali and Low [4]. The properties such as elastic modulus and flexural strength, impact strength and fracture toughness are analysed according to ASTM D790-86, ASTM D 256-06 and ASTM D5045-99 standards, respectively. The addition of n-SiC

TABLE 9.3
Mechanical Properties of Nanosilica-Vinyl Ester Nanocomposites [8]

Nanosilica Content (wt.%)	Elongation (%)	Toughness (J/m³)	Tensile Strength (MPa)	Young's Modulus (GPa)
0	5.69	$350{\cdot}46 \times 10^{-3}$	84.2	3·215
0.3	5.39	$498{\cdot}93 \times 10^{-3}$	80.4	3·223
0.75	2.19	$88{\cdot}985 \times 10^{-3}$	60.8	3·446
1	1.75	$57{\cdot}818 \times 10^{-3}$	52.2	3·984

particles to VE resin reduces the toughness and increases the elastic modulus and strength since n-SiC (457 GPa) has greater elastic modulus than VE resin (2.9 GPa) [36 The n-SiC particles have nanosize of 50–150 nm, with 100 nm average size with higher specific surface area and spherical shape, which provides the interfacial bonding between polymer matrix and filler, thus attributing to good degree of dispersion and interfacial adhesion, and thereby enhances the composite strength [37].

The thermal property of polymer matrix is enhanced with the addition of nanosilica. The addition of 1, 2 and 3 wt.% silica loading decreases the onset temperature to 339°C, 331°C and 332°C, respectively. It clearly shows that smaller amount of silicate loading increases the thermal properties where the dispersion should be achieved and the barrier property effect will dominate. Conversely, the higher amount of silicate led to reduction in the thermal stability where the catalysing outcome will dominate [38]. Overall, it can be concluded that VESiNPs composite provides good mechanical and thermal property.

9.4 EFFECT OF SiNPs INCORPORATION ON THERMAL AND MECHANICAL PROPERTIES IN VESiNPs COMPOSITES

Despite extensive research and commercialization of several inorganic fillers and polymer composites, there exist some failure modes as well. Delamination, breakage of fibre, cracking of matrix and shrinkage issues caused by thermal and/or mechanical treatment to the composites are the major failure modes. A natural fibre, hemp-sisal-reinforced hybrid of epoxy ester was made composite with Si NPs of varied concentrations [39]. With the increased concentration of Si NPs, the composites density increased and the void content decreased. Maximum tensile, impact strength and hardness were observed for the one with 2 wt.% Si NPs. Maximum flexural strength was observed for the one with 3 wt.%. The Taguchi-based L_{16} orthogonal array design revealed that the wear performance of the prepared composites was increased due to the 32.61% contribution of Si NPs at its 2 wt.%. Using the same model, the contribution of each parameter was made into a chart and is given in Figure 9.2. Another group prepared composites of sisal natural fibre reinforced with polyester [40]. The addition of Si NPs to the composite increased tensile strength and modulus for about 1.5 and 1.08 times, respectively.

A nanosilica-filled sisal fibre reinforced with phenol formaldehyde was investigated for the effect of nanosilica filling on the thermal and mechanical features on composite [41]. Added nanosilica could relive the heat fade observed during the friction, and it also increases the interaction between fibres and resin at their interfaces. Along with silica filling, treatment with alkali has contributed to the alterations in density of the composite as well as mechanical features. Silica filling is found to increase the resistance against the failure of the interlamination made while fabricating the composites [42]. 2 wt.% fillers contributed to increase of 12% in the flexural modulus, and this was attributed to the positive effect provided by the unidirectional fillers during lamination process. This has improved the compressive stiffness and strengths. Besides, the bidirectionally oriented nanosilica fillers reduced the strength by 13%, wherein, the combination of 0° and 90° oriented fibres reduced the stiffness

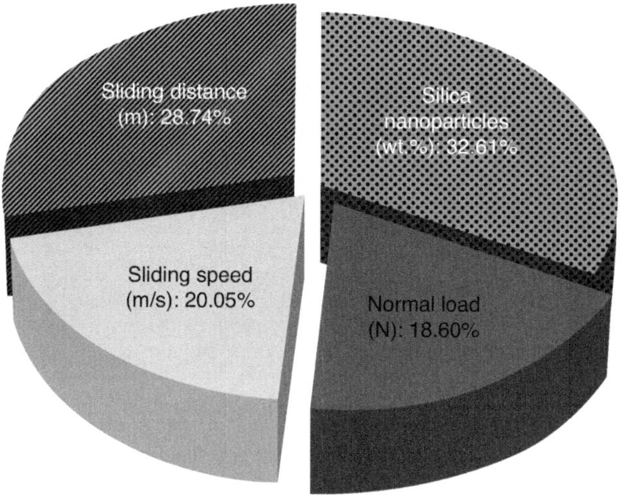

FIGURE 9.2 Contribution of each control factor to volumetric wear of the composites [39].

in line to perpendicular direction. Si NPs are known to increase the fracture energy in the matrix phase of polymers. The toughening process in the matrix is due to the plastic shear-bands localized and originated by concentrated stress around the side-line of Si NPs.

A composite of polylactic acid (PLA) and sisal with varied concentrations of silica aerogel were studied by Venkata Prasad et al. [43]. PLA is the aliphatic thermoplastic polyester and is derived from renewable sources. This has been made composite with sisal fibres and the aerogel of silica. The obtained composite showed a tensile strength of 53.86 MPa and a flexural strength of 90.50 MPa. Sisal palm reinforced with glass fibre was fabricated by hand lay-up method, followed by hot press moulding. Likewise, in other composites, in this case also, 2 wt.% nanosilica played a major role in the improvization of mechanical strength and 1 wt.% nanosilica significantly affected the charpy impact feature of the prepared composite [44].

In a previous study, sisal/Kevlar-reinforced polyesters were studied for its mechanical properties with the introduction of nanosilica [45]. In another study, composite made of mat fibre with altered stacking sequence of sisal and Kevlar was fabricated, and the influence of distribution pattern of nanosilica in the mat fibre was explored in detail. The mechanical properties of the obtained composite were greater and positive when Kevlar was the outer layer. The distribution of nanofiller silica by double-stirring process has improved the performance in terms of tensile strength, and performance decreased beyond 4 wt.% loading. The decreased tensile strength is due to poor dispersion of silica in the matrix or its agglomeration at higher concentration. This would cause the homogeneity, and therefore required extra energy for the mixing process due to the uncured resins. Zhou et al. [46] have also proved that nanosilica would increase the mechanical strength of the polymer, but at higher concentration, the mechanical strength would worsen. The ultrasonic stirring process in the

FIGURE 9.3 SEM images polyester fibre (a), which is without nanosilica and (b) with 4 wt.% of nanosilica showing less crack formation [45].

composites fabrication helped the uniform/feasible distribution of nanosilica in the matrix. Improper wetting of filler in the matrix reduced the mechanical strength and posed the fibre handling issues. Double-stirring method, which includes ultrasonic and chemical process, increased the flexural strength by 37% and tensile strength by 23%. The SEM analysis clearly showed the failure of the fibre in the absence of nanosilica and the increased flexural strength with the addition of silica, as shown in Figure 9.3.

Not only the nanosilica, but also the surface functional groups, attached to the nanosilica, has affected the mechanical features. A uniform nanosized silica was subjected to four modifications, wherein the first one is thermal treatment and the other three are inclusion of epoxy ring, amine group and isocyanate ring on its surface [47]. The scheme for the same is provided in Figure 9.4. Measured T_g values for all the functionalities showed reduced damping and increased T_g value for S-NH_2. The results are attributed to the well dispersion of fillers, and increased T_g is due the restricted chain movement in the polymer matrix, which could also enhance the mechanical aspects of the composite. However, the thermal-treated one with a system of weak interfaces possessed lower T_g values. The suppression in aggregation of nanofillers was observed in case of epoxy ring and amine group-introduced composites.

Epoxy resin was hybridized with carbon fibres and made composite with silica micro and NPs functionalized with polydiallul dimethyl ammonium chloride (PDDA) [48]. This polymer was used to reduce the agglomeration of silica in the composite by decreasing the apparent density. Due to the presence of polar bonds in the epoxy resin, PDDA was able to completely soluble in epoxy resin and improved the homogenization to the system, which also enhanced the difference between silica and hydroxyl groups on the surface of nanosilica. Interfacial Transition Zone (ITZ) is the key feature influencing the mechanical features of composites, wherein the adhesion ability between the polymer matrix and the filler material supporting the distribution of filler material is analysed. Elastic characteristics and durability of the laminates

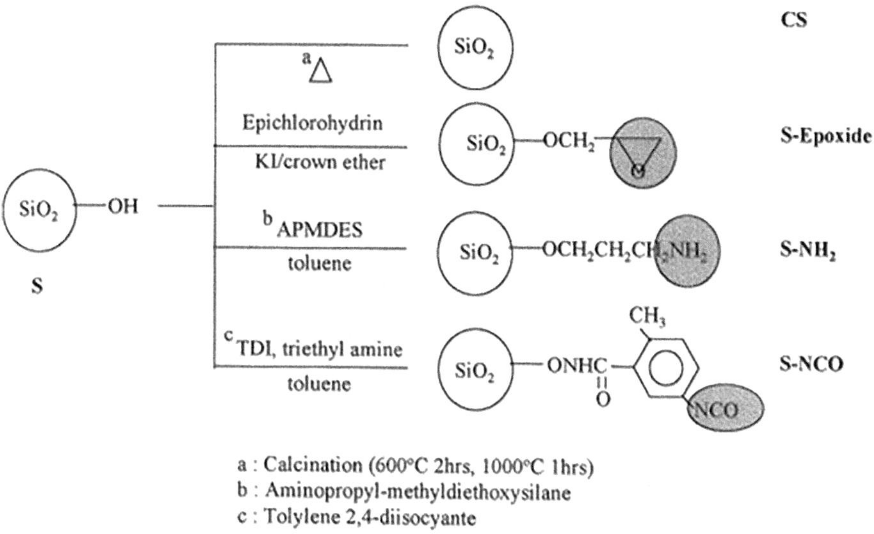

FIGURE 9.4 Scheme for modification of surface of nanosized silica [47].

are evaluated in ITZ. The correlation existing between the strength and the modulus is an effective factor that indirectly decides the condition of ITZ [49,50].

Two kinds of fibre-reinforced polymers were reinforced epoxy and VEs which are revised by NPs were studied by Su et al. [51]. Specimens with epoxy resins and VE behaved differently with the addition of silica NPs. VE resin showed improved breaking stress compared to the epoxy resin. Epoxy resins reinforced with basalt fibre showed 15% greater breaking stress than the one with our basalt fibre. VEs made with the lower amount of fibre impregnated roving and the made of which improved with silica NPs insertion has significantly increased the breaking stress. VE671, epoxy VE resin was made composite with silica NPs using ultrasonic instrument to ensure the feasible distribution of silica NPs in the resin matrix [10]. The con centration of Si NPs guided the dynamic curing procedure and hence, with an aid of Sun method, the curing rate and degree of curing were modelled. The modelled data were in well agreement with the experimental results and showed the performance of nanosilica as catalyst. Introduction of 4% of silica decreased the activation energy of VE671 by 5 kJ/mol and decreased the yield of char.

Silicon carbide (SiC) is another fascinating material, with pure carbon and silicon. Nano-silicon carbide was made composite with VE. Like Si NPs, silicon carbide also provides a good adhesion and offers feasible degree of dispersion in composites, which further alters the mechanical features of the fabricated specimen [4]. A pineapple leaf fibre was hybridized with epoxy resin and SiC mix [52]. Reuss model, an analytical method, was used to study the mechanical features and also for the development of MATLAB code. With an increase in particles concentration, Young's modulus of the composite also increased. On an average, 5.7% of Young's modulus was increased for every volume fraction. Nevertheless, the Poisson ratio of epoxy/SiC decreased with increase in particle volume ration. Longitudinal Young's modulus for the prepared hybrids with both PALF and SiC reinforcement, and the concentration was of 25% volume fraction of SiC in the epoxy matrix. Due to clustering effect, the particulate features of composite are expected to reduce, and hence, 25% was considered as an ideal one.

Similarly, a combination of SiC and rice husk was introduced into the tin-lead alloy for their engineering applications, with an idea of increasing the hardness of alloy [53]. However, the results were not as expected; the composite material possessed greater ductile property than the alloy of tin and lead with 60/40 solder. The results were observed due to the lack of uniformity of the particulate in the mould. Nonetheless, the specimen with rice husk fibre served the purpose by achieving lesser elastic limit than the alloy. The lowered elastic feature is feasible enough to withstand the applied force while clamping the strap clamp.

Silica fume is another category, which has contributed enormously for the concrete-based structural requests [54,55]. Coconut fibre-reinforced concrete is studied for its mechanical properties in the presence and absence of silica fume [56]. With several optimization parameters, 10% silica fume and 2% coconut fibre in the concrete are said to be the best one. The researchers have observed the increased modulus of elasticity, compressive strength and total energy adsorption; however, the durability of coconut fibre poses durability issue of the composite. Steel and polypropylene fibres were reinforced with concrete, and addition of silica fume to them increased the tensile and flexural strengths of the composite [57]. Silica fume was also introduced in the composites made of waste rubber fibre [58] and waste rubber tyre [59], where it partially replaced the cement in the composite.

Silica sol, a colloidal form of silica with a particle size of 5–40 nm and the distribution of which is uniform in aqueous water, is another interesting category considered as grouted silica. An enormous number of active silanol moieties in silica sol interconnected by hydrogen bonds offer good heat resistance. A formation of conductive material with greater mechanical features when silica sol is subjected to heating and dehydration process to form siloxane–silicon bonds is another important feature. This would form a hard and thin inorganic film contributing to enhanced mechanical aspects [60]. Recently, an ultra-lightweight material derived from cellulose fibre was reinforced with silica sol by foam technology in a study [61]. The work justified the deposition of silisol in the cellulose fibre matrix through Si–O–Si bonds (IR analysis) and is expected as a result of dehydration condensation process of silisol with cellulose. A camera picture of prepared composite indicating the macromorphology with the simulated model of its structure and the simulated cross-sectional representation of the composite are provided in Figure 9.5. Due to 3D dispersed fibre structure, very low dense, high porous and good sound adsorption nature of the prepared composite, enhanced thermal and mechanical properties were observed. Some of the studies reported using various forms of silica in the fibre/polymer matrix are tabulated in Table 9.4.

Layered silicates are another ideal class of reinforcing material for organic-inorganic hybrid composites. The intercalation chemistry of silicates has been explored enormously. The dispersion of these nanolayers in the polymer matrix is quite difficult due to their face-to-face stacking feature observed in agglomerated phases [62]. The hydrophilicity of layered silicates and hydrophobicity of the

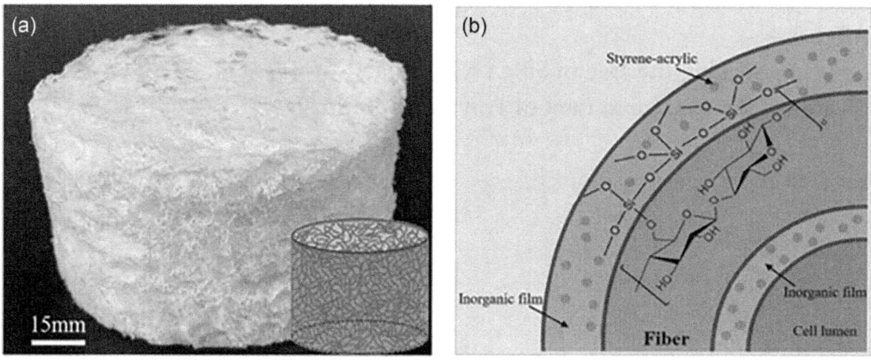

FIGURE 9.5 (a) Macromorphological structure of ultra-lightweight cellulose/silica sol composite; (b) its simulated cross-sectional image.

polymers further possess intrinsic compatibility issues. However, the optimization of their content will promote desirable positive features.

In conclusion, Si content of any form either present in natural fibres itself or made composite externally is found to (1) maximize the interlocking system at the interface of matrix and fibre, (2) provide feasible dispersion of the filler at optimum concentration and (3) reduce the requirement of matrix content. These features aid in successful improvement of mechanical properties, as well as lower the density of the composite.

9.5 CONCLUSION AND FUTURE PROSPECTIVE

VE resins, the thermosetting polymers, are greatly used in various industrial applications because of its anti-corrosion coating property and also show great resistance against the hydrothermal aging. Since the VE resin is brittle, it is very much essential to stabilize the resin with the help of fillers (such as silica and its derivatives). The incorporation of silica nanofillers has accelerated the durability, resistance and coating performance against hydrothermal coating. The silica fillers help in reducing void fraction, prevent the crack formation throughout resin composite by lessening its shrinkage, provide greater stability at higher temperatures and in strong acidic and basic environments and exhibit a strong adhesion between VE resin and silica particles. The mechanical and thermal properties of the resin enhanced drastically upon incorporation of silica NPs. Enhancement in tensile and flexural strength, short-beam shear and hardness is noticed, and this is because of the strong interaction between the silica particles and the VE resins. The silica particles with higher surface area provides the strong interfacial bonding between VE and filler, and this attributes to a great degree of dispersion and interfacial adhesion and thus enhances the composite strength. It is clearly observed that only the desired amount of silica loading helps in enhancing the thermal and mechanical properties because of its thorough dispersion. In comparison to polyester resins, VEs are more reliable and decisive because of their low permeability, abrasion resistance, greater strength and robust mechanical properties. Since VE resins are in significant demand across various industrial sectors because of its outstanding weathering resistance properties,

TABLE 9.4
Literature Serving the Use of Silica NPs, Silica Carbide, Fumed Silica and Silicates in the Reinforcement of Polymer Composites

Form of Silica	Polymer Matrix	Key Alterations at Optimized Concentrations	References
Silica NPs	Epoxy resin	Increase in fracture toughness and deformation resistance	[63]
	Epoxy resin	Increase in storage and loss modulus	[64]
	Epoxy resin	Enhanced Young's modulus	[65]
	Polyurethane	T_g values were constant at varied NPs size and concentrations	[66]
	Polypropylene	Greater elastic modulus and thermal degradation	[67]
	Polystyrene	Enhanced polymer and silane group interaction	[68]
SiC	Epoxy matrix	Enhanced flexural features and work of fracture	[69]
	Polyurethane	Increased tensile strength and thermal resistance	[70]
	Polyvinyl alcohol/MgO	Enhanced electrical, thermal and mechanical features	[71]
	Polystyrene	Increased dielectric constant, tensile strength and elongation at break	[71]
Silica fume	PEODME	Increased energy conversion in dye-sensitized solar cells	[72]
	Nylon 66	Increased flexural, impact and compression strength	[74]
Silicates	Polyamide-6	Improved in-plane shear modulus and strength	[75]
	Polyamide-6	Longer critical aspect ratio	[76]
	Epoxy	Increased mechanical features	[77]

intense research is required in future to design suitable fillers to enhance its thermal and mechanical properties, to make them highly waterproof, and to manifest strong tolerance to structural deformations and to corrosion by chemical agents.

9.6 ACKNOWLEDGEMENT

The authors acknowledge the financial support from SERB-TARE project, Govt. of India (Project Sanction No: TAR/2019/000042).

REFERENCES

1. Mohamed, S. A. N.; Zainudin, E. S.; Sapuan, S. M.; Azaman, M. D.; Arifin, A. M. T. Chapter 1: Introduction to natural fiber reinforced vinyl ester and vinyl polymer composites. In *Natural Fibre Reinforced Vinyl Ester and Vinyl Polymer Composites*; Sapuan, S. M.; Ismail, H.; Zainudin, E. S., Eds.; Cambridge: Woodhead Publishing: 2018; pp. 1–25.

2. Mohammed, L.; Ansari, M. N. M.; Pua, G.; Jawaid, M.; Islam, M. S. A review on natural fiber reinforced polymer composite and its applications. *International Journal of Polymer Science* **2015**, *2015*, 243947, DOI: 10.1155/2015/243947.
3. Hasan, K. M. F.; Horvath, P. G.; Alpar, T. Potential natural fiber polymeric nanobiocomposites: A review. *Polymers* **2020**, *12* (5), 1072, DOI: 10.3390/polym12051072.
4. Alhuthali, A. M.; Low, I.-M. Characterization of mechanical and fracture behaviour in nano-silicon carbide-reinforced vinyl-ester nanocomposites. *Polymer-Plastics Technology and Engineering* **2013**, *52* (9), 921–930, DOI: 10.1080/03602559.2013.763372.
5. Launikitis, M. B. *Handbook of Composites: Vinyl Ester Resins*, Van Nostrand Reinhold Company Inc., New York 1982: pp. 38–49.
6. Labani Motlagh, K.; Seyfi, J.; Khonakdar, H. A.; Mortazavi, S. Vinyl ester/silica aerogel nanocomposite coatings with enhanced hydrophobicity and corrosion protection properties. *Polymers for Advanced Technologies* **2021**, *32* (5), 2176–2184, DOI: 10.1002/pat.5249.
7. Alateyah, A. I.; Dhakal, H. N.; Z. Y. Zhang. Mechanical and thermal properties characterization of vinyl ester matrix nanocomposites based on layered silicate. *International Journal of Mechanical and Mechatronics Engineering* **2013**, *7* (9), 1770–1777.
8. Dehghan Baniani, D.; Jahromi, S. A. J.; Zebarjad, S. M. A study on role of nanosized SiO_2 on deformation mechanism of vinyl ester. *Bulletin Material Science* **2014**, *37* (7), 1677–1683.
9. Shaker, K.; Nawab, Y.; Saouab, A. Influence of silica fillers on failure modes of glass/vinyl ester composites under different mechanical loadings. *Engineering Fracture Mechanics* **2019**, *218*, DOI: 10.1016/j.engfracmech.2019.106605.
10. Arabli, V.; Aghili, A. The effect of silica nanoparticles, thermal stability, and modeling of the curing kinetics of epoxy/silica nanocomposite. *Advanced Composite Materials* **2014**, *24* (6), 561–577, DOI: 10.1080/09243046.2014.944254.
11. Gu, H.; Guo, J.; He, Q.; Tadakamalla, S.; Zhang, X.; Yan, X.; Huang, Y.; Colorado, H. A.; Wei, S.; Guo, Z. Flame-retardant epoxy resin nanocomposites reinforced with polyaniline-stabilized silica nanoparticles. *Industrial & Engineering Chemistry Research* **2013**, *52* (23), 7718–7728, DOI: 10.1021/ie400275n.
12. Gauvin, F.; Robert, M. Durability study of vinylester/silicate nanocomposites for civil engineering applications. *Polymer Degradation and Stability* **2015**, *121*, 359–368, DOI: 10.1016/j.polymdegradstab.2015.09.010.
13. Rejón, L.; Saldivar-Guerrero, R.; Castillo-Ocampo, P. Characterization of nanocomposites produced with thermosetting resins and silica nanoparticles. In *Polymer Processing Society 24th Annual Meeting*, Italy, 2008.
14. Yong, V.; Thomas Hahn, H. Monodisperse SiC/vinyl ester nanocomposites: Dispersant formulation, synthesis, and characterization. *Journal of Materials Research* **2011**, *24* (4), 1553–1558, DOI: 10.1557/jmr.2009.0176.
15. Yong, V.; Hahn, H. T. Processing and properties of SiC/vinyl ester nanocomposites. *Nanotechnology* **2004**, *15* (9), 1338–1343, DOI: 10.1088/0957-4484/15/9/038.
16. Pun, A. K.; Siddhartha. A comparative study for the leverage of micro and nano SiC fillers on thermo-mechanical and erosion wear peculiarity of woven glass fibre-based vinyl ester composite. *Internal journal of advance research in science and engineering* **2018**, *12* (5–6), 467–495.
17. Faruk, O.; Bledzki, A. K.; Fink, H.-P.; Sain, M. Biocomposites reinforced with natural fibers: 2000–2010. *Progress in Polymer Science* **2012**, *37* (11), 1552–1596, DOI: 10.1016/j.progpolymsci.2012.04.003.
18. El-Shekeil, Y. A.; Sapuan, S. M.; Abdan, K.; Zainudin, E. S. Influence of fiber content on the mechanical and thermal properties of Kenaf fiber reinforced thermoplastic polyurethane composites. *Materials & Design* **2012**, *40*, 299–303, DOI: 10.1016/j.matdes.2012.04.003.

19. Wang, C.; Ying, S. A novel strategy for the preparation of bamboo fiber reinforced polypropylene composites. *Fibers and Polymers* **2014**, *15* (1), 117–125, DOI: 10.1007/s12221-014-0117-z.
20. Aprilia, N. S.; Khalil, H. A.; Bhat, A. H.; Dungani, R.; Hossain, M. S. Exploring materials properties of vinyl ester biocomposites filled carbonized jatropha seed shell. *Bioresources* **2014**, *9* (3), 4888–4898.
21. Nadlene, R.; Sapuan, S. M.; Mohammad, J.; Mohamad, R. I.; Yusriah, L. Mechanical and Thermal properties of Roselle fibre reinforced vinyl ester composites. *Bioresources* **2016**, *11* (4), 9325–9339.
22. Manshor, M. R.; Anuar, H.; Nur Aimi, M. N.; Ahmad Fitrie, M. I.; Wan Nazri, W. B.; Sapuan, S. M.; El-Shekeil, Y. A.; Wahit, M. U. Mechanical, thermal and morphological properties of durian skin fibre reinforced PLA biocomposites. *Materials & Design* **2014**, *59*, 279–286, DOI: 10.1016/j.matdes.2014.02.062.
23. Razali, N.; Salit, M. S.; Jawaid, M.; Ishak, M. R.; Lazim, Y. A study on chemical composition, physical, tensile, morphological, and thermal properties of Roselle fibre: Effect of fibre maturity. *BioResources* **2015**, *10* (1), DOI: 10.15376/biores.10.1.1803-1824.
24. Nadlene, R.; Nadlene, R.; Sapuan, S. M.; Sapuan, S. M.; Jawaid, M.; Ishak, M. R.; Ishak, M. R.; Ishak, M. R.; Yusriah, L. Material characterization of Roselle fibre (*Hibiscus sabdariffa* L.) as potential reinforcement material for polymer composites. *Fibres and Textiles in Eastern Europe* **2015**, *23* (6(114)), 23–30, DOI: 10.5604/12303666.1167413.
25. Huzaifah, M. R. M.; Sapuan, S. M.; Leman, Z.; Ishak, M. R. Effect of fibre loading on the physical, mechanical and thermal properties of sugar palm fibre reinforced vinyl ester composites. *Fibers and Polymers* **2019**, *20* (5), 1077–1084, DOI: 10.1007/s12221-019-1040-0.
26. Huzaifah, M. R. M.; Sapuan, S. M.; Leman, Z.; Ishak, M. R. Effect of soil burial on physical, mechanical and thermal properties of sugar palm fibre reinforced vinyl ester composites. *Fibers and Polymers* **2019**, *20* (9), 1893–1899, DOI: 10.1007/s12221-019-9159-6.
27. Yusriah, L.; Sapuan, S. M. Properties of betel nut husk reinforced vinyl ester composites. In *Natural Fibre Reinforced Vinyl Ester and Vinyl Polymer Composites*; Series in Composites Science and Engineering: Woodhead Publishing: 2018; pp. 129–155.
28. Mahato, K.; Goswami, S.; Ambarkar, A. Morphology and mechanical properties of sisal fibre/vinyl ester composites. *Fibers and Polymers* **2014**, *15* (6), 1310–1320, DOI: 10.1007/s12221-014-1310-9.
29. Rahman, I. A.; Padavettan, V. Synthesis of silica nanoparticles by sol-gel: Size-dependent properties, surface modification, and applications in silica-polymer nanocomposites—A review. *Journal of Nanomaterials* **2012**, *2012*, 132424, DOI: 10.1155/2012/132424.
30. Vallet-Regí, M.; Balas, F.; Arcos, D. Mesoporous materials for drug delivery. *Angewandte Chemie (International ed. in English)* **2007**, *46* (40), 7548–7558, DOI: 10.1002/anie.200604488.
31. Slowing, I. I.; Vivero-Escoto, J. L.; Trewyn, B. G.; Lin, V. S. Y. Mesoporous silica nanoparticles: Structural design and applications. *Journal of Materials Chemistry* **2010**, *20* (37), 7924–7937, DOI: 10.1039/C0JM00554A.
32. Suthabanditpong, W.; Takai, C.; Fuji, M.; Buntem, R.; Shirai, T. Improved optical properties of silica/UV-cured polymer composite films made of hollow silica nanoparticles with a hierarchical structure for light diffuser film applications. *Physical Chemistry Chemical Physics* **2016**, *18* (24), 16293–16301, DOI: 10.1039/C6CP01005A.
33. Yong-Taeg, O.; Fujino, S.; Morinag, K. Fabrication of transparent silica glass by powder sintering. *Science and Technology of Advanced Materials* **2002**, *3* (4), 297–301, DOI: 10.1016/s1468-6996(02)00030-x.
34. Mittal, V. Functional polymer nanocomposites with graphene: A review. *Macromolecular Materials and Engineering* **2014**, *299* (8), 906–931, DOI: 10.1002/mame.201300394.

35. Jayavani, S.; Deka, H.; Varghese, T. O.; Nayak, S. K. Recent development and future trends in coir fiber-reinforced green polymer composites: Review and evaluation. *Polymer Composites* **2016**, *37* (11), 3296–3309, DOI: 10.1002/pc.23529.
36. Zhao, L.-D.; Zhang, B.-P.; Li, J.-F.; Zhou, M.; Liu, W.-S.; Liu, J. Thermoelectric and mechanical properties of nano-SiC-dispersed Bi_2Te_3 fabricated by mechanical alloying and spark plasma sintering. *Journal of Alloys and Compounds* **2008**, *455* (1), 259–264, DOI: 10.1016/j.jallcom.2007.01.015.
37. Rooj, S.; Das, A.; Thakur, V.; Mahaling, R. N.; Bhowmick, A. K.; Heinrich, G. Preparation and properties of natural nanocomposites based on natural rubber and naturally occurring halloysite nanotubes. *Materials & Design* **2010**, *31* (4), 2151–2156, DOI: 10.1016/j.matdes.2009.11.009.
38. Alateyah, A. I.; Dhakal, H. N.; Zhang, Z. Y. Water absorption behavior, mechanical and thermal properties of vinyl ester matrix nanocomposites based on layered silicate. *Polymer-Plastics Technology and Engineering* **2014**, *53* (4), 327–343, DOI: 10.1080/03602559.2013.844246.
39. Singh, T.; Gangil, B.; Ranakoti, L.; Joshi, A. Effect of silica nanoparticles on physical, mechanical, and wear properties of natural fiber reinforced polymer composites. *Polymer Composites* **2021**, *42* (5), 2396–2407, DOI: 10.1002/pc.25986.
40. Gowthami, A.; Ramanaiah, K.; Prasad, A. R.; Reddy, K. H. C.; Rao, K. M.; Babu, G. S. Effect of silica on thermal and mechanical properties of sisal fiber rein-forced polyester composites. *Journal of Materials and Environmental Science* **2013**, *4*, 199–204.
41. Wei, C.; Zeng, M.; Xiong, X.; Liu, H.; Luo, K.; Liu, T. Friction properties of sisal fiber/nano-silica reinforced phenol formaldehyde composites. *Polymer Composites* **2015**, *36* (3), 433–438.
42. Vieira, L. M. G.; Santos, J. C. D.; Panzera, T. H.; Christoforo, A. L.; Mano, V.; Campos Rubio, J. C.; Scarpa, F. Hybrid composites based on sisal fibers and silica nanoparticles. *Polymer Composites* **2018**, *39* (1), 146–156.
43. Venkata Prasad, C.; Sudhakara, P.; Prabhakar, M. N.; Ur Rehman Shah, A.; Song, J.-I. An Investigation on the effect of silica aerogel content on thermal and mechanical properties of sisal/PLA nanocomposites. *Polymer Composites* **2018**, *39* (3), 835–840, DOI: 10.1002/pc.24005.
44. Raghul, K. S.; Logesh, M.; Kiran Kisshore, R.; Muhila Ramanan, P.; Muralitharan, G. Mechanical behaviour of sisal palm glass fiber reinforced composite with addition of nano silica. *Materials Today: Proceedings* **2021**, *37*, 1427–1431, DOI: 10.1016/j.matpr.2020.07.063.
45. Chowdary, M. S.; Raghavendra, G.; Kumar, M. N.; Ojha, S.; Boggarapu, V. Influence of nano-silica on enhancing the mechanical properties of sisal/Kevlar fiber reinforced polyester hybrid composites. *Silicon* **2022**, *14*, 539–546.
46. Zhou, S. X.; Wu, L. M.; Sun, J.; Shen, W. D. Effect of nanosilica on the properties of polyester-based polyurethane. *Journal of Applied Polymer Science* **2003**, *88* (1), 189–193.
47. Kang, S.; Hong, S. I.; Choe, C. R.; Park, M.; Rim, S.; Kim, J. Preparation and characterization of epoxy composites filled with functionalized nanosilica particles obtained via sol–gel process. *Polymer* **2001**, *42* (3), 879–887.
48. Santos, J. C.; Vieira, L. M.; Panzera, T. H.; Christoforo, A. L.; Schiavon, M. A.; Scarpa, F. Hybrid silica micro and PDDA/nanoparticles-reinforced carbon fibre composites. *Journal of Composite Materials* **2017**, *51* (6), 783–795.
49. Fu, S.-Y.; Feng, X.-Q.; Lauke, B.; Mai, Y.-W. Effects of particle size, particle/matrix interface adhesion and particle loading on mechanical properties of particulate–polymer composites. *Composites Part B: Engineering* **2008**, *39* (6), 933–961.
50. Cao, Y.; Cameron, J. Flexural and shear properties of silica particle modified glass fiber reinforced epoxy composite. *Journal of Reinforced Plastics and Composites* **2006**, *25* (4), 347–359.

51. Su, C.; Wang, X.; Ding, L.; Wu, Z. Enhancement of mechanical behavior of FRP composites modified by silica nanoparticles. *Construction and Building Materials* **2020**, *262*, 120769.
52. Potluri, R. Mechanical properties of pineapple leaf fiber reinforced epoxy infused with silicon carbide micro particles. *Journal of Natural Fibers* **2019**, *16* (1), 137–151, DOI: 10.1080/15440478.2017.1410511.
53. Taufik, R.; Muhamad, M.; Hasib, H. Feasibility study of natural fiber composite material for engineering application. *Journal of Mechanical Engineering and Sciences* **2014**, *6*, 940.
54. Shannag, M. High strength concrete containing natural pozzolan and silica fume. *Cement and Concrete Composites* **2000**, *22* (6), 399–406.
55. Toutanji, H. A. Properties of polypropylene fiber reinforced silica fume expansive-cement concrete. *Construction and Building Materials* **1999**, *13* (4), 171–177.
56. Khan, M.; Rehman, A.; Ali, M. Efficiency of silica-fume content in plain and natural fiber reinforced concrete for concrete road. *Construction and Building Materials* **2020**, *244*, 118382, DOI: 10.1016/j.conbuildmat.2020.118382.
57. Afroughsabet, V.; Ozbakkaloglu, T. Mechanical and durability properties of high-strength concrete containing steel and polypropylene fibers. *Construction and Building Materials* **2015**, *94*, 73–82, DOI: 10.1016/j.conbuildmat.2015.06.051.
58. Gupta, T.; Sharma, R. K.; Chaudhary, S. Impact resistance of concrete containing waste rubber fiber and silica fume. *International Journal of Impact Engineering* **2015**, *83*, 76–87, DOI: 10.1016/j.ijimpeng.2015.05.002.
59. Gupta, T.; Chaudhary, S.; Sharma, R. K. Mechanical and durability properties of waste rubber fiber concrete with and without silica fume. *Journal of Cleaner Production* **2016**, *112*, 702–711, DOI: 10.1016/j.jclepro.2015.07.081.
60. Morris, C. A.; Anderson, M. L.; Stroud, R. M.; Merzbacher, C. I.; Rolison, D. R. Silica sol as a nanoglue: Flexible synthesis of composite aerogels. *Science* **1999**, *284* (5414), 622–624.
61. Dong, K.; Wang, X. Development of cost effective ultra-lightweight cellulose-based sound absorbing material over silica sol/natural fiber blended substrate. *Carbohydrate Polymers* **2021**, *255*, 117369.
62. Chandradass, J.; Ramesh Kumar, M.; Velmurugan, R. Effect of clay dispersion on mechanical, thermal and vibration properties of glass fiber-reinforced vinyl ester composites. *Journal of Reinforced Plastics and Composites* **2008**, *27* (15), 1585–1601.
63. Yao, X.; Zhou, D.; Yeh, H. Macro/microscopic fracture characterizations of SiO_2/epoxy nanocomposites. *Aerospace Science and Technology* **2008**, *12* (3), 223–230.
64. Kwon, S.-C.; Adachi, T.; Araki, W.; Yamaji, A. Thermo-viscoelastic properties of silica particulate-reinforced epoxy composites: Considered in terms of the particle packing model. *Acta Materialia* **2006**, *54* (12), 3369–3374.
65. Wang, H.; Bai, Y.; Liu, S.; Wu, J.; Wong, C. Combined effects of silica filler and its interface in epoxy resin. *Acta Materialia* **2002**, *50* (17), 4369–4377.
66. González-Irún Rodríguez, J.; Carreira, P.; García-Diez, A.; Hui, D.; Artiaga, R.; Liz-Marzán, L. Nanofiller effect on the glass transition of a polyurethane. *Journal of Thermal Analysis and Calorimetry* **2007**, *87* (1), 45–47.
67. Palza, H.; Vergara, R.; Zapata, P. Composites of polypropylene melt blended with synthesized silica nanoparticles. *Composites Science and Technology* **2011**, *71* (4), 535–540.
68. Naka, Y.; Komori, Y.; Yoshitake, H. One-pot synthesis of organo-functionalized monodisperse silica particles in W/O microemulsion and the effect of functional groups on addition into polystyrene. *Colloids and Surfaces A: Physicochemical and Engineering Aspects* **2010**, *361* (1–3), 162–168.

69. Davies, I. J.; Hamada, H. Flexural properties of a hybrid polymer matrix composite containing carbon and silicon carbide fibres. *Advanced Composite Materials* **2001**, *10* (1), 77–96, DOI: 10.1163/15685510152546376.
70. Guo, Z.; Kim, T. Y.; Lei, K.; Pereira, T.; Sugar, J. G.; Hahn, H. T. Strengthening and thermal stabilization of polyurethane nanocomposites with silicon carbide nanoparticles by a surface-initiated-polymerization approach. *Composites Science and Technology* **2008**, *68* (1), 164–170, DOI: 10.1016/j.compscitech.2007.05.031.
71. Ahmed, H.; Hashim, A.; Abduljalil, H. Analysis of structural, electrical and electronic properties of (polymer nanocomposites/ silicon carbide) for antibacterial application. *Egyptian Journal of Chemistry* **2019**, *62* (4), 767–776, DOI: 10.21608/ejchem.2019.6241.1522.
72. Cao, J.-P.; Zhao, J.; Zhao, X.; Hu, G.-H.; Dang, Z.-M. Preparation and characterization of surface modified silicon carbide/polystyrene nanocomposites. *Journal of Applied Polymer Science* **2013**, *130* (1), 638–644, DOI: 10.1002/app.39186.
73. Kim, J. H.; Kang, M.-S.; Kim, Y. J.; Won, J.; Park, N.-G.; Kang, Y. S. Dye-sensitized nanocrystalline solar cells based on composite polymer electrolytes containing fumed silica nanoparticles. *Chemical Communications* **2004**, (14), 1662–1663, DOI: 10.1039/B405215C.
74. Raja, V. L.; Kumaravel, A. Studies on physical and mechanical properties of silica fume-filled nylon 66 polymer composites for mechanical components. *Polymers and Polymer Composites* **2015**, *23* (6), 427–434, DOI: 10.1177/096739111502300608.
75. Daud, W.; Bersee, H. E. N.; Picken, S. J.; Beukers, A. Layered silicates nanocomposite matrix for improved fiber reinforced composites properties. *Composites Science and Technology* **2009**, *69* (14), 2285–2292, DOI: 10.1016/j.compscitech.2009.01.009.
76. Vlasveld, D. P. N.; Parlevliet, P. P.; Bersee, H. E. N.; Picken, S. J. Fibre–matrix adhesion in glass-fibre reinforced polyamide-6 silicate nanocomposites. *Composites Part A: Applied Science and Manufacturing* **2005**, *36* (1), 1–11, DOI: 10.1016/j.compositesa.2004.06.035.
77. Lin, L.-Y.; Lee, J.-H.; Hong, C.-E.; Yoo, G.-H.; Advani, S. G. Preparation and characterization of layered silicate/glass fiber/epoxy hybrid nanocomposites via vacuum-assisted resin transfer molding (VARTM). *Composites Science and Technology* **2006**, *66* (13), 2116–2125, DOI: 10.1016/j.compscitech.2005.12.025.

10 Natural Fiber–Reinforced Vinyl Ester Composites
Influence of Moisture Absorption on the Physical, Thermal and Mechanical Properties

Le Duong Hung Anh and Pásztory Zoltán
University of Sopron

CONTENTS

10.1 INTRODUCTION

In recent decades, natural fibers–based composites have attracted researchers, engineers, and businesses due to a wide range of applications as a potential replacement for synthetic fiber in polymer composites [1]. Some outstanding benefits since replacing conventional fibers are mechanical capabilities, material strength, being environment-friendly, biological degradation, and the low cost [2]. They showed excellent performance because of the combination of the fibers and matrix polymers; for example, a strong damage endurance or better break elongation when the load applied. However, some drawbacks were found since the natural fibers were used to

DOI: 10.1201/9781003270997-10

make the composites, such as the mechanical properties were reduced because the low interfacial bonding between the fibers and matrix/voids has transformed into a concentration of stress [3]. Another disadvantage is the hydrophilicity of natural fibers resulting in incompatibility with hydrophobic polymers, thus resulting in a drop in mechanical performances, and thermophysical properties of the composites due to the swelling at the fiber–matrix interphase [4].

Numerous practical studies have investigated the physical and mechanical properties of vinyl ester composites derived from renewable resources. The most common natural fibers sourced from plants were used as reinforcement in vinyl ester composites (NFRVC), namely, flax [5], woven hemp fabric [6], jute [7], green coconut husk [8], sugar palm [9,10], rice husk/coir [11], kenaf [12,13], roselle [14–16], date palm seed [17], tamarind seed [18], bamboo [19], pineapple leaves [20], bagasse [21,22], betel nut husk [23], cellulose [24–26], polyalthia longifolia seed [27], and vetiver/banana [28]. These studies followed the method of manipulating the vinyl ester system to produce fiber-reinforced composites and conducted water absorption tests with different types of solvents (i.e., sea water, groundwater, and distilled water). They also studied mechanical properties, thermal behavior at different fiber contents as well as observed the outer surface of fibers and their composites by using the scanning electronic microscopy (SEM). According to experimental results, the mechanical properties increased with a rise in fiber content loading, but started to drop since fiber content reached a maximum value. Besides, these studies also exhibited the thermophysical properties of the composites at different volume fractions using the thermal gravimetric analysis and derivative thermogravimetric (TGA and DTG), in order to observe the variation of the mass under the impact of high temperature over time. In addition, the thermal behavior of composites made from these abovementioned natural fibers was observed when the fiber content increased.

NFRVC are interesting because of low density, high strength, biodegradability, and being environment-friendly. However, because the cell wall polymer of fibers contains hydroxyl groups, it increases significantly the hydrophilicity of the fiber-based composites and affects dimensional stability, density, porosity, biodegradation, and thickness swelling caused by the water uptake [29]. It is apparent that a low interfacial connection between the fibers and polymer matrix has reduced the composite's properties [3]. Hence, it is interesting to explore the influence of water absorption on the mechanical and thermophysical properties of the natural fiber–reinforced vinyl ester composites.

10.2 NATURAL FIBERS REINFORCEMENT

Natural fibrous materials have been currently used with polymers due to recent advancement in technology. They are renewable resources cultivated from plants, which are mainly available in tropical regions of developing countries, and they became more attractive because of their high strength, lightweight materials, flexibility, biodegradability, and being eco-friendly. The properties of some common natural fibers used to fabricate composites are shown in Table 10.1. One of the advantages is the low density than other conventional fibers; for example, the value normally

TABLE 10.1
Properties of Some Commonly Used Natural Fibers

Name of Fibers	Density (g/cm³)	Moisture (%)	Tensile Strength (MPa)	Elongation at Break (%)	Thermal Conductivity (W/(m.K))	Reference
Flax	1.4–1.5	8–12	345–1500	0.2–3.2	0.037–0.045	[34,35]
Hemp	1.48	6.2–12	550–900	2–4	0.04–0.05	[35,36]
Jute	1.3–1.46	12.5–13.7	393–773	1.5–1.8	0.038–0.055	[35–37]
Coir	1.15	11.36	500	20	0.04–0.05	[38]
Sugar palm	1.21	9.16	15.5	5.75	-	[39,40]
Rice husk	0.5	9	19–135	-	0.046–0.056	[37,41]
Kenaf	1.31	6.2–12	427–519	2.5–3.5	0.04–0.065	[42,43]
Roselle	1.33–1.4	3.7–5.8	147–184	5–8	-	[29,44]
Date palm	1.2	24–25	97–196	2–4.5	0.083	[45–47]
Bamboo	0.6–1.1	9.16	503	1.4	-	[45,48]
Pineapple leaf	1.32	11.8	400–700	0.8–1.6	0.07	[42,49, 50]
Bagasse	1.25	45–55	290	-	0.046–0.055	[37,45,51]
Betel nut husk	0.19	12–15	123.93	22.56	-	[52,53]
Banana	-	12.57	600–720	1.9–3.2	0.044	[28,54]

ranges from 1.2 to 1.6 g/cm³ when compared to the density of glass fibers (2.4 g/cm³), as seen in the Table 10.1 [1,30]. The lower density has more volume or number of fibers for the same weight.

Natural fibers are commonly comprised of 5%–20% lignin, 60%–80% cellulose, and moisture up to 20%. They are essentially hygroscopic due to the polar characteristic of their main component. More specifically, the abundant presence of free hydroxyl groups in the amorphous domains of hemicellulose in the cell wall of the fibers [31]. These groups basically tend to establish hydrogen bonding with water molecules when the fibers are exposed to water. The more the existence of hydroxyl groups on natural fibrous materials the more the hydrophilic characteristic. It is apparent that hydrogen bonds are more formed since more water molecules are strapped onto the cellulose fibers. Besides, the more additional cellulose fibers to the polymer composites raise the dimensions of the interfacial region, thereby enabling more water to be attracted. Hence, the higher amount of fiber content in the polymer matrix composites has led to strengthening the large amount of water absorption [32]. Consequently, the whole properties of composites will be affected markedly [33].

10.3 VINYL ESTER

Epoxy is normally used as the thermoset resin for polymer matrixes in composites. Vinyl ester (VE) resins, on the other hand, are the additional product of an epoxy resin and an unsaturated carboxyl acid with a molecular structure quite similar to

TABLE 10.2
Physical and Mechanical Properties of Vinyl Ester Resin

Properties	Viscosity (Cps)	Density (g/cm^3)	Tensile Strength (MPa)	Flexural Strength (MPa)	Flexural Modulus (GPa)	Thermal Conductivity (W/(m.K))
Value	450 ± 100	1.04	80–90, 73–81	130–140	3	0.2–0.22

that of the polyester resin [55]. The water absorption of vinyl ester resin was investigated and compared to other resins. It is observed that the relationship between the water uptake and time of polyester and vinyl ester resin follows the Fickian mechanism with the initial liner portion. Visco et al. [56] concluded that the maximum water absorption value is lower than 0.7%, where the value of polyester resin is 1.2%. Another study by Alia et al. [57] revealed that the diffusion coefficient of vinyl ester is smaller than polyurethane resin, while the maximum water uptake is around 0.55% compared to 0.6% for polyurethane resin. Table 10.2 shows the physical, mechanical properties and thermal conductivity of vinyl ester resin [13,58,59].

VE composite has more application because it's good mechanically when reinforced with natural fibers. The use of VE may enhance the strength, the stiffness, dimensional stability, and other properties of a composite. Besides, VE reveals better resistance to moisture absorption and water penetration than polyester or epoxy resins due to a smaller number of ester groups in the structure. However, the process of vinyl ester is more complicated than polyester resin, requiring thorough surface penetration during cure to achieve acceptable levels of adhesion between fiber and vinyl ester matrix [5].

The mechanical properties of the composites using vinyl ester are relatively similar to those of epoxy resins; however, VE resin is breakable, and its performance needs to be strengthened by reinforcing with the fillers [60]. Nevertheless, for brittle thermosetting resins such as VE, the expansion of the natural fibers occurs by the extension of microcracking since the composites are exposed to extreme moisture environments for long, resulting in the composite's failure. Because the capillarity originated from the damage of composite cracks, the flow of water molecules and the diffusion through the bulk matrix attack the interface, causing the degradation of the fiber/matrix interface areas [24,44,61]. It is compatible with the investigation of carbon fiber–reinforced vinyl ester composites by Gargano et al. [62]. They stated that the composite degradation was strongly influenced by hydrolysis and cracks caused by swelling.

10.4 FIBER REINFORCEMENT ON VINYL ESTER COMPOSITES

10.4.1 Preparation of Composites

To obtain the natural fiber–reinforced vinyl ester composites, a technique named wet hand lay-up is commonly used. Firstly, the raw fibers were moderately added to the vinyl ester resin and stirred using a mechanical stirrer until the mixture was uniformly distributed. Then, a hardener called methyl ethyl ketone peroxide (MEKP) was mixed in a different ratio by weight for curing (i.e., in the ratio of 1:44 [6]). Finally, the mixture (as wet fabrics) was poured into the metal mold and was pressed by using a compressive machine and left to cure in room temperature for 24 h.

10.4.2 Properties of Natural Fiber–Reinforced Vinyl Ester Composites

The physical and mechanical properties (mostly of tensile and flexural strength) of natural fiber–reinforced composites were investigated under the influence of filler loading, volume fraction, chemical treatment, manufacturing process, interfacial adhesion, etc. Numerous studies have also indicated the water damage of natural fiber vinyl ester composites [6,24,63,64]. Table 10.3 shows the properties of natural fibers reinforced vinyl ester composites according to the published data. In general, the composite properties are concluded as a result of the nature of the bonding between the fibers/polymer matrix, and the interfacial adhesion quality of the finalized composite [26]. Nevertheless, moisture was revealed as an essential key to further research due to degrading mechanical and thermophysical properties in wet conditions. The mechanisms of the water transport process mainly include the infiltration of water molecules through the microcracks in the matrix structure, the diffusion of water vapor in the gaps between polymer chains, and capillary action at the interface of fibers and polymer matrix [65,66]. It creates obstacles in maintaining good adhesion between fiber/polymer matrix composites and contributes to higher water absorption, which weakens the composite product in applications. One of the negative consequences due to varying water absorption are unexpected dimensional changes of the composites. Because of shrinking and swelling of the material, visual and structural problems occur, including splitting, development of decay and stain fungi, and therefore loss of mechanical strength.

10.5 WATER UPTAKE EXPERIMENTS

Water uptake tests have been experimentally conducted by soaking the composites in water. The weight of tested samples was measured first (noted as W_0), and then they were immersed in various liquids at room temperature. Next, the samples were taken out and reweighted (noted as W_1) after being dried. Theoretically, the moisture absorption is computed by the weight difference using the below equation (10.1). Moreover, the percentage of water uptake was plotted versus the square root of immersion time [70].

$$\text{Water absorption}(\%) = \frac{W_1 - W_0}{W_0} \times 100\% \quad (10.1)$$

In essence, the water uptake level increased with an increase in fiber content in all cases [44]. This is the basis of enhancement of micro-void formation in the matrix resin [71]. Some studies were investigated to highlight the maximum moisture absorption percent before evaluating its effect on the properties of the composites [10,15,19,63], as seen in Table 10.3. Manickam et al. observed a linear increase in

TABLE 10.3
Physical and Mechanical Properties of Some Natural Fiber–Reinforced Vinyl Ester Composites

Name of Fibers	Density (g/cm³)	Maximum Water Absorption (%)	Tensile Strength (MPa)	Impact Strength (kJ/m²)	Flexural Strength (MPa)	Reference
Jute	1.3	-	11.89	22.1	199.1	[7,67]
Woven hemp fabric	-	3.43	46.47	-	70.23	[6]
Flax	1.19–1.32	5.92	216.13	-	334.5	[5]
Rice husk/Coir	-	-	31	43.4	38	[8,11]
Kenaf	1.24	-	147.5	-	180	[13]
Roselle	1.6	2.2 7.2	17–24	5.43–2.5	58–110	[14–16,68]
Bagasse	1.25	6.95	76.63	5.89	112.47	[21]
Betel nut husk	1.15–1.28	8	39–100	2–5.7	60–80	[23]
Cellulose	1.18–1.27	1.29	-	2.6–4.45	42–56.5	[24]
Tamarind seed	-	-	34.1	14.02	121	[18]
Bamboo	1.158	-	90	-	150	[69]
Polyalthia longifolia seed	-	7.5	32.5	31.09	125	[27]
Vetiver/banana	-	-	25–47	23–53	45–86	[28]
Sugar palm	-	9	6.1–25.1	5.9	2.5–48.5	[10]

water absorption with immersion time and reached a specified value at a saturation level [15]. Practical tests were implemented with roselle fiber–reinforced vinyl ester composites by immersing in three types of liquids. It is identified that the higher volume fraction of composites that were soaked in sea water showed a higher water absorption percentage than that of other water environments. Furthermore, the maximum water absorption was pointed out at 45% of the volume fraction of all cases and the water absorption proportion increased with increased fiber content. The moisture penetration behavior for different relative humidity levels of the composites made from bamboo strips was observed in the study by Chen et al. [19]. Water absorption behavior was investigated for untreated and fire retardant treated woven hem fiber–reinforced vinyl ester composites, which were immersed in distilled water at room temperature [6]. According to the curves, the fire retardant treated hem fiber reached a maximum water uptake after only 552 h, and the weight started decreasing after 840 h up to 2700 h immersion, whereas the untreated fiber was observed to continue increasing from the starting point to the maximum water uptake of 3.43% after 1848 h and reached saturation level. The results of water absorption percentages of the composites made from sugar palm fiber/vinyl ester collected since being immersed for 7 days agreed with the conclusion mentioned above [10].

As the moisture can critically damage the fibers and matrix adhesion, especially at high humidity, the tensile and the flexural strength tend to drop, which was caused by the poor interfacial bonding [4,60]. From these viewpoints, the moisture absorption must be taken into consideration while investigating the natural fiber–reinforced composites.

10.6 INFLUENCE OF MOISTURE ABSORPTION

10.6.1 Physical Properties

Natural fibers are influenced by moisture absorption in two ways—it changes the fiber density through absorbed water molecules and the fiber swelling [72]. Therefore, the physical properties (i.e., density and thickness swelling) are also influenced. Figure 10.1 shows how the fiber density changes with moisture content.

The theoretical and experimental composites' densities were found to increase when the weight fraction of the fibers was adjusted to be larger. This is because of the existence of fiber content within the composites, which was denser than the polymer matrix. Besides, an increase in water uptake is always related to a rise in fiber content due to the hydrophilic nature of cellulose fibers, associated with an increased density of the composites [61,68]. The physical density of natural fibers determines the weight of the natural fiber–reinforced composites. Practical results showed that since the percentage of water uptake increased from 0.76% to 14.62%, the experimental densities of cellulose fiber–reinforced vinyl ester composites increased from 1.18 to 1.27 g/cm^3 [61]. The same result was concluded in the study of betel nut husk fiber–reinforced vinyl ester composites at four different levels of fiber content [23], in which a linear increase in the beginning phase slowly fluctuates until reaching the saturation point during the water absorption process.

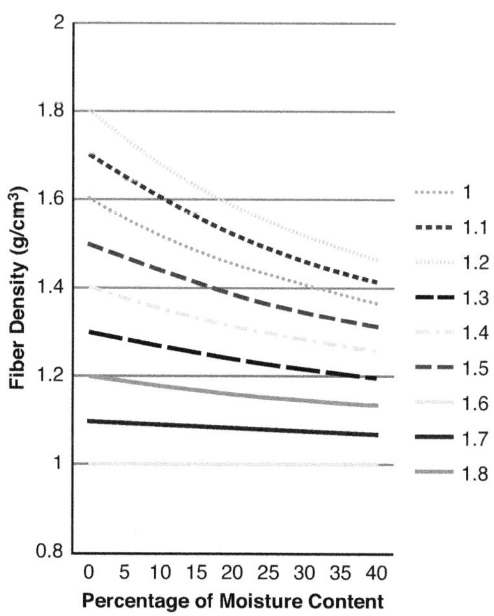

FIGURE 10.1 Effect of moisture content on the natural fiber density [72].

Because of hydrophilicity, the fibers relatively expand when the composites are immersed in water, resulting in changes in the bulk density due to the incorporation of raw fibers in the vinyl ester matrix. In addition, an increased density could be related to water diffusion through the open-cavity of fibers' structure and the penetration of ester resin liquid, which is then entrapped into the hollow lumen of fibers. Hence, the more presence of fibers in the vinyl ester composites has caused a high percentage of moisture content level, consequently leading to an increase in the weight of composites [68].

An increase in water absorption and thickness swelling was reported in the study on sugar palm fiber–reinforced vinyl ester composites at different fiber contents [10]. As a result, the composite containing higher fiber content showed a higher water uptake and swelling percentage. It is evident that the more presence of absorbed water caused more hydrogen bonding in the cell wall of fibers, and then the composites were expanded and swollen.

10.6.2 Thermal Properties

It is believed that the moisture content significantly affects the thermal properties of the natural fiber–reinforced composites because of the hydrophilicity of plant-based fibrous materials. Some published experiments have indicated that higher moisture content is always associated with higher thermal conductivity of natural fibrous materials (i.e., bagasse, cotton, hemp, flax, and cellulose fibers) [73–75]. The low thermal conductivity value of natural fiber–reinforced composites is normally governed by the low heat conductivity of fibrous materials (ranging from 0.03 to 0.06 W/(m.K)) [76] and the resin is often insulated, according to Bavan and Kumar [77]. Besides, the lumen structure of natural fibers is a hollow filled with air, which creates a barrier to heat transmission because of the low thermal conductivity of air (0.026 W/(m.K)) at 25°C; hence, the thermal conductivity value of the whole composites is affected as well [78]. It is compatible with a report of betel nut husk–reinforced vinyl ester composites [23]; they found that the thermal conductivity decreased slightly from 0.22 to 0.14 W/(m.K) compared to a neat sample when the fiber loading raised from 10% to 40%. There are two influencing factors mentioned in this study—the rise in the number of hollow lumen structures and the compact fiber composition in additional fibers as the fiber content increased. Another explanation supported by Mangal et al. [79] was the declined conductive paths for the dissipation of heat energy at high fiber loading. However, this report did not investigate the increased water uptake when the fiber content increased and how the moisture absorption influenced thermal properties because once the water molecules penetrate markedly inside the lumen structures, they may change the heat conduction of the natural fiber as more moisture absorption causes high heat conduction because the thermal conductivity of water is about 20 times higher than that of stationary air (0.598 W/(m.K) at 20°C [35]).

Thermal stability and behavior of some fiber reinforcement composites were evaluated using TGA and DTG. It has been reported that the degradation of composites is more affected by water than temperature [64]. In general, there are three main stages of degradation introduced in which the first phase is governed by the moisture occurrence, and the neat VE showed one phase during the heating process [16,23,69,80].

Razali et al. concluded that the first stage of degradation process of roselle–reinforced VE composites was introduced by the moisture loss (30°C–110°C) [14]. In addition, 10% fiber content showed the highest thermal degradation compared to other fiber loadings, and the percentage of moisture loss has risen with an increase in the fiber loading [16]. The thermal degradation of neat VE/VE combined with lowest and highest betel nut husk fiber content (10% and 40%) was investigated [23]. Results showed that the increased moisture content with the presence of more fiber content decreased the thermal stability of composites, whereas the neat VE resin showed a very low moisture content. According to the TGA curves, the initial thermal degradation phase was associated with moisture reduction when the temperature increased from 75°C to 225°C. All the above investigations indicated that the effect of the moisture presence increased together with the growth of more fiber loading into the VE resin (10–40 wt.%) in the thermal stability of the composites. In conclusion, the initial weight loss of the fiber reinforcement composites was mainly attributed to the evaporation of the absorbed moisture between fibers/matrix, and the other phases of degradation were commonly caused by the decomposition of hemicellulose, lignin components at high temperatures.

10.6.3 Mechanical Properties

Water absorption produces a negative impact on mechanical properties, for example, reducing the dimensional stability, creating the microcracking, and speeding up the biodegradation process [81]. The water molecules that penetrate the composite's structure from cracks and voids have declined the bonding between the fiber and matrix phase through the hydrolysis reaction of unsaturated groups within the resin [82]. The tensile strengths of natural fiber–reinforced composites are mainly influenced by the interfacial adhesion between the fibers and the matrix, and together with Young's modulus, they will increase with increased fiber loading [60,83]. The lignin in the fibers also contributes to the strength of the composite's structure [68]. As the fiber content increases, the mechanical properties of composites increase due to the interaction between the fiber and the composite matrix [15]. However, when the composites are exposed to water, the hydrophilicity of the natural fibers tends to increase the fiber swelling at the fiber–matrix interphase. Prolonged exposure to water absorption may cause a reduction in mechanical performance of the natural fiber–reinforced composite [24]. In addition, the longtime duration of immersing in water also changes glass transition temperature and may lead to obtaining of short chain molecules, influencing the flexural modulus [55].

The effect of water absorption on the mechanical properties of the natural fiber–reinforced vinyl ester composites has been studied. Practical results displayed a decrease in flexural and modulus strength relating to a decrease in interfacial bonding, which was formed at relatively high humidity conditions [6,24]. Misnon et al. [6] investigated the influence of water absorption on the mechanical properties of woven hemp fabric–reinforced vinyl ester composites and found that the existence of water reduced the tensile strength and tensile modulus up to 24% and 39%, respectively, as water penetration has weakened the adhesion between the fibers and vinyl resin. The same behavior was also revealed for flexural properties in the case of both untreated and fire-treated fibers after being exposed to water for up to 168 h. Interestingly, since the composites were immersed in water after 2688 h, an increase in flexural properties was found because a high amount of water filled the gaps between the fiber and the matrix, causing an increase in friction between them, and creating the resistance to resist the flexural load.

The effect of different types of water (sea, ground, and drilled) in the mechanical properties of roselle fiber–reinforced vinyl ester composites was investigated [14,15]. Results showed that the tensile and flexural strength increased when the fiber loading increased from 15% to 45% at a dry condition, while the impact strength values were found to decrease. After being exposed to water, mechanical properties showed a considerable decrease in three cases of fiber loading (15%, 30%, and 45%). This result agreed with the investigation of water absorption on elastic modulus, flexural strength in cellulose fiber–reinforced vinyl ester composites [24]. It is obvious that the penetration of water damages interfacial adhesion, and then creates the debonding of the fiber and the matrix, contributing to a decrease in the mechanical properties of the composites. Alkali solution is a common treatment that reduces the water absorption to obtain the better mechanical properties of the natural fiber–reinforced composites. Results showed that the flexural and modulus strength of jute fiber–reinforced vinyl ester composites improved significantly since the composites at 35% fiber loading were treated by 5% NaOH after 4 h [7]. The same result was revealed with coir fiber–reinforced polyester composite [84].

The mechanical properties of betel nut husk–reinforced vinyl ester composites were carried out with an increase in water uptake from low to high fiber content [23]. As shown in the results, it is considered that further increases in the water uptake at high fiber content (at 40 wt.%) resulted in a significant decrease in the flexural properties. The same conclusion was also evident in the study of date palm seed and cellulose fiber–reinforced vinyl ester composites in wet conditions [17,25]. The reduction in flexural strength could be explained by the weakening of the fiber–matrix interface due to the formation of intermolecular hydrogen bonding between the absorbed water caused by increased fiber content and the fibers, leading to a reduction of adhesion between the fibers and matrix; therefore, the performance of the whole composite was significantly decreased. Consequently, a further increase in the water absorption resulted in a negative effect on the flexural strength of composites [20,24]. On the other hand, investigation of mechanical properties of pineapple leaf fiber–reinforced composites revealed that storing the composites in hot and humid conditions does not significantly reduce their reinforcing capability [20]. However, this paper also concluded that the presence of more fibers may cause greater moisture absorption and result in higher weight loss in a high temperature.

The effect of moisture absorption on the interfacial shear strength of bamboo vinyl ester composites was investigated [19]. It is concluded that the exposure of post-fabrication composites to moisture was less damaged than the moisture uptake during the composite fabrication. Experimental data indicated that the interfacial strength reduced by 40% when immersing in water for 9 days, but no further reduction happened even when immersed up to 100 days. A small increase of tensile strength was also observed in this study; however, there was no effect on flexural strength since the relative humidity increased up to 90%. The same conclusion with an increase of the tensile strength of flax fiber reinforced bioepoxy composites was also noted in the study by Muñoz et al. [63]. In this study, the tensile strength of the biocomposites immersed in water for 768 h showed a higher value than that of dry samples. It is because the swelling of the fibers caused by high absorption quantities and possibly resulted in the gaps between the fiber and the matrix caused by the manufacturing process could be filled up, therefore, can lead to an increase of approximately 10%–35% of the tensile strength. In addition, it is also found that the flexural strength decreased as moisture absorption content increased due to the water absorption at a higher percentage has led to the formation of higher number of microcracks as a result of fiber swelling leading to the weak interfacial adhesion between fiber and matrix.

10.7 CONCLUSIONS

The specific objective of this chapter was to discuss the influence of water absorption on the mechanical, thermal, and physical properties of natural fiber–reinforced vinyl ester composites. According to previous studies, an increase in fiber content loading can enhance the endurance of reinforcement of fiber-based composites. However, it is also evident that the water uptake increases when the fiber content increases, leading to the restriction in obtaining good bonding in fiber–matrix adhesion, resulting in poor performance in the composite's products. Specifically, it caused a decrease in the mechanical properties, affecting the thermal degradation, and influencing the heat conductivity of the whole composite.

The potential of plant-based fibers is one of the viable solutions to solve the costly, wasteful, and nonenvironment-friendly methods of disposing of nondegradable composites. Many efforts were made to investigate the outstanding performances of natural fiber–reinforced composites but still experimental investigations are lacking on the effect of moisture on the mechanical and thermophysical properties. Therefore, this present chapter can be useful to provide a source of literature for further research concerning the use of natural fiber–reinforced composites in various applications.

REFERENCES

1. Nguong, C., S. Lee, and D. Sujan, A review on natural fibre reinforced polymer composites. *International Journal of Materials and Metallurgical Engineering*, 2013. **7**(1): pp. 52–59.
2. Malkapuram, R., V. Kumar, and Y.S. Negi, Recent development in natural fiber reinforced polypropylene composites. *Journal of Reinforced Plastics and Composites*, 2009. **28**(10): pp. 1169–1189.
3. Milanese, A.C., M.O.H. Cioffi, and H.J.C. Voorwald, Mechanical behavior of natural fiber composites. *Procedia Engineering*, 2011. **10**: pp. 2022–2027.
4. Mehta, G., et al., Effect of novel sizing on the mechanical and morphological characteristics of natural fiber reinforced unsaturated polyester resin-based bio-composites. *Journal of Materials Science*, 2004. **39**(8): pp. 2961–2964.
5. Huo, S., V.S. Chevali, and C.A. Ulven, Study on interfacial properties of unidirectional flax/vinyl ester composites: resin manipulation on vinyl ester system. *Journal of Applied Polymer Science*, 2013. **128**(5): pp. 3490–3500.

6. Misnon, M.I., et al., Woven hemp fabric reinforced vinyl ester composite: effect of water absorption on the mechanical properties degradation. *International Journal of Advances in Science, Engineering and Technology*, 2016. **4**(3): pp. 96–101.
7. Ray, D., et al., The mechanical properties of vinylester resin matrix composites reinforced with alkali-treated jute fibres. *Composites Part A: Applied Science and Manufacturing*, 2001. **32**(1): pp. 119–127.
8. Jayabal, S., et al., Mechanical performance of bio particulated natural green husk coir fiber-vinyl ester composites. *Cellulose*, 2016. **28**(35.7): pp. 31–33.
9. Ammar, I., et al., Development of sugar palm fiber reinforced vinyl ester composites, in: S.M. Sapuan, H. Ismail, E.S. Zainudin (eds), *Natural Fibre Reinforced Vinyl Ester and Vinyl Polymer Composites*. 2018, Elsevier. pp. 211–224.
10. Huzaifah, M., et al., Effect of fibre loading on the physical, mechanical and thermal properties of sugar palm fibre reinforced vinyl ester composites. *Fibers and Polymers*, 2019. **20**(5): pp. 1077–1084.
11. Ramprasath, R., et al., Investigation on impact behavior of rice husk impregnated coir-vinyl ester composites, in: *Third International Conference on Polymer Processing and Characterization*, Kottayam, India, October 2014. *Macromolecular Symposia*. 2016. Wiley Online Library.
12. Fairuz, A., et al., Optimization of pultrusion process for kenaf fibre reinforced vinyl ester composites, in: M.J.M. Nor, M.E.A.Manaf, L.K.Tee (eds), *International Conference on Design and Concurrent Engineering, Malacca, Malaysia (iDECON 2014)*, September 22–23, 2014. *Applied Mechanics and Materials* (761), Switzerland. 2015. Trans Tech Publication Ltd. p. 499–503.
13. Fairuz, A., et al., Effect of filler loading on mechanical properties of pultruded kenaf fibre reinforced vinyl ester composites. *Journal of Mechanical Engineering and Sciences*, 2016. **10**(1): pp. 1931–1942.
14. Nadlene, R., et al., The effects of chemical treatment on the structural and thermal, physical, and mechanical and morphological properties of roselle fiber-reinforced vinyl ester composites. *Polymer Composites*, 2018. **39**(1): pp. 274–287.
15. Manickam, C., et al., Effect of various water immersions on mechanical properties of roselle fiber–vinyl ester composites. *Polymer Composites*, 2015. **36**(9): pp. 1638–1646.
16. Razali, N., et al., Mechanical and thermal properties of Roselle fibre reinforced vinyl ester composites. *BioResources*, 2016. **11**(4): pp. 9325–9339.
17. Nagaraj, N., et al., Effect of cellulosic filler loading on mechanical and thermal properties of date palm seed/vinyl ester composites. *International journal of biological macromolecules*, 2020. **147**: pp. 53–66.
18. Stalin, B., et al., Evaluation of mechanical and thermal properties of tamarind seed filler reinforced vinyl ester composites. *Journal of Vinyl and Additive Technology*, 2019. **25**(s2): pp. E114–E128.
19. Chen, H., M. Miao, and X. Ding, Influence of moisture absorption on the interfacial strength of bamboo/vinyl ester composites. *Composites Part A: Applied Science and Manufacturing*, 2009. **40**(12): pp. 2013–2019.
20. Mohamed, A., S. Sapuan, and A. Khalina, Mechanical and thermal properties of josapine pineapple leaf fiber (PALF) and PALF-reinforced vinyl ester composites. *Fibers and Polymers*, 2014. **15**(5): pp. 1035–1041.
21. Athijayamani, A., et al., Mechanical properties of unidirectional aligned bagasse fibers/vinyl ester composite. *Journal of Polymer Engineering*, 2016. **36**(2): pp. 157–163.
22. Athijayamani, A., et al., Parametric analysis of mechanical properties of bagasse fiber-reinforced vinyl ester composites. *Journal of Composite Materials*, 2016. **50**(4): pp. 481–493.

23. Yusriah, L., et al., Thermo-physical, thermal degradation, and flexural properties of betel nut husk fiber reinforced vinyl ester composites. *Polymer Composites*, 2016. **37**(7): pp. 2008–2017.
24. Alhuthali, A.M. and I.M. Low, Effect of prolonged water absorption on mechanical properties in cellulose fiber reinforced vinyl-ester composites. *Polymer Engineering & Science*, 2015. **55**(12): pp. 2685–2697.
25. Alhuthali, A. and I.M. Low, Mechanical properties of cellulose fibre reinforced vinyl-ester composites in wet conditions. *Journal of Materials Science*, 2013. **48**(18): pp. 6331–6340.
26. Alhuthali, A., I.M. Low, and C. Dong, Characterisation of the water absorption, mechanical and thermal properties of recycled cellulose fibre reinforced vinyl-ester eco-nanocomposites. *Composites Part B: Engineering*, 2012. **43**(7): pp. 2772–2781.
27. Stalin, B., et al., Evaluation of mechanical, thermal and water absorption behaviors of Polyalthia longifolia seed reinforced vinyl ester composites. *Carbohydrate Polymers*, 2020. **248**: p. 116748.
28. Stalin, A., et al., Mechanical properties of hybrid vetiver/banana fiber mat reinforced vinyl ester composites. *Journal of Industrial Textiles*, 2020. 51(4S): p. 5869S–5886S.
29. Razali, N., et al., A study on chemical composition, physical, tensile, morphological, and thermal properties of roselle fibre: effect of fibre maturity. *BioResources*, 2015. **10**(1): pp. 1803–1824.
30. Huda, M.S., et al., Chopped glass and recycled newspaper as reinforcement fibers in injection molded poly (lactic acid)(PLA) composites: a comparative study. *Composites science and technology*, 2006. **66**(11–12): pp. 1813–1824.
31. Neagu, R.C., E.K. Gamstedt, and M. Lindström, Influence of wood-fibre hygroexpansion on the dimensional instability of fibre mats and composites. *Composites Part A: Applied Science and Manufacturing*, 2005. **36**(6): pp. 772–788.
32. Adhikary, K.B., S. Pang, and M.P. Staiger, Long term moisture absorption and thickness swelling behaviour of recycled thermoplastics reinforced with Pinus radiata sawdust. *Chemical Engineering Journal*, 2008. **142**(2): pp. 190–198.
33. Sathishkumar, T., et al., Characterization of natural fiber and composites–A review. *Journal of Reinforced Plastics and Composites*, 2013. **32**(19): pp. 1457–1476.
34. De Rosa, I.M., C. Santulli, and F. Sarasini, Mechanical and thermal characterization of epoxy composites reinforced with random and quasi-unidirectional untreated Phormium tenax leaf fibers. *Materials & Design (1980–2015)*, 2010. **31**(5): pp. 2397–2405.
35. Pfundstein, M., et al., *Insulating Materials: Principles, Materials, Applications*. 2012. Walter de Gruyter, Birkhäuser, Basel.
36. Li, X., L.G. Tabil, and S. Panigrahi, Chemical treatments of natural fiber for use in natural fiber-reinforced composites: a review. *Journal of Polymers and the Environment*, 2007. **15**(1): pp. 25–33.
37. D'Alessandro, F., et al., Insulation materials for the building sector: a review and comparative analysis. *Renewable and Sustainable Energy Reviews*, 2016. **62**: pp. 988–1011.
38. Joseph, K., et al., A review on sisal fiber reinforced polymer composites. *Revista Brasileira de Engenharia Agrícola e Ambiental*, 1999. **3**: pp. 367–379.
39. Afzaluddin, A., et al., Physical and mechanical properties of sugar palm/glass fiber reinforced thermoplastic polyurethane hybrid composites. *Journal of Materials Research and Technology*, 2019. **8**(1): pp. 950–959.
40. Leman, Z., et al., Moisture absorption behavior of sugar palm fiber reinforced epoxy composites. *Materials & Design*, 2008. **29**(8): pp. 1666–1670.
41. Chen, Z., Y. Xu, and S. Shivkumar, Microstructure and tensile properties of various varieties of rice husk. *Journal of the Science of Food and Agriculture*, 2018. **98**(3): pp. 1061–1070.

42. Munawar, S.S., K. Umemura, and S. Kawai, Characterization of the morphological, physical, and mechanical properties of seven nonwood plant fiber bundles. *Journal of Wood Science*, 2007. **53**(2): pp. 108–113.
43. Rassmann, S., R. Reid, and R. Paskaramoorthy, Effects of processing conditions on the mechanical and water absorption properties of resin transfer moulded kenaf fibre reinforced polyester composite laminates. *Composites Part A: Applied Science and Manufacturing*, 2010. **41**(11): pp. 1612–1619.
44. Athijayamani, A., et al., Effect of moisture absorption on the mechanical properties of randomly oriented natural fibers/polyester hybrid composite. *Materials Science and Engineering: A*, 2009. **517**(1–2): pp. 344–353.
45. John, M.J. and R.D. Anandjiwala, Recent developments in chemical modification and characterization of natural fiber-reinforced composites. *Polymer composites*, 2008. **29**(2): pp. 187–207.
46. Al-Oqla, F.M. and S. Sapuan, Natural fiber reinforced polymer composites in industrial applications: feasibility of date palm fibers for sustainable automotive industry. *Journal of Cleaner Production*, 2014. **66**: pp. 347–354.
47. Falade, K.O. and E.S. Abbo, Air-drying and rehydration characteristics of date palm (Phoenix dactylifera L.) fruits. *Journal of Food Engineering*, 2007. **79**(2): pp. 724–730.
48. Rao, K.M.M. and K.M. Rao, Extraction and tensile properties of natural fibers: vakka, date and bamboo. *Composite structures*, 2007. **77**(3): pp. 288–295.
49. Mittal, M. and R. Chaudhary, Experimental investigation on the thermal behavior of untreated and alkali-treated pineapple leaf and coconut husk fibers. *International Journal of Applied Science and Engineering*, 2019. **7**(1): pp. 1–16.
50. Anonaba, A.U., E.C. Mbamala, and U.S. Mbamara, *Thermal Properties of Pineapple Leaf Composite and its Suitability as a Viable Alternative for Efficient Roofing Material*. 2019.
51. Panyakaew, S. and S. Fotios, New thermal insulation boards made from coconut husk and bagasse. *Energy and buildings*, 2011. **43**(7): pp. 1732–1739.
52. Yusriah, L., et al., Exploring the potential of betel nut husk fiber as reinforcement in polymer composites: effect of fiber maturity. *Procedia Chemistry*, 2012. **4**: pp. 87–94.
53. Srisang, N., S. Srisang, and T. Chungcharoen, The study of the drying kinetic and the cutting performance in dried betel nut production machine. in *MATEC Web of Conferences*. 2018. EDP Sciences.
54. Manohar, K., A comparison of banana fiber insulation with biodegradable fibrous thermal insulation. *American Journal of Engineering Research (AJER)*, 2016. **5**(8): pp. 249–255.
55. Lima Sobrinho, L., M. Ferreira, and F.L. Bastian, The effects of water absorption on an ester vinyl resin system. *Materials Research*, 2009. **12**: pp. 353–361.
56. Visco, A., N. Campo, and P. Cianciafara, Comparison of seawater absorption properties of thermoset resins based composites. *Composites Part A: Applied Science and Manufacturing*, 2011. **42**(2): pp. 123–130.
57. Alia, C., et al., Degradation in seawater of structural adhesives for hybrid fibre-metal laminated materials. *Advances in Materials Science and Engineering*. **2013**: p. 869075.
58. Crosky, A., et al., Thermoset matrix natural fibre-reinforced composites. In: A. Hodzic, R. Shanks (eds), *Natural Fibre Composites*. Woodhead Publishing, Cambridge, MA. 2014: pp. 233–270.
59. Mutnuri, B., *Thermal Conductivity Characterization of Composite Materials*. (MSc Thesis), Department of Mechanical Engineering, Morgantown, West Virginia. 2006: West Virginia University.
60. Ku, H., et al., A review on the tensile properties of natural fiber reinforced polymer composites. *Composites Part B: Engineering*, 2011. **42**(4): pp. 856–873.

61. Kim, H.J., Effect of water absorption fatigue on mechanical properties of sisal textile-reinforced composites. *International Journal of Fatigue*, 2006. **28**(10): pp. 1307–1314.
62. Gargano, A., et al., Effect of seawater immersion on the explosive blast response of a carbon fibre-polymer laminate. *Composites Part A: Applied Science and Manufacturing*, 2018. **109**: pp. 382–391.
63. Muñoz, E. and J.A. García-Manrique, Water absorption behaviour and its effect on the mechanical properties of flax fibre reinforced bioepoxy composites. *International Journal of Polymer Science*. **2015**: p. 390275.
64. Abdurohman, K. and M. Adhitya. Effect of water and seawater on mechanical properties of fiber reinforced polymer composites: a review for amphibious aircraft float development. in *IOP Conference Series: Materials Science and Engineering*. 2019. IOP Publishing.
65. Lin, Q., X. Zhou, and G. Dai, Effect of hydrothermal environment on moisture absorption and mechanical properties of wood flour–filled polypropylene composites. *Journal of Applied Polymer Science*, 2002. **85**(14): pp. 2824–2832.
66. Barsberg, S. and L.G. Thygesen, Nonequilibrium phenomena influencing the wetting behavior of plant fibers. *Journal of colloid and interface science*, 2001. **234**(1): pp. 59–67.
67. Wambua, P., J. Ivens, and I. Verpoest, Natural fibres: can they replace glass in fibre reinforced plastics? *Composites Science and Technology*, 2003. **63**(9): pp. 1259–1264.
68. Sapuan, S., H. Ismail, and E. Zainudin, *Natural Fiber Reinforced Vinyl Ester and Vinyl Polymer Composites: Development, Characterization and Applications*. Woodhead Publishing, Cambridge, US. 2018
69. Chin, S.C., et al., Thermal and mechanical properties of bamboo fiber reinforced composites. *Materials Today Communications*, 2020. **23**: p. 100876.
70. Dhakal, H.N., Z.A. Zhang, and M.O. Richardson, Effect of water absorption on the mechanical properties of hemp fibre reinforced unsaturated polyester composites. *Composites Science and Technology*, 2007. **67**(7–8): pp. 1674–1683.
71. Zhang, Z. and S. Zhu, Microvoids in unsaturated polyester resins containing poly (vinyl acetate) and composites with calcium carbonate and glass fibers. *Polymer*, 2000. **41**(10): pp. 3861–3870.
72. El Messiry, M., Theoretical analysis of natural fiber volume fraction of reinforced composites. *Alexandria Engineering Journal*, 2013. **52**(3): pp. 301–306.
73. Abdou, A. and I. Budaiwi, The variation of thermal conductivity of fibrous insulation materials under different levels of moisture content. *Construction and Building Materials*, 2013. **43**: pp. 533–544.
74. Manohar, K., et al., Biodegradable fibrous thermal insulation. *Journal of the Brazilian Society of Mechanical Sciences and Engineering*, 2006. **28**(1): pp. 45–47.
75. Mahapatra, A.K., Thermal properties of sweet sorghum bagasse as a function of moisture content. *Agricultural Engineering International: CIGR Journal*, 2018. **19**(4): pp. 108–113.
76. Hung Anh, L.D. and Z. Pásztory, An overview of factors influencing thermal conductivity of building insulation materials. *Journal of Building Engineering*, 2021. **44**: p. 102604.
77. Bavan, D.S. and G.M. Kumar, Thermal properties of maize fiber reinforced unsaturated polyester resin composites, in: *Proceedings of the World Congress on Engineering*, London, UK. 2013. pp. 2091–2096.
78. Osugi, R., et al., Thermal conductivity behavior of natural fiber-reinforced composites. in *Proceedings of the Asian pacific conference for materials and mechanics*, Yokohama, Japan. 2009.
79. Mangal, R., et al., Thermal properties of pineapple leaf fiber reinforced composites. *Materials Science and Engineering: A*, 2003. **339**(1–2): pp. 281–285.

80. Fraga, A.N., et al., Relationship between dynamic mechanical properties and water absorption of unsaturated polyester and vinyl ester glass fiber composites. *Journal of Composite Materials*, 2003. **37**(17): pp. 1553–1574.
81. Wang, W., M. Sain, and P. Cooper, Study of moisture absorption in natural fiber plastic composites. *Composites science and technology*, 2006. **66**(3–4): pp. 379–386.
82. Tezara, C., et al., Factors that affect the mechanical properties of kenaf fiber reinforced polymer: a review. *Journal of Mechanical Engineering and Sciences*, 2016. **10**(2): pp. 2159–2175.
83. Ahmad, I., A. Baharum, and I. Abdullah, Effect of extrusion rate and fiber loading on mechanical properties of Twaron fiber-thermoplastic natural rubber (TPNR) composites. *Journal of Reinforced Plastics and Composites*, 2006. **25**(9): pp. 957–965.
84. Prasad, S., C. Pavithran, and P. Rohatgi, Alkali treatment of coir fibres for coir-polyester composites. *Journal of Materials Science*, 1983. **18**(5): pp. 1443–1454.

11 Natural Fiber-Reinforced Vinyl Ester Composites

Influence of Soil Burial on Physico-Chemical, Thermal and Mechanical Properties

Theivasanthi Thirugnanasambandan and Senthil Muthu Kumar Thiagamani
Kalasalingam Academy of Research and Education

CONTENTS

11.1 INTRODUCTION

Vinyl ester resins are made by the esterification of epoxy resin and an unsaturated monocarboxylic acid. They are used in transportation, buildings, energy and marine applications for making products like pipelines and storage tanks. Vinyl ester, a thermoset polymer, is used for manufacturing automotive parts, cables, and roller systems. These resins are susceptible to acids, alkalis, solvents, hypochlorites, and peroxides. The performance of these polymers is superior to other polymers because they offer high corrosion resistance and high elongation to failure. Natural fibers are added as reinforcements in thermoset composites since they are more economical and biodegradable. They also provide mechanical properties that are comparable to the synthetic fibers [1–5].

DOI: 10.1201/9781003270997-11

FIGURE 11.1 Image analyzer photographs of PLA and PLA/30% coir fiber and PLA/ 30% pineapple leaf fiber before and after degradation in the soil [6].

Polymer matrix plays an important role in the degradation of fiber-reinforced polymer composites. The factors like water absorption, sunlight, heat and microorganisms decide the rate at which the composite degrades. These factors become less important when the composite is buried under the soil for degradation. Figure 11.1 shows that the neat polylactic acid (PLA) retains its original shape even after degradation. But, the surface of the composites becomes rougher and many fractures and cracks are formed on their surface after soil burial degradation. When the samples are buried under the soil, they are subjected to various biotic and abiotic factors. The rate of degradation depends on the thickness of the film and the composting conditions in the soil. The degradation is also found to be fast when compared to the degradation due to accelerated weathering [6–9].

The natural fibers used as fillers in composites are responsible for rapid degradation of the composite. The degradation of the composite is more rapid when compared to the pure polymer matrices. Normally, the degradation of polyethylene starts to degrade after 60 days of soil burial only. The soil microbes are able to perform biochemical reactions on the composite and thereby breaking down of the composite occurs. The degradation of polymer composites in soil may happen aerobically or anaerobically. The aerobic process produces H_2O and CO_2 and the anaerobic process produces CH_4. The degradation of the composite depends on the type of soil microbe and the factors that affect the condition of the soil such as pH, moisture content, temperature and oxygen. The properties of the polymer matrix of a composite such as molecular weight, its constituents and crystallinity also determine the degradability of the composite. Since the pure polymers are hydrophobic in nature, the soil microbes do not attach easily on their surface. But the hydrophilic nature of the natural fibers as fillers in the composites facilitates the degradation in a quick manner through the acceleration of the growth of the microbes on the surface of the composites [10–13].

PVC shows a high degradation of 68 wt.% upon soil burial for 6 months when blended with linseed oil epoxy. A high linseed oil epoxy content in the blend results in a faster degradation. The films fabricated out of such blends can be used for packaging applications. Microorganisms like bacteria and fungi degrade this natural linseed oil epoxy [14–17]. The starch-grafted polyethylene is degraded under soil burial of the sample. Some biochemical tests are performed to identify the microorganism responsible for the degradation. It is found that the bacterium is *Rhizobium meliloti* [18–21]. The polymer composites mostly consist of acrylate, esters and bisphenol which can be easily broken down by soil microbes [22].

11.2 SOIL BURIAL TESTS

Recently, various biodegradation mechanisms of plastics are extensively studied to analyze the problems related to the environment [23]. The degradation of the polymer composites is usually studied by the soil burial test. The degradation of effect of sugar palm fiber-reinforced vinyl ester composites is tested by burial of the composite in the soil for 0–1000 h. Water is poured on the samples for every 2 days. The samples with dimension of 10 mm × 10 mm × 3 mm are tested for every 200 h by collecting, washing and drying [24,25]. For soil burial degradation tests, the approximate dimension of the composite may be 10 cm × 5 cm with the soil composition as manure:sand:soil in 2:1:1 ratio. The samples are buried at a height of 15 cm for a period of 10–80 days. The samples are collected and analyzed for their weight loss, tensile strength, surface morphology and structural properties like composition and crystallinity [26]. The composite is subjected to modified temperature and humidity conditions. Natural fiber-reinforced composites are preferable in structural strengthening over synthetic fiber-reinforced composites because they are more lightweight, economical, renewable and biodegradable. Natural fibers such as kenaf, sisal, coir, banana, jute, flax, oil palm, pineapple leaf and bamboo are used as fillers in polymer matrices. Their applications lie in automotive, sporting goods, electrical, household and construction. A loading of 40% fiber volume fraction is preferable to achieve high mechanical properties. When buried in the soil, the tensile and flexural properties of the composite decrease up to 6%–19%. The durability of the composite can be found by performing these soil burial tests. The soil burial decreases the compatibility between the polymer and the filler through the formation of hydrogen bond between the hydroxyl group of fibers and water molecules. The fiber swells by the water absorption that forms the microcracks. Finally, the fibers are detached from the composite. The tensile modulus and the flexural strength are not affected by soil burial [27].

11.3 WATER ABSORPTION

The absorption of water in the soil by the sugar palm fiber is analyzed. The hydrophobic nature of the composite is analyzed by contact angle and water absorption studies. If the composites are buried for a longer time, a large amount of water is absorbed. The water absorption test is performed by immersing in fresh water for 24h. The water absorption is calculated using the formula given below:

$$\text{Water absorption (\%)} = M_1 - M_0/M \times 100,$$

where M_0 and M_1 are the masses of the samples before and after immersing in water, respectively.

A high water uptake is observed when the natural fiber-reinforced vinyl esters are buried for 200 hours. The natural fibers absorb more water due to their hydrophilic property. They consist of celluloses, hemicelluloses and lignin and more hydroxyl groups on their surface which offer this hydrophilicity [24,28].

The vinyl ester composites when reinforced with sugar palm fibers exhibit more water uptake upon soil burial for a longer time [29]. The natural fibers show more water absorption in the soil. But, when the fibers are treated with alkali (NaOH/H_2O_2), they show a reduced absorption of water [30]. The natural fiber-reinforced composites are degraded more easily than the pure polymers. This occurs due to the degradation of the cellulosic chains by the enzymes in natural fibers. Since the composites are more amorphous in nature, the weight is lost at a greater rate. It is also reported that cellulose absorbs more water, and the fibers like kenaf fibers with more cellulose content are susceptible to more degradation. At the same time, the fibers with more lignin content degrade slowly as lignin is hydrophobic [31].

The absorption of water reduces the mechanical properties to a larger extent when woven hemp fabric is reinforced with vinyl ester resin. The hemp fibers possess 65%–68% of cellulose and so absorb more water [32]. The water diffusion coefficient is calculated using Fick's law. A graph is plotted with the water uptake versus square root of time and from the slope k gives the value of diffusion coefficient D. If h is the thickness of the composite and M_m is the maximum water uptake, then the diffusion coefficient is given by

$$D = \pi\left[\frac{kh}{4M_m}\right]^2.$$

11.4 MASS LOSS

Thermoplastic starch is reinforced with 30% jute, and a strength of 27.3 MPa is achieved. The soil burial test shows a complete degradation of this composite. So, they can be used to prepare manures for agriculture applications. The jute fiber-reinforced thermoplastic starch exhibit increased weight loss. This is because pure thermoplastic starch consists of aliphatic polyester that prevents the microorganism's attack. The degradation upon soil burial is shown in terms of weight loss of the material. The weight loss of the pure thermoplastic starch is very less and it occurs within fifteen days. This is because of the aliphatic polyesters that resist the attack of soil microorganisms. Jute and the composites show mass loss for 60 days. The degradation of the composite is very high when the jute content is high. The jute is able to create a large number of micro-voids in the composite that leads to penetration of water and attack of microorganisms [26,33].

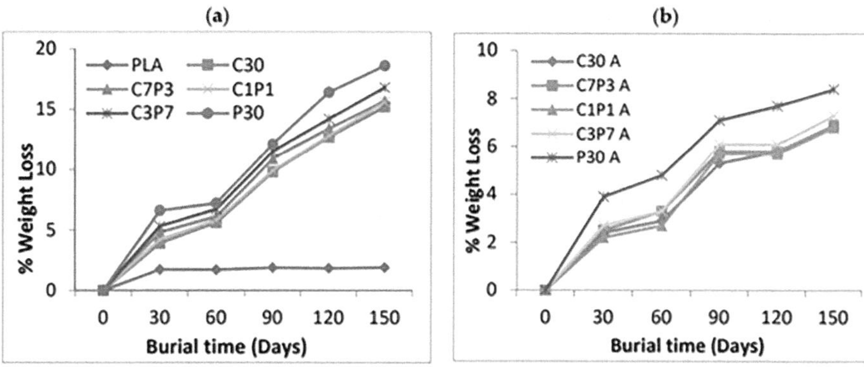

FIGURE 11.2 (a and b) Weight loss curves for biocomposites with untreated and treated fibers [6].

The aliphatic biodegradable polymers degrade easily in the soil and there is no exception for the polymer PLA. PLA degrades in such a way that it is hydrolyzed and then the lactic acid is transformed into gas and water. The molecular weight of the polymer is lowered with the help of acids or bases. Microorganisms such as bacteria and fungi participate in this process as catalysts in the reaction. Figure 11.2 shows the graphs of percentage weight loss of the films versus soil burial time in days. The biodegradability level is analyzed from 30 to 150 days of soil burial. The analysis is performed for untreated and treated coir fiber/pineapple leaf fiber/PLA films with alkali. There is no loss of weight for neat PLA as seen from the curve. As the number of days of soil burial increases, the percentage weight loss of the films also increases. This is high for the untreated fibers when compared with the alkali-treated fibers. A weight loss of 15% is obtained for Polyvinyl alcohol (PVA) films with 30% of coir or pineapple leaf fibers. But, the weight loss for the same composites when treated with alkali is found to be only 8%. The same results are observed from the curves even for hybrid fibers with both coir and pineapple leaf fibers. The untreated fibers offer less adhesion with the matrix that leads to more weight loss [6].

The degradation of bamboo/kenaf/epoxy composites due to accelerated weathering and soil burial is analyzed. A high degradation in terms of weight loss is obtained for the composite with larger content of kenaf fiber. Kenaf shows a high degradability when compared to the other constituents of the composite. This is due to the hygroscopic feature of the fiber that promotes the microbial activity, and the weight loss is achieved. The presence of a large number of hemicelluloses is responsible for this hygroscopic feature. The bamboo fiber is found to possess less amount of hemicellulose content as compared with kenaf fiber [34].

Figure 11.3 shows the weight loss of the pure and photo-oxidized polymer films after soil burial. In the soil burial, the polymers are firstly converted into monomers. They are transformed into biomass, which finally produces CO_2 and H_2O. The photo-oxidized films exhibit an increased weight loss when compared to the pure polymer. More weight loss confirms the complete disintegration of the photo-oxidized films in the soil. The photo-oxidation affects not only the surface of the film but also the entire matrix of the composite [35].

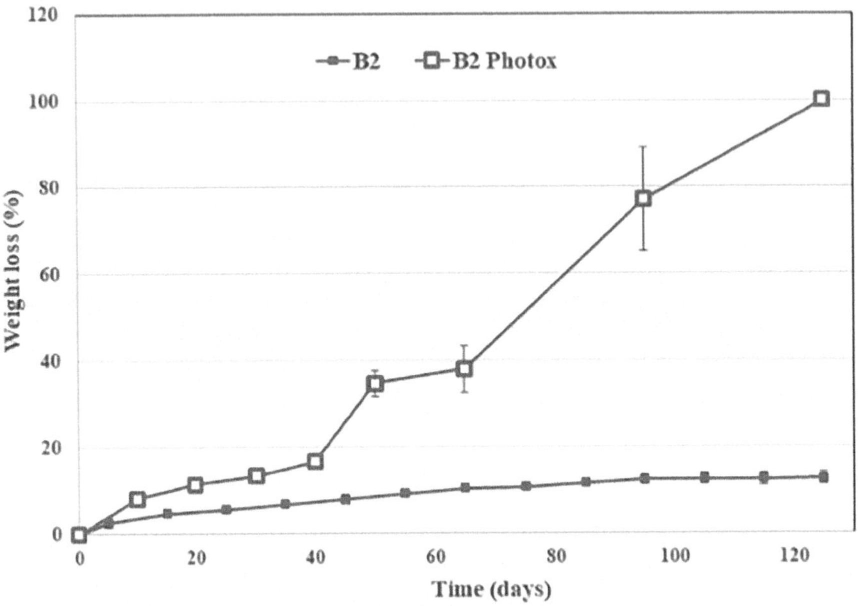

FIGURE 11.3 Weight loss versus degradation time curves of pure and photo-oxidized films [35].

11.5 MECHANICAL PROPERTIES

A composite of banana fiber-reinforced vinyl ester resin at a loading of 35% of fiber volume fraction results in an enhancement of 38.6% tensile strength [36]. The tensile and flexural strength of vinyl ester is improved with 40 wt.% of bagasse fiber [37]. The mechanical properties are greatly reduced after soil burial for 200 h. The temperature onset of the composite increases after soil burial [29]. The mechanical properties are decided by the interaction between the fillers and the matrix. When this interaction is affected by soil burial degradation, the mechanical properties are seriously affected.

The mechanical properties like tensile and flexural measurements are made before and after the soil burial. The tensile strength of kenaf fiber-reinforced thermoplastic polyurethane composites decreased after soil burial. At the same time, the flexural properties of the composite are not affected by soil burial. The decrease in tensile strength of the composite may be due to the polymer matrix, natural fiber filler and their interaction. The tensile strength may also be reduced because of accelerated weathering and soil burial effects [38].

The mechanical properties such as tensile strength, tensile modulus, elongation at break and toughness are analyzed for oil palm trunk fiber/polypropylene composite after soil burial degradation. They decrease with increase of soil burial time [39].

Polypropylene/plasticized cassava starch/polypropylene-graft-maleic anhydride composite is prepared by melt blending and the degradation is studied through soil burial tests. The mechanical properties such as the tensile strength, elongation at break and Young's modulus decrease with increase of burial time. The addition of polypropylene-graft-maleic anhydride in the composite greatly enhances the degradation rate through the deterioration of the mechanical properties [40]. The soil burial tests are performed for windmill palm fiber to study its degradation properties. The fiber is treated with alkali to remove the hemicellulose. The fiber is then bleached to eliminate the lignin content. These three samples are buried in the soil for 90 days. The parenchyma cells and silica bodies of the fibers are lost during soil burial. Normally, hemicellulose shows a high degradation rate than lignin in natural fibers. As the burial time increases, more inorganic substances present in the soil are adhered to the fibers. So, a thick layer covers the composite after 90 days. The X-ray Diffraction (XRD) spectrum (Figure 11.4) of windmill palm fiber shows two reflections at $2\theta = 18°$ (amorphous phase) and 22° (crystalline phase) of cellulose with the crystallinity index of 34%. The amorphous portion of the composite is completely degraded upon soil burial and so the crystallinity increases after 90 days. The XRD patterns of the samples after 60–90 days of soil burial exhibit a narrow peak at 26°. This is due to the impurities that arise after degradation in the soil. As the soil burial time increases, the intensity of the peak increases. The impurities are identified as some inorganic particles present in the soil that adhered to the samples. The inorganic impurities may be potassium, calcium, chlorides and phosphors [41].

The tensile properties of the pure fiber, alkali-treated fiber and bleached fiber before and after soil burial are shown in Figure 11.5. The alkali treatment increases the strength of the fibers. This is because of the fact that the impurities and hemicellulose are removed from the fiber and the diameter of the fiber is greatly decreased by alkali treatment. The soil degradation decreases both the tensile strength and elongation at break. It is also said that even though the alkali treatment improves the tensile strength, the durability of the fiber is not enhanced to a higher level. The soil burial increases the microfibrillar angle that results in a decrease in elongation at break.

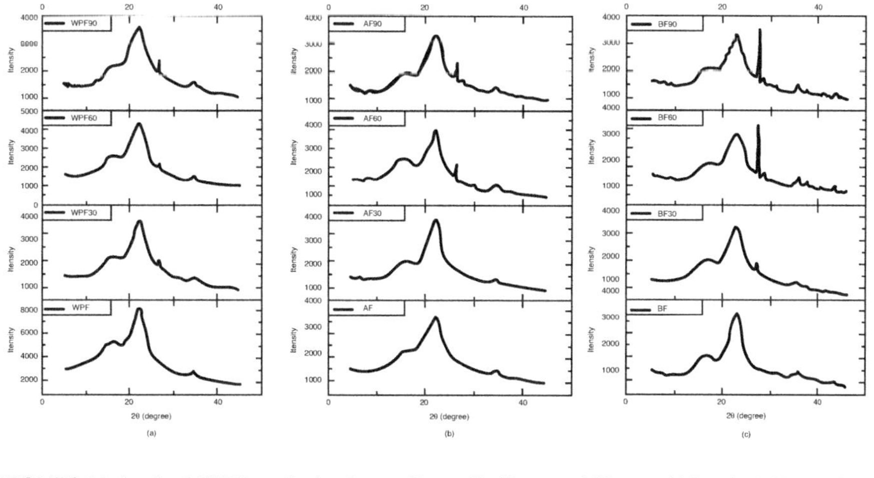

FIGURE 11.4 (a-c) XRD analysis of pure fiber, alkali-treated fiber and bleached fiber before and after soil burial [41].

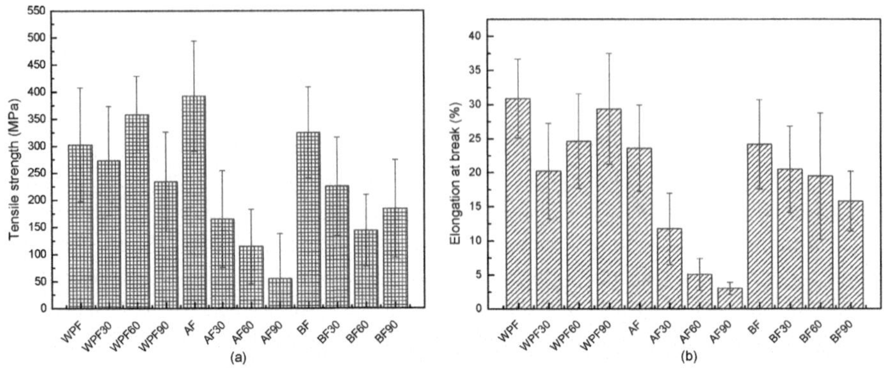

FIGURE 11.5 Tensile strength and elongation at break of the pure fiber, alkali-treated fiber and bleached fiber before and after soil burial [41].

The tensile strength and percentage of elongation of polypropylene/starch blend decrease as the starch content and burial time increase. The starch molecules increase the chain spacing in polypropylene which weakens the strength of the secondary bonding. More stress is concentrated over the starch that leads to the formation of cracks. Water absorption and the attack of soil microbes are the two major reasons for the reduction of mechanical properties of the composite [42].

The polymers degrade in the soil in such a way that they are divided into monomers which are finally degraded. This can be understood with the help of viscosity analysis. The crystallinity of the composite is increased after soil burial because the amorphous part of the composite is degraded quickly in the soil The viscosity analysis showed that the degradation of the composite was achieved by the simultaneous processes of breaking big molecules into small molecules and by the degradation of small molecules [43].

11.6 THERMAL PROPERTIES

Thermogravimetric analysis (TGA) gives the weight loss as a function of temperature. Since polymers decompose at high temperatures, TGA can provide the detailed information regarding their thermal stability. Normally pure polymers decompose from 350°C to 470°C due to the breakage of bonds in the polymer chains. The constituents of natural fibers include cellulose, lignin and hemicellulose that decompose at various temperatures. The decomposition temperatures of cellulose, lignin and hemicellulose are reported as 300°C–420°C, 200°C–900°C and 200°C–420°C [10, 44].

The alkali treatment also decreases the thermal stability of the composite. This is clearly shown by an increase in the percentage weight loss in TGA analysis. The soil burial degradation is less in the alkali-treated fibers when compared with the untreated fibers. During degradation, the weight loss is more in the elephant grass fiber/PLA composite, whereas the composite shows a decrease in weight loss for the alkali-treated elephant grass fiber. The weight loss is linear with the number of days of soil burial in both treated and untreated fibers. This happens because the treated fibers adhere strongly with the polymer. The interaction between the untreated fiber and the polymer is less. Also, the treated fibers attain hydrophobic nature which leads to a less absorption of water. The degradation of the untreated fibers/PLA composite is seen with the cracks on its surface whereas the erosion is found to be less in the case of the composite with the alkali-treated fibers [30].

The degradation is enhanced when the temperature in the soil environment increases. A high rate of degradation is reported at 60°C [45]. In TGA, the parameters like onset temperature, maximum rate degradation temperature and activation energy decrease when the sample is buried in the soil for a long time [46]. The decrease in the molecular weight of the polymers can be measured from gel permeation chromatography. The degradation of polymers occurs through decomposition of the materials. The thermal stability of the composites is very less when they are reinforced with natural materials that is confirmed with TGA analysis. In the same manner, the synthetic polymers possess high thermal stability. After degradation in the soil, the size of the composite becomes small and its shape becomes irregular. This is visible on seeing the composite and also through Scanning Electron Microscopy (SEM) analysis. The degradation can also be analyzed by the variations in the Fourier Transform Infrared Spectroscopy (FTIR) spectra. The degradation rate of the composites is fast in the presence of enzymes and microorganisms. It is also enhanced by the temperature and the presence of water content. The water absorption of the composite and thereby the mass loss can be measured to study about the degradation process [43].

11.7 FTIR ANALYSIS

The effects of degradation by soil burial and microorganism are understood using the FTIR spectra. A reduction in the peak intensity is observed after degradation. Due to degradation of the composite, the cellulosic hydroxyl group breaks, which decreases the height of the OH stretching peak. The loss of cellulose and hemicellulose after degradation reduces the peak intensity of C H stretching. The soil microbes degrade the composite to a larger extent, which is shown by C=O stretching and CH_2 bending. The loss of resin is indicated by the peaks corresponding to C–O stretching, C–C and C–O–H stretching. The intensities of all these vibrations are found to be decreased after degradation [26].

Polyurethane is modified using *Mesua ferrea* L. seed oil and is blended with epoxy and melamine formaldehyde resins. The degradation effects of these samples are studied using the FTIR technique. The soil burial of the blends results in the decrease of the intensity of ester linkages that are caused by the enzyme esterase and the peaks due to hydrocarbon moieties are also reduced by microbes in the soil [47].

The degradation degree is much greater for the natural materials than the synthetic polymers even though the polymers are biodegradable. The FTIR spectra of fiber/starch, fiber/PVA and fiber/PA (polyacrylate) are compared before and after degradation. The changes are observed both in the intensity and position of the peak. The intensity of the peaks in the FTIR spectrum for fiber/starch and fiber/PVA films shows more variations before and after degradation of the films. But the FTIR

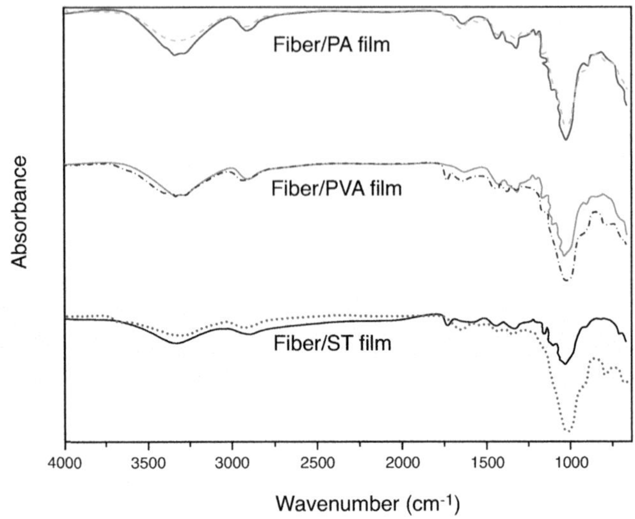

FIGURE 11.6 FTIR spectra of fiber/starch, fiber/PVA and fiber/PA films [48].

spectrum of fiber/PA film is not showing much variation for before and after degradation of the films. This confirms the fast and slow degradation of fiber/starch, fiber/PVA films and fiber/PA film, respectively [48] (Figure 11.6).

The treatment of kenaf fiber with silane decreases the degradation rate of the kenaf fiber/low-density polyethylene/polyvinyl alcohol composites. The silane treatment helps for the enhanced adhesion of kenaf fiber with the polymer matrices. And the moisture absorption of kenaf is reduced by the silane treatment. The soil burial causes the microcracks on the composite through which the moisture can diffuse into the composite. A burial time of up to 180days results in a greater number of such pores with increased size. Soil burial for a long duration degrades the sample more effectively. The FTIR peaks of kenaf fiber that are assigned to O–H bending, C–O and C–O–C stretching and C–H and CH_2 stretching are found to possess reduced absorption after soil burial. These results confirm the leaching of kenaf out of the composite after soil burial. The crystallinity of the composites is decreased when the kenaf fibers are treated with silane. A high loading of kenaf causes more weight loss and improved crystallinity of the composite during soil burial. At the same time, the silane treatment reduces these two properties [49].

11.8 SEM ANALYSIS

The effect of degradation can be analyzed through SEM analysis. The smooth surface is modified into rough surface because of degradation of the composite as viewed from the SEM images. The jute fibers are completely covered by the thermoplastic starch in the composite. After degradation of the composite, the debonding of resin occurs, which appears like a rough surface [26]. The soil burial creates more voids and holes on the surface of the composite as viewed from SEM images. This is attributed to the fact that during soil burial, the microorganisms uptake the moisture content in the composite and then remove the cellulose content from the composite [38]. The formation of ripples on the surface shows the partial degradation of the films. A high degradation rate is visible from the SEM images of the fiber/starch and fiber/PVA films. So, these films can be used as mulching films in agriculture. But the fiber/PA degradation is very slow and so the polyacrylate films are termed as non-biodegradable polymers [48] (Figure 11.7).

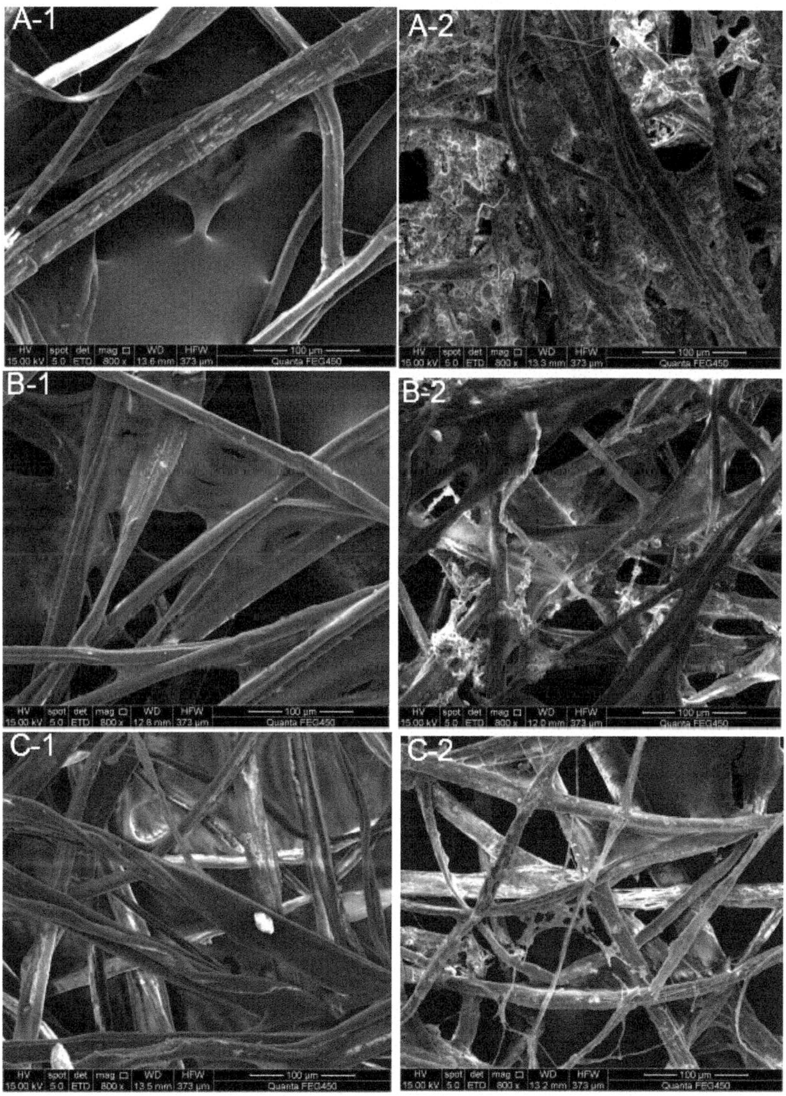

FIGURE 11.7 SEM images of fiber/starch film, original fiber/PVA film and fiber/polyacrylate film before and after degradation [48].

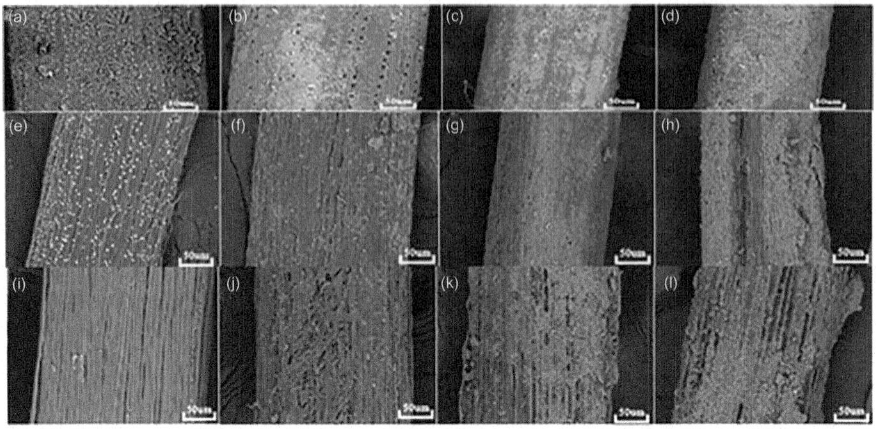

FIGURE 11.8 (a–l) SEM images of untreated, alkali-treated and bleached windmill palm fibers before and after soil burial [41].

The alkali treatment provides more droplets like structures that are having lignin on the surface of the fiber. This is clearly visible in the SEM images as shown in Figure 11.8. After degradation in the soil, these droplets disappear and inorganic layers are coated on the surface of the fiber, which is similar to what happens in the untreated fiber. This inorganic layer becomes thick as the duration of burial increases. Bleaching of fibers causes removal of lignin that results in a striated surface. All the substances like hemicellulose, silica bodies and parenchyma cells are removed from the fibers. As the time proceeds, the cell wall becomes degraded and some grooves are found on the surface of the fiber. The presence of a greater number of grooves is an indication of the complete degradation of the fiber [41].

11.9 X-RAY PHOTOELECTRON SPECTROSCOPY (XPS)

The PLA-blended starch is tested for its degradation using soil burial tests. The starch supplies biological fuel to the soil microbes that accelerate the degradation of the composite. The weight loss after soil burial is more for high starch content of the composite. The SEM images exhibit gullies and cavities on the surface of the composite. Stress will be concentrated on these gullies and cracks will start to grow from these gullies. Finally, the composite loses its mechanical properties. After degradation, the oxygen/carbon ratio is found out using XPS. The addition of starch not only increases the degradation rate but also reduces the amount of PLA needed, thus becoming more economical.

Before degradation, the morphology appears like a smooth surface which then becomes as a fractured surface after degradation. PLA covers completely all the starch and wood particles. Since all the three constituents are biodegradable, they are easily affected by water and soil microbes. They get loosened out of the matrix because of degradation for a long time. The molecular chains of PLA are broken, which results in the decrease of mechanical properties. The degradation mechanism can be clearly explained using XPS technique since it provides idea on the composition of the composites upon degradation. In the hydrolysis process, the molecular weight of PLA is reduced, which is then degraded by microbes. Thus, the polymer composite is finally converted into water, carbon dioxide and biomass [50].

11.10 CONCLUSION

The degradation of polymers is more important to avoid environmental pollution. The rate of degradation of the composite increases as the polymer matrix is filled with natural fibers. The natural fibers are more susceptible to degradation upon soil burial. The soil burial test is performed by burying the composite under the soil to analyze the level of degradation. When the composite is buried in the soil, the composite is subjected to soil moisture absorption, mass loss and attack of soil microbes. The strength and the thermal stability of the composite are reduced after soil burial degradation. The various compositional changes are analyzed using FTIR and XPS techniques. The SEM analysis of the composite provides the information about the morphological changes before and after soil burial degradation.

REFERENCES

1. S. M. K. Thiagamani, N. Rajini, S. Siengchin, A. V. Rajulu, N. Hariram, and N. Ayrilmis, "Influence of silver nanoparticles on the mechanical, thermal and antimicrobial properties of cellulose-based hybrid nanocomposites," *Compos. Part B Eng.*, vol. 165, pp. 516–525, 2019.
2. T. S. M. Kumar, K. S. Kumar, N. Rajini, S. Siengchin, N. Ayrilmis, and A. V. Rajulu, "A comprehensive review of electrospun nanofibers: Food and packaging perspective," *Compos. Part B Eng.*, vol. 175, p. 107074, 2019.
3. S. M. K. Thiagamani et al., "Investigation into mechanical, absorption and swelling behaviour of hemp/sisal fibre reinforced bioepoxy hybrid composites: Effects of stacking sequences," *Int. J. Biol. Macromol.*, vol. 140, pp. 637–646, 2019.
4. S. Krishnasamy et al., "Recent advances in thermal properties of hybrid cellulosic fiber reinforced polymer composites," *Int. J. Biol. Macromol.*, vol. 141, pp. 1–13, 2019.
5. T. Senthil Muthu Kumar, N. Rajini, S. Siengchin, A. Varada Rajulu, and N. Ayrilmis, "Influence of *Musa acuminata* bio-filler on the thermal, mechanical and visco-elastic behavior of poly (propylene) carbonate biocomposites," *Int. J. Polym. Anal. Charact.*, vol. 24, no. 5, pp. 439–446, 2019.
6. R. Siakeng et al., "Accelerated weathering and soil burial effect on biodegradability, colour and texture of coir/pineapple leaf fibres/PLA biocomposites," *Polymers*, vol. 12, no. 2, pp. 1–15, 2020.https://doi.org/10.3390/polym12020458
7. I. D. MP and S. Siengchin, "Antimicrobial properties of poly (propylene) carbonate/Ag nanoparticle-modified tamarind seed polysaccharide with composite films," *Ionics (Kiel)*, vol. 25, no. 7, pp. 3461–3471, 2019.
8. M. P. I. Devi et al., "Biodegradable poly (propylene) carbonate using in-situ generated CuNPs coated *Tamarindus indica* filler for biomedical applications," *Mater. Today Commun.*, vol. 19, pp. 106–113, 2019.
9. R. M. Kumar, N. Rajini, T. S. M. Kumar, K. Mayandi, S. Siengchin, and S. O. Ismail, "Thermal and structural characterization of acrylonitrile butadiene styrene (ABS) copolymer blended with polytetrafluoroethylene (PTFE) particulate composite," *Mater. Res. Express*, vol. 6, no. 8, p. 85330, 2019.

10. M. Alshabanat, "Morphological, thermal, and biodegradation properties of LLDPE/treated date palm waste composite buried in a soil environment," *J. Saudi Chem. Soc.*, vol. 23, no. 3, pp. 355–364, 2019.
11. S. Krishnasamy et al., "Effect of fibre loading and Ca (OH) 2 treatment on thermal, mechanical, and physical properties of pineapple leaf fibre/polyester reinforced composites," *Mater. Res. Express*, vol. 6, no. 8, p. 85545, 2019.
12. M. Chandrasekar et al., "Flax and sugar palm reinforced epoxy composites: Effect of hybridization on physical, mechanical, morphological and dynamic mechanical properties.," *Mater. Res. Express*, vol. 6, no. 10, pp. 1–11, 2019.
13. S. Krishnasamy et al., "Effects of stacking sequences on static, dynamic mechanical and thermal properties of completely biodegradable green epoxy hybrid composites," *Mater. Res. Express*, vol. 6, no. 10, p. 105351, 2019.
14. U. Riaz, A. Vashist, S. A. Ahmad, S. Ahmad, and S. M. Ashraf, "Compatibility and biodegradability studies of linseed oil epoxy and PVC blends," *Biomass and Bioenergy*, vol. 34, no. 3, pp. 396–401, 2010.
15. K. Senthilkumar, M. Chandrasekar, N. Rajini, S. Siengchin, and V. Rajulu, "Characterization, thermal and dynamic mechanical properties of poly (propylene carbonate) lignocellulosic *Cocos nucifera* shell particulate biocomposites," *Mater. Res. Express*, vol. 6, no. 9, p. 96426, 2019.
16. T. S. M. Kumar, N. Rajini, T. Huafeng, A. V. Rajulu, N. Ayrilmis, and S. Siengchin, "Improved mechanical and thermal properties of spent coffee bean particulate reinforced poly (propylene carbonate) composites," *Part. Sci. Technol.*, vol. 37, no. 5, pp. 643–650, 2019.
17. K. Yorseng, N. Rajini, S. Siengchin, N. Ayrilmis, and V. Rajulu, "Mechanical and thermal properties of spent coffee bean filler/poly (3-hydroxybutyrate-co-3-hydroxyvalerate) biocomposites: Effect of recycling," *Process Saf. Environ. Prot.*, vol. 124, pp. 187–195, 2019.
18. G. Neena and K. Inderjeet, "Soil burial biodegradation studies of starch grafted polyethylene and identification of *Rhizobium meliloti* therefrom," *J. Environ. Chem. Ecotoxicol.*, vol. 5, no. 6, pp. 147–158, 2013.
19. K. Senthilkumar et al., "Dual cantilever creep and recovery behavior of sisal/hemp fibre reinforced hybrid biocomposites: Effects of layering sequence, accelerated weathering and temperature," *J. Ind. Text.*, p. 1528083720961416, vol. 51, no. 2S, pp. 2372S–2390S, 2022.
20. J. Sangilimuthukumar, T. S. M. Kumar, C. Santulli, M. Chandrasekar, K. Senthilkumar, and S. Siengchin, "The use of pineapple fiber composites for automotive applications: A short review," *J. Mater. Sci. Res. Rev.*, pp. 39–45, 2020.
21. T. Kumar et al., "Characterization, thermal and antimicrobial properties of hybrid cellulose nanocomposite films with in-situ generated copper nanoparticles in Tamarindus indica Nut Powder," *J. Polym. Environ.*, vol. 29, no. 4, pp. 1134–1142, 2021.
22. A. M. Breister et al., "Soil microbiomes mediate degradation of vinyl ester-based polymer composites," *Commun. Mater.*, vol. 1, no. 1, pp. 1–15, 2020.
23. T. Fakhrul and M. A. Islam, "Degradation behavior of natural fiber reinforced polymer matrix composites," *Procedia Eng.*, vol. 56, pp. 795–800, 2013.
24. M. R. M. Huzaifah, S. M. Sapuan, Z. Leman, M. R. Ishak, and R. A. Ilyas, "Effect of soil burial on water absorption of sugar palm fibre reinforced vinyl ester composites," in *6th Postgraduate Seminar on Natural Fiber Reinforced Polymer Composites*, 2018, pp. 52–54.
25. K. Senthilkumar et al., "Performance of sisal/hemp bio-based epoxy composites under accelerated weathering," *J. Polym. Environ.*, vol. 29, no. 2, pp. 624–636, 2021.
26. A. K. Behera, C. Mohanty, S. K. Pradhan, and N. Das, "Assessment of soil and fungal degradability of thermoplastic starch reinforced natural fiber composite," *J. Polym. Environ.*, vol. 29, no. 4, pp. 1031–1039, 2021.

27. S. C. Chin, F. S. Tong, S. I. Doh, K. S. Lim, and J. Gimbun, "Effect of soil burial on mechanical properties of bamboo fiber reinforced epoxy composites," in *IOP Conf. Ser.: Mater. Sci. Eng.*, 2020, vol. 736, no. 5, p. 52016.
28. S. Jeyaguru, S. M. K. Thiagamani, A. G. Rajkumar, S. M. Rangappa, and S. Siengchin, "Solid particle erosion, water absorption and thickness swelling behavior of intra ply Kevlar/PALF fiber epoxy hybrid composites," *Polym. Compos.*, vol. 43, no. 6, pp. 3929–3943, 2022.
29. M. R. M. Huzaifah, S. M. Sapuan, Z. Leman, and M. R. Ishak, "Effect of soil burial on physical, mechanical and thermal properties of sugar palm fibre reinforced vinyl ester composites," *Fibers Polym.*, vol. 20, no. 9, pp. 1893–1899, 2019.
30. R. Gunti, A. V Ratna Prasad, and A. Gupta, "Mechanical and degradation properties of natural fiber-reinforced PLA composites: Jute, sisal, and elephant grass," *Polym. Compos.*, vol. 39, no. 4, pp. 1125–1136, 2018.
31. A. A. Yussuf, I. Massoumi, and A. Hassan, "Comparison of polylactic acid/kenaf and polylactic acid/rise husk composites: The influence of the natural fibers on the mechanical, thermal and biodegradability properties," *J. Polym. Environ.*, vol. 18, no. 3, pp. 422–429, 2010.
32. M. I. Misnon, M. M. Islam, J. A. Epaarachchi, H. Wang, and K. Lau, "Woven hemp fabric reinforced vinyl ester composite: Effect of water absorption on the mechanical properties degradation," *Int. J. Adv. Sci. Eng. Technol.*, vol. 4, no. 3, pp. 96–101, 2016.
33. N. A. Nasimudeen et al., "Mechanical, absorption and swelling properties of vinyl ester based natural fibre hybrid composites," *Appl. Sci. Eng. Prog.*, vol. 14, no. 4, pp. 680–688, 2021.
34. S. S. Chee, M. Jawaid, M. T. H. Sultan, O. Y. Alothman, and L. C. Abdullah, "Thermomechanical and dynamic mechanical properties of bamboo/woven kenaf mat reinforced epoxy hybrid composites," *Compos. Part B Eng.*, vol. 163, pp. 165–174, 2019.
35. F. P. La Mantia, L. Ascione, M. C. Mistretta, M. Rapisarda, and P. Rizzarelli, "Comparative investigation on the soil burial degradation behaviour of polymer films for agriculture before and after photo-oxidation," *Polymers (Basel).*, vol. 12, no. 4, p. 753, 2020.
36. R. Ghosh, G. Reena, A. R. Krishna, and B. H. L. Raju, "Effect of fibre volume fraction on the tensile strength of Banana fibre reinforced vinyl ester resin composites," *Int J adv Eng Sci Technol*, vol. 4, no. 1, pp. 89–91, 2011.
37. A. Athijayamani, B. Stalin, S. Sidhardhan, and C. Boopathi, "Parametric analysis of mechanical properties of bagasse fiber-reinforced vinyl ester composites," *J. Compos. Mater.*, vol. 50, no. 4, pp. 481–493, 2016.
38. S. M. Sapuan, F. Pua, Y. A. El-Shekeil, and F. M. AL-Oqla, "Mechanical properties of soil buried kenaf fibre reinforced thermoplastic polyurethane composites," *Mater. Des.*, vol. 50, pp. 467–470, 2013.
39. H. P. S. Abdul Khalil et al., "The effect of soil burial degradation of oil palm trunk fiber-filled recycled polypropylene composites," *J. Reinf. Plast. Compos.*, vol. 29, no. 11, pp. 1653–1663, 2010.
40. H. C. Obasi, I. O. Igwe, and I. C. Madufor, "Effect of soil burial on tensile properties of polypropylene/plasticized cassava starch blends," *Adv. Mater. Sci. Eng.*, vol. 2013, pp. 1–5, 2013. http://dx.doi.org/10.1155/2013/326538
41. C. Chen, W. Yin, G. Chen, G. Sun, and G. Wang, "Effects of biodegradation on the structure and properties of windmill palm (Trachycarpus fortunei) fibers using different chemical treatments," *Materials (Basel).*, vol. 10, no. 5, p. 514, 2017.
42. Z. J. A. Amer and A. Q. Saeed, "Soil burial degradation of polypropylene/starch blend," *Int. J. Tech. Res. Appl.*, vol. 3, no. 1, pp. 91–96, 2015.
43. O. Wilfred et al., "Biodegradation of Polylactic Acid and starch composites in compost and soil," *Int. J. Nano Res.*, vol. 1, no. 2, pp. 1–11, 2018.

44. T. S. M. Kumar, N. Rajini, K. O. Reddy, A. V. Rajulu, S. Siengchin, and N. Ayrilmis, "All-cellulose composite films with cellulose matrix and Napier grass cellulose fibril fillers," *Int. J. Biol. Macromol.*, vol. 112, pp. 1310–1315, 2018.
45. N. S. Yatigala, "Thermal, Physico-Mechanical and Degradation Characteristics of Compatibilized Biodegradable Biopolymers and Composites." North Dakota State University, 2017.
46. S. Martelli, E. Fernandes, and E. Chiellini, "Thermal analysis of soil-buried oxo-biodegradable polyethylene based blends," *J. Therm. Anal. Calorim.*, vol. 97, no. 3, pp. 853–858, 2009.
47. S. Dutta, N. Karak, J. P. Saikia, and B. K. Konwar, "Biodegradation of epoxy and MF modified polyurethane films derived from a sustainable resource," *J. Polym. Environ.*, vol. 18, no. 3, pp. 167–176, 2010.
48. Z. Tan, Y. Yi, H. Wang, W. Zhou, Y. Yang, and C. Wang, "Physical and degradable properties of mulching films prepared from natural fibers and biodegradable polymers," *Appl. Sci.*, vol. 6, no. 5, p. 147, 2016.
49. A. L. Pang, A. Arsad, M. Ahmadipour, H. Ismail, and A. A. Bakar, "Effect of soil burial on silane treated and untreated kenaf fiber filled linear low-density polyethylene/polyvinyl alcohol composites," *BioResources*, vol. 15, no. 4, p. 8648, 2020.
50. S. Lv, Y. Zhang, J. Gu, and H. Tan, "Biodegradation behavior and modelling of soil burial effect on degradation rate of PLA blended with starch and wood flour," *Colloids Surf. B Biointerfaces*, vol. 159, pp. 800–808, 2017.

12 Vinyl Ester-Based Biocomposites for Various Applications

Xiaoan Nie and Jie Chen
Chinese Academy of Forestry

CONTENTS

12.1 INTRODUCTION

Because of high toughness, low volume shrinkage, low exothermal heat, low weight and low cost [1–6], vinyl ester resins (VERs) are applied to various commercial fields, such as coatings, adhesives, automobile part, molding compounds, structural laminates, and composites.

Biocomposite materials usually mainly contain a matrix phase and a reinforcement phase, as well as one or more biomass-derived phase(s). In terms of the bio-based matrix, a growing number of studies have been carried out on thermosets recently, including VERs from biomass such as plant oils, rosins, cardanol, saccharides, terpenes, and lignin. Furthermore, the use of bio-based filler reinforcements in thermosets has been accepted in many applications recently. Plant fillers such as banana ribbon, bagasse, kusha grass, Indian mallow, nendran, red banana peduncle, vetiver, flax, banana peduncle, jute, oil palm, sisal, kenaf, *Cissus quadrangularis*, coconut shell, powder and roselle fiber [7–13] are widely used as reinforcements for polymer materials. The fillers are added to improve the processing properties or to reduce the production cost. About 170 billion tons of biomass are generated from the nature annually, of which only about 3.5% is utilized [14,15]. Luckily, nearly 10% of global agricultural and forestry residues can be used as biochemicals and biofuels

DOI: 10.1201/9781003270997-12

[16–18]. The use of biomass as a feedstock for resins, monomers, and polymers is important for economic competitiveness [19–22]. Recently, agricultural and forestry byproducts have gained increasing attention as alternative resins and fillers on account of their low costs and abundance. The following part will focus on the major advances in preparing bio-based VER composites and their potential applications.

12.2 POTENTIAL APPLICATIONS

Diverse plant biomass can be used as feedstocks to produce alternative chemicals and precursors for novel VERs with promising properties suitable for various structural uses. Such biomass mainly includes lignocellulosics, plant oils, and other less-voluminous streams, including myrcene, tung oils, dimer fatty acids, and cashew nut shell liquids. The main recent progress in preparing and application of biocomposite materials containing bio-based VERs will be highlighted below.

12.2.1 Composite

12.2.1.1 Bio-based Vinyl Ester Resins

Novel VERs derived from diverse renewable biomass have been used in various applications. These biomass feedstocks include plant oils, lignocellulosics, terpenes, rosin, and dimer fatty acids.

12.2.1.1.1 Plant Oils-Derived VERs Composite

Plant oils are often acquired from the nuts, nut shells, leaves, flowers, seeds, fruits, or buds of plants. Plant oils can be used as crucial renewable resources to synthesize unique building blocks for polymer resins. Plant oils or plant oil-based fatty acids can be used to prepare VERs [23]. The chemical molecular structures, such as unsaturation degree, number and stereochemistry of double bonds, and alkyl chain length, can impact the physiochemical properties of plant oil-based resins [24]. Chemical reactions, such as acrylation, epoxidation, and methacrylation, can be conducted to prepare the precursors of thermoset resins [25–27]. These resins are appropriate for producing composite materials in molding processes.

Epoxy soybean oil (ESO) can be obtained by functionalizing the unsaturated double bonds of soybean oil [28,29], which is used and widely commercialized as an additive [30,31]. Numerous chemical functional groups can be grafted onto the ESO backbone, such as fatty acids, diester, ketals, acrylates, allyl group, diamine, and azide [32–39]. ESO can be modified through epoxy ring-opening reaction to VERs. The soybean oil derivations are reported, such as vinyl esters of acrylated epoxidized soybean oil (AESO) [40], of maleinized hydroxylated soybean oil (HO/MA), and of maleinized soybean oil monoglyceride (SOMG/MA) [40]. AESO is prepared via epoxy ring-opening reaction of ESO with acrylic acid [40–42]. AESO products are commercially available [43,44]. HO/MA is produced by transforming double bonds to hydroxyl groups, which are then functionalized by maleic anhydride [40]. SOMG/MA can be synthesized via epoxy ring-opening and esterification reactions [40]. These precursors can be used as main components of thermoset resins, which also show mechanical properties identical to common composites. AESO can be mixed with styrene and cured via free radical polymerization. The tensile strength, modulus, storage modulus, and glass transition temperature (T_g) of pure AESO at room temperature are 6 MPa, 440 MPa, 1.3 GPa, and 65°C, respectively. When mixed with 40% styrene, however, the AESO/styrene composite showed a much higher tensile strength of 21 MPa and modulus of 1.6 GPa. The AESO/glass composites with 35 wt.% and 50% fibers exhibited moduli of 5.2 and 24.8 GPa, respectively [40]. Furthermore, the maleic acid-modified AESO has larger T_g of 105°C and storage modulus of 1.9 GPa. The AESO modified with cyclohexane dicarboxylic anhydride has T_g of 65°C and storage modulus of 1.6 GPa. The improvement in relevant properties of the modified AESO can be ascribed to the larger crosslinking density of the VER matrix in the composite. The SOMG/MA has better thermomechanical properties, with T_g of 133°C and storage modulus of 0.92 GPa. The HO/MA has T_g of 116°C and storage modulus of 1.55 GPa. These results are all better than the pure AESO composite.

12.2.1.1.2 Cellulose-Derived VERs Composite

Cellulose is one of the most abundant biomass resources, and is estimated to be produced by almost 830 million tons every year [45]. In recent years, there is much research on VER composites derived from cellulose and its derivatives, such as furanyl and isosorbide. Because of the unique heterocycle structure and properties, furanyl-based VERs demonstrate various features. The heterocycle structure endows furanyl-derived VER composites with high stiffness and functionality. Furthermore, the furanyl-based VERs are applicable into self-healing composites via Diels–Alder reactivity [46].

Isosorbide, another derivate of cellulose, is a good candidate for thermosets [47]. Isosorbide has excellent chemical and UV stability because of its chemical structure [48,49]. Sadler et al. [47] studied the productions of several isosorbide-based VERs, and prepared isosorbide dimethacrylate (IM) from esterification reaction of methacrylic anhydride or acrylic anhydride with the hydroxyl groups of isosorbide. The T_g and storage modulus of IM polymers are 240°C and 2.9 GPa, respectively [47]. Furthermore, for a vinyl ester, the T_g of IM-based VERs is the highest known. However, the cured IM-based composite has T_g of nearly 212°C and storage modulus of 3.4 GPa after the addition of 35 wt.% styrene. Mechanical tests show that the cured IM has strength of 85 MPa and tensile modulus of 4 GPa. Palmese et al. [50] also prepared an IM-based composite with free hydroxyl groups via glycidylation and esterification reactions. When blended with 35 wt.% styrene, the cured glycidyl methacrylate isosorbide had T_g of 110°C and storage modulus of 3 GPa at 25°C, which changed to 130°C and 3.3 GPa after the addition with 35% styrene.

12.2.1.1.3 Sucrose-Derived VERs Composite

Sucrose, the disaccharide of fructose and glucose, is a bio-based polyol [51,52]. Recently, studies on sucrose-derived VER composites are reported. Yan and Webster [51] prepared a suite of methacrylated epoxidized sucrose soyate (MESS) and sucrose soyate-B6 (MESSB6), which are commercially named Sefose® 1618U and Sefose® 1618U-B6, respectively. MESS and MEEEB6 can be copolymerized with styrene via free radical polymerization. The T_g of the cured resin increased with the addition of styrene into MESS and MESSB6, which may be on account of the intermolecular propagation and the rigid benzene rings promoted by styrene. However, the blend of styrene into the thermoset resins will lead to two opposite effects on the crosslinking density. Stryene as a functional monomer can form a linear polymer, which decreases the crosslinking density of the thermosets. On the contrary, stryene can increase the crosslinking density due to the low viscosity in the system, enhancing the molecular mobility and thereby the intermolecular propagation during curing. The mechanical properties of the MESS- or MESSB6-based composites lie in the sucrose-derived oligomers. The modulus and tensile strength of the composites are enhanced with the styrene content.

12.2.1.2 Natural Fibers-Reinforced VERs Composite

VER composites reinforced with natural fibers or synthetic fibers have various applications owing to their excellent mechanical properties. Natural and synthetic fibers are the two main types of reinforcing materials of VER composites. Natural fibers are naturally extracted from plants and animals, whereas synthetic fibers are some artificial materials such as nylon fibers, glass fibers, and carbon fibers. Both natural fibers and synthetic fibers have merits. Fibers are often treated with chemicals to improve the mechanical properties and toughness of vinyl ester composites [53–55]. Furthermore, the mechanical properties can be enhanced significantly by adding secondary reinforcement [56]. Owing to the high strength and stiffness, VER composites can also be used in structural applications [56,57].

Walnut shell powder (WSP) can be used as reinforcement materials for green vinyl ester composites, which were also tested by sliding wear studies [58]. Compared with pure VERs, the WSP-filled composite shows lower wear rate and higher friction coefficient. The wear mechanism was studied by scanning electron microscopy. The mechanical properties of WSP-filled vinyl ester composites were investigated, such as flexural, tensile, and impact strengths.

Subramanian et al. [59] investigated the thermal stability of basalt-filled vinyl ester green composites coated or not coated with polytetrafluoroethylene. The storage modulus, loss modulus, and damping properties of the coated vinyl ester composite increased by 18%, 14%, and 13%, respectively. The friction coefficient and wear resistance of the coated composite are significantly enhanced in both dynamic and static conditions. Clearly, the polytetrafluoroethylene-coated basalt-filled vinyl ester composite has friction coefficients of 0.22 and 0.12 under dynamic and static conditions, respectively, and a wear rate of $4.87484 \times 10^{-9} m^3/Nm$.

The machining and drilling properties of vinyl ester composite reinforced with basalt fiber and nanocarbon at different contents were tested. The drilling experiments on the VER composites were conducted to optimize the drilling parameters (e.g. spindle speed, feed rate, drill diameter) and material parameters (e.g. drill material, weight percent). Nagaraja et al. [60] designed the drilling experiments on Minitab and took L16 orthogonal array for drilling parameter operation. The delamination factor and thrust force were outputted to optimize the drilling and material parameters. The minimum thrust force was determined under the optimum parameters of 3 wt.% nanocarbon powder, high-speed steel (HSS) drill, hole diameter of 3 mm, 0.3 mm/rev, and 600 rpm. Furthermore, minimum delamination factor was obtained under the optimum parameters of 9 wt.% nanocarbon powder, cobalt HSS drill, hole diameter of 3 mm, 0.3 mm/rev, and 1260 rpm. All the results suggest carbon plays a major role in decreasing the delamination factor and thrust force of VER composites.

Coconut shell-derived nanofillers such as the kenaf/coconut fiber can be used to improve the properties of reinforced vinyl ester composites. The coconut-based nanofiller was ground over 30 h under high-energy ball milling. Particle size analysis showed that mostly 90% of the reinforce nanofiller fell between 15 and 140 nm. In addition, water absorption was enhanced with the presence of the filler. When up to 3% of the filler was loaded, mechanical properties such as tensile, impact and flexural strengths were improved. However, the mechanical properties were weakened when the filler concentration exceeded 3%. Reportedly, morphological analysis demonstrated that only minimum fiber pull-outs and voids were observed after adding 3% of the filler [61].

Manickam et al. [62] studied the roselle fiber-reinforced VER composite, especially the effect of water absorption on the mechanical properties. Samples were prepared using wet hand lay-up with the same fiber length and fiber weight percents of 15%, 30%, and 45%. The composites soaked in seawater absorbed much more water than groundwater and distilled water. Mechanical properties were reduced as the moisture content rose. Furthermore, SEM showed that the composites were brittle in dry condition and soaked in distilled water, but were ductile in sea and ground water. Nicolai et al. [63] characterized the hybrid filler-reinforced VER composite. For VER composites reinforced with glass fiber or cellulose fiber, the addition of 3 wt.% clay nanofiller improved the mechanical and physical properties of the green hybrid composites.

12.2.2 Coating

There is much research about novel green VERs with beneficial properties designed for coating applications. In the VERs system, the matrix derived from renewable biomass resources potentially substituted the petroleum-derived matrix in common use. These biomass resources include plant oils, cardanol, dimer fatty acids, and myrcene.

Zhang et al. [64] prepared and characterized the vinyl ester of acrylated soybean oil (ASO). The ASO was copolymerized with styrene in the presence of a free radical initiator – benzoyl peroxide and an accelerating agent – dimethyl aniline. The composite was initially cured at 140°C for 2 h and post-cured at 180°C for 12 h. The vinyl ester of ASO at less content (59.3%) had T_g of 55.5°C and storage modulus of 892 MPa at 25°C. Contrarily, the vinyl ester of ASO with higher acrylation (75.7%) had larger T_g of 63.7°C and storage modulus at 25°C of 1247 MPa. These results indicate the vinyl ester of ASO with high acrylation shows higher stiffness and T_g because of the high cross-linking density. Furthermore, the vinyl ester of ASO with high acrylation has higher elastic modulus and yield stress but lower yield strain. However, the rheological aspect and viscosity of the resin were not explored.

Sultania and Garg et al. [65–69] prepared a vinyl ester of cardanol-based novolac resin from the formaldehyde and cardanol reaction, epoxidation reaction, and esterification reaction. Firstly, the first reaction occurred in the presence of a suitable catalyst to form a cardan ol novolac resin. Then, the epoxidation reaction proceeded between phenolic hydroxyl and epichlorohydrin in the presence of a basic catalyst, forming cardanol novolac epoxidized resin (CNER). Finally, the esterification reaction between the methacrylic acid and CNER happened in the presence of a hydroquinone inhibitor and a triphenylphosphine catalyst, forming the vinyl ester of cardanol-based novolac resin (CNEVER). In the presence of 2 wt.% benzoyl peroxide and 40 wt.% styrene diluent, the CNEVER was cured under heating. Kinetic analysis indicated that the curing reaction obeyed the Sestak–Berggren model. Due to the unique chemical structure of the C15 alkyd chain, the CNEVER showed significantly lower T_g than bisphenol A-derived VERs. In addition, the elongation at break of the CNEVER-based composites was slightly enhanced. However, at the same styrene content, the cured systems of CNEVER-based composites showed worse properties such as impact strength and tensile stress than the bisphenol A-derived VERs. On account of its higher reactivity and lower viscosity, this CNEVER is mainly used as an anti-corrosive coating product, and was patented by Cardolite® [70,71].

Jaillet et al. [72] studied the preparation and properties of cardanol-based di-phenyl di-epoxy resin NC514 vinyl ester. Owing to the byproducts of cardanol phenolation reaction, the NC514 vinyl ester contains many impurities, such as the open epoxies and oligomers with large molecular weight [73,74]. Moreover, the modified resin was blended with several vinyl-type reactive diluents, such as dibutyl itaconate, styrene, 1,6-hexanediol dimethacrylate, isobornyl methacrylate, 1,4-butanediol dimethacrylate, and divinylbenzene. The composite mixed with 40 wt.% reactive diluent and 3% tert-butyl peroxybenzoatewas was first thermally cured at 120°C for 12 h, and thermally post-cured at 150°C for 2 h. A Bisphenol A (BPA)-based resin was also methacrylated and cured using the same reaction condition with the same process. Thermal Gravimetric Analyzer (TGA) of the BPA-based or cardanol-based NC514 composites displayed single-stage degradation, without significant difference. In addition, the cardanol-based NC514 composites had lower T_g than BPA-based composites prepared with the same reactive diluents. The reason may be the effect of the long flexible alkyd chain on the backbone of cardanol. Tetrahydrofuran was chosen as a diffusing solvent to explore the insoluble part of different vinyl ester formulations for 24 h. Results showed the insoluble contents agreed with their crosslinking densities. The BPA-based composites had higher insoluble contents than cardanol-based NC514 composites. However, this research did not report the rheological stress of the system.

Li et al. prepared two kinds of dimer fatty acids (DA)-based VERs, DA-polymerized glycidyl methacrylate resin (DA-p-GMA) and maleic anhydride-modified DA-p-GMA (MA-m-DA-p-GMA) through ring-opening and esterification reactions [75]. The DA-based VERs were copolymerized with 1.5 wt.% benzoyl peroxide as a free radical initiator and 30–50 wt.% styrene as a reactive diluent. The samples were first thermally cured at 100°C for 3 h and post-cured at 150°C for 5 h [75]. The tensile strengths of the DA-based VER composites were enhanced with the styrene content. The reason can be that the homopolymers were formed from the high-content styrene, which influenced the tensile properties of the cured system. Compared to the DA-p-GMA-based composite, the MA-m-DA-p-GMA/styrene-based composite had a higher tensile strength, because of the different chemical structures and the high degree of resin curing.

Myrcene is mainly obtained from conifers and pine trees [76]. Yang et al. [76] synthesized ultraviolet (UV)-curable myrcene-based VERs (ADGEMYMs), which were copolymerized with AESO. ADGEMYM was formed from the maleic anhydride and myrcene Diels–Alder reaction, the glycidylation reaction, and the acrylation reaction. Furthermore, the acrylic acid-modified myrcene-based diglycidyl ester (ADMM) can be obtained via dimerization reactions [77]. ADGEMYM was blended with various AESO and dimethoxybenzoin photoinitiator to obtain ADGEMYM-based composites, which were cured by UV light (400 nm) for 20 min. Additionally, the performance of the curing process was studied by FTIR, which showed that above 83% of double bonds in the VERs were converted after 4 min of UV exposure. The tensile strength and breaking elongation of the UV-cured ADGEMYM were 49.04 MPa and 3.2%, respectively. With the increase of AESO content, only the tensile strength decreased. The ADGEMYM/AESO composite at a 1:1 weight ratio had tensile strength of 17.57 MPa and elongation at break of 9.19%. However, T_g was the same between the dimer ADMM and the ADGEMYM. TGA under nitrogen was done to detect the thermal resistance of the composites. TGA suggests that the compositions are thermally stable at 200°C. However, at 220°C, 5% weight loss occurred in the AESO-based sample. In the second step, the ADGEMYM-based composite showed the highest thermal resistance of 402.9°C (with 5% weight loss), whereas the ADGEMYM/AESO-based composite had lower second degradation temperature below 360°C. Moreover, Yang et al. [77] prepared another novel UV-curable VER composite from myrcene and tung oil. Tung oil was extracted from the seeds of tung tree, which consist of nearly 80% of oleic acid, α-eleostearic acids, and linoleic acid. The tung oil-based VERs were synthesized via Diels–Alder and ring-opening esterification reactions. Firstly, acrylic acid reacted with tung oil, with hydroquinone as the polymerization inhibitor. Then the intermediate reacted with glycidyl methacrylate to form tung oil-based VERs (TOA-GMA). The TOA-GMA VERs had modulus of 154.3 MPa, tensile strength of 4.8 MPa, elongation at break of 14.9%, and T_g of 34.8°C. The possible reason for these results is that the tung oil-based VERs have unique flexible alky chains, high free volume, and low crosslinking density in the network. The T_g of the TOA-GMA/ADMM copolymer was enhanced to 77.1°C after curing with ADMM. All the composites were thermally stable below 250°C.

12.2.3 Fire Retardant

Eren and Küsefoğlu [78] prepared bromoacrylated triglyceride oils starting from soybean oil, acrylic acid, and N-bromosuccinimide. Then the bromoacrylated soybean oil radically copolymerized with styrene to form a flame-retardant rigid thermoset polymer, which had T_g of 55°C–65°C and ignition response index of 5 B at 2 mm. Furthermore, other triglycerides, such as sunflower, show the same chemistry [78] and can be used as fire retardants.

Recently, Mao et al. [79] reported a novel UV-curable dipentene-based VER from multi-step reactions of Diels–Alder, glycidylation, epoxy ring-opening and esterification reactions. Dipentene is a plant-derived terpene liquid and a by-product of the pine oil and camphor process. Firstly, dipentene and maleic anhydride underwent the Diels–Alder reaction to form terpinene maleic anhydride. Then dipentene-derived VERs were formed from epoxidation and methacrylation reactions and named MDDMD. Then MDDMD and poly(ethylene glycol) dimethacrylate (PEGDMA)-200 were blended at different contents to form various UV-cured VERs. Application tests indicated that the pure dipentene-based MDDMD VER is brittle with the lowest tensile strength of 0.63 MPa and elongation at break of 1.86%. With the addition of PEGDMA-200, however, the tensile strength and elongation at break of the MDDMD VERs were enhanced significantly, which may be due to the flexible structure of PEGDMA-200. Contrarily, the shore hardening of the MDDMD VERs decreased from nearly 50 HD to about 23 HD. The pure MDDMD VERs and the sample with 10 wt.% PEGDMA-200 showed no transition in DMA at above 0°C. However, the T_g values of the samples with 20, 30, and 50 wt.% PEGDMA-200 were 35.5°C, 44.6°C, and 51.9°C, respectively. The TGA results indicate that the MDDMD VERs with more PEGDMA-200 have higher thermal stability. Moreover, for all the cured resins, the initial decomposition temperature is mainly above 260°C.

12.2.4 Adhesive

Rosin-based VERs can be synthesized by Diels–Alder and epoxy reactions with the rosin-based epoxy monomer, maleic anhydride, acrylic acid, methacrylic acid, and epichlorohydrin [80]. Four rosin-based VERs were prepared. The T_g values of the pre-cured and post-cured VERs were explored using DSC. Probably due to aliphaticity, all of the rosin-based VERs have pre-cure T_g between 84°C and 95°C. All the acrylopimaric acid-based VERs have high post-cure T_g of 94°C–105°C, which can be because of the highly crosslinked network formed in the styrene homopolymerization. However, for the maleopimaric acid-based VERs, the probability in styrene homopolymerization decreased. Compared with the acrylopimaric acid-based VERs, the maleopimaric acid-based VERs had lower vitrification temperature, higher monomer mobility, and higher cure rate. The mechanical properties of the novel VERs were studied by the T-bend and adhesion pull tests for adhesion application. Due to the polar groups, all the rosin-based VERs performed well in adhesion with steel materials. As reported, maleopimaric acid-based VERs have higher pull-off resistance and impact values than acrylopimaric acid-based VERs. Furthermore, compared to the methacrylic-based rosin monomers, the acrylic acid-based rosin monomers display higher adhesive performance. The reason may be due to the high crosslinking density of the acrylic-based composite. The maleopimaric acid-based VERs have better mechanical properties than acrylopimaric acid-based VERs.

Atta et al. [81] reported furan-based VERs, including photo-polymerizable furanic VERs, furan di-2-hydroxypropyl methacrylate (FdHPM) and the furan mono-2-hydroxypropyl methacrylate (FmHPM), from furan-based monoreaction with methacrylic acid. The photo-polymerizing behaviors such as shrinkage and photo-polymerizing rate were studied. FmHPM demonstrated lower photo-polymerizing rate and higher conversion in photo-polymerization than FdHPM, which were attributed to the two methacrylate groups of FdHPM. The mechanical properties of the photo-polymerized composites such as surface hardness and bonding properties were also explored, which were similar to those of petroleum-based bisphenol A-based VER composites. The mechanical tests also suggest that the tensile shear strength of glass falls within the range 0.2–0.6 MPa.

Goiti et al. [82] studied the effects of diene on furan-based VERs, and they also found that furfuryl methacrylate can take place the post-polymerization alone or with styrene, forming an insoluble crosslinked network. The T_g of the furfuryl methacrylate/styrene composites decreased with the increase of furfuryl methacrylate content. Compared with the composite with pure stryene (99°C), the T_g of the composite with 55.6 wt.% furfuryl methacrylate decreased to 78°C.

12.3 SUMMARY

The synthesis and application of biocomposite materials containing bio-based VERs are of growing interest. Novel bio-based VERs, which have better properties, lower toxicity, and sustainability, are prepared from biomass resources, including plant oils, lignocellulosics, and rosins. Many works have been done on the potential applications of VERs, such as composites, coatings, fire retardants, and adhesives. Bio-based VERs derived from plant oils, cellulose, or sucrose can be used as potential composite matrices. Bio-based fillers are widely used as reinforcements for VERs composites. The fillers can be added to improve the processing properties or to reduce the production cost. There is much research about bio-based VERs with beneficial properties designed for coating applications. Plant oil- or dipentene-based VERs can be used as fire retardants. Rosin- and furan-based VERs are reportedly used as adhesives with enhanced adhesive performance. However, there is a largely underexploited field in VER applications, and large gap shall be filled before the commercialization of VER biocomposites. Furthermore, the competitiveness and the substantial commercial utilization of bio-based VERs relative to petroleum-derived VERs still need to be improved.

REFERENCES

1. F.A. Cassis, R.C. Talbot, Polyester and vinyl ester resins, in: S.T. Peters (Ed.), *Handbook of Composites*, Springer, Boston, MA, 1998, pp. 34–47.
2. Y. Lu, R.C. Larock, Novel polymeric materials from vegetable oils and vinyl monomers: preparation properties, and applications, *ChemSusChem* 2 (2009) 136–147.
3. N.A. St John, C.P. Gardiner, L.A. Dunlop, *Characterisation of the Thermo-mechanical Behaviour of a Glass Reinforced Vinyl Ester Composite, Composite Technologies for 2020*, Woodhead Publishing, Sawston, Cambridge, UK, 2004, pp. 131–136.
4. S. Dua, R.L. McCullough, G.R. Palmese, Copolymerization kinetics of styrene/vinyl-ester systms: low temperature reactions, *Polym. Compos.* 20 (1999) 379–391.
5. G.R.P.S. Ziaee, Effects of temperature on cure kinetics and mechanical properties of vinyl–ester resins, *J. Polym. Sci. Part B* 37 (1999) 725–744.
6. G.R.P. Russell, P. Brill, An investigation of vinyl–ester–styrene bulk copolymerization cure kinetics using Fourier transform infrared spectroscopy, *J. Appl. Polym. Sci.* 76 (2000) 1572–1582.
7. J.J. La Scala, J.M. Sands, J.A. Orlicki, E.J. Robinette, G.R. Palmese, Fatty acid-based monomers as styrene replacements for liquid molding resins, *Polymer* 45 (2004) 7729–7737.

8. H. Hong, B. Harvey, G. Palmese, J. Stanzione, H. Ng, S. Sakkiah, W. Tong, J. Sadler, Experimental data extraction and in silico prediction of the estrogenic activity of renewable replacements for bisphenol A, *Int. J. Environ. Res. Publ. Health.* 13 (2016) 705.
9. NCSL Policy Update: State Restriction on Bisphenol A (BPA) in Consumer Products, National Conference of State Legislatures, 2015.
10. M.D. Shelby, NTP-CERHR monograph on the potential human reproductive and developmental effects of bisphenol A, *Ntp cerhr mon*, 2008 v, vii-ix, 1–64 passim.
11. Bisphenol A (BPA): use in food contact application, Food and Drug Administration (FDA), 2010.
12. P. Penczek, P. Czub, J. Pielichowski, *Unsaturated Polyester Resins: Chemistry and Technology, Crosslinking in Materials Science*, Springer, Berlin, Heidelberg, 2005, pp. 1–95.
13. S. Jaswal, B. Gaur, New trends in vinyl ester resins, *Rev. Chem. Eng.* 567 (2014).
14. D.S. Roland Ulber, *Thomas Hirth Biobased Products – Market Needs and Opportunities, Renewable Raw Materials: New Feedstocks for the Chemical Industry*, Wiley-VCH Verlag GmbH & Co. KGaA, Weinheim, Germany, 2011.
15. R. Ulber, K. Muffler, N. Tippkötter, T. Hirth, D. Sell, *Introduction to Renewable Resources in the Chemical Industry, Renewable Raw Materials*, Wiley-VCH Verlag GmbH & Co. KGaA, Weinheim, Germany, 2011, pp. 1–5.
16. J.R. Mauck, S.K. Yadav, J.M. Sadler, J.J. La Scala, G.R. Palmese, K.M. Schmalbach, J.F. Stanzione, Preparation and characterization of highly bio-based epoxy amine thermosets derived from lignocellulosics, *Macromol. Chem. Phys.* 218 (2017) 1700013.
17. J.F. Stanzione III, *Lignin-based Monomers: Utilization in High-Performance Polymers and the Effects of Their Structures on Polymer Properties*, University of Delaware, Delaware, OH, 2013.
18. R. Auvergne, S. Caillol, G. David, B. Boutevin, J.-P. Pascault, Biobased thermosetting epoxy: present and future, *Chem. Rev.* 114 (2014) 1082–1115.
19. P.N. Shah, N. Kim, Z. Huang, M. Jayamanna, A. Kokil, A. Pine, J. Kaltsas, E. Jahngen, D.K. Ryan, S. Yoon, R.F. Kovar, Y. Lee, Environmentally benign synthesis of vinyl ester resin from biowaste glycerin, *RSC Adv.* 5 (2015) 38673–38679.
20. I. Delidovich, P.J.C. Hausoul, L. Deng, R. Pfützenreuter, M. Rose, R. Palkovits, Alternative monomers based on lignocellulose and their use for polymer production, *Chem. Rev.* 116 (2016) 1540–1599.
21. A. Gandini, T.M. Lacerda, A.J.F. Carvalho, E. Trovatti, Progress of polymers from renewable resources: furans, vegetable oils, and polysaccharides, *Chem. Rev.* 116 (2016) 1637–1669.
22. N. Hosseini, D.C. Webster, C. Ulven, Advanced biocomposite from highly functional methacrylated epoxidized sucrose soyate (MAESS) resin derived from vegetable oil and fiberglass fabric for composite applications, *Eur. Polym. J.* 79 (2016) 63–71.
23. L. Montero de Espinosa, M.A.R. Meier, Plant oils: the perfect renewable resource for polymer science, *Eur. Polym. J.* 47 (2011) 837–852.
24. J.M. Raquez, M. Deléglise, M.F. Lacrampe, P. Krawczak, Thermosetting (bio)materials derived from renewable resources: a critical review, *Prog. Polym. Sci.* 35 (2010) 487–509.
25. R. Davenport, POINT OF VIEW: chemicals & polymers from biomass, *Ind. Biotechnol.* 4 (2008) 59–63.
26. M.R. Islam, M.D.H. Beg, S.S. Jamari, Development of vegetable-oil-based polymers, *J. Appl. Polym. Sci.* 131 (2014) 40787.
27. G. Lligadas, J.C. Ronda, M. Galià, V. Cádiz, Renewable polymeric materials from vegetable oils: a perspective, *Mater. Today* 16 (2013) 337–343.
28. J.S. Bashar Mudhaffar Abdullah, Epoxidation of vegetable oils and fatty acids: catalysts, methods and advantages, *J. Appl. Sci.* 10 (2010) 1545–1553.

29. P.M. Tayde Saurabh, S.L. Bhagt, V.C. Renge, Epoxidation of vegetable oils: a review, *Int. J. Adv. Eng. Technol.* II (IV) (2011) 491–501.
30. J. Zhu, K. Chandrashekhara, V. Flanigan, S. Kapila, Curing and mechanical characterization of a soy-based epoxy resin system, *J Appl. Polym. Sci.* 91 (2004) 3513–3518.
31. V. Flanigan, S. Kapila, K. Chandrashekhara, R. Seemamahannop, A. Garg, S. Misra, Soybean based epoxy resin and methods of making and use, US Patents US8481622 B2, 2013.
32. Q. Luo, M. Liu, Y. Xu, M. Ionescu, Z.S. Petrović, Thermosetting allyl resins derived from soybean oil, *Macromolecules* 44 (2011) 7149–7157.
33. B.K. Sharma, Z. Liu, A. Adhvaryu, S.Z. Erhan, One-pot synthesis of chemically modified vegetable oils, *J. Agric. Food Chem.* 56 (2008) 3049–3056.
34. A. Biswas, A. Adhvaryu, S.H. Gordon, S.Z. Erhan, J.L. Willett, Synthesis of diethylamine-functionalized soybean oil, *J. Agric. Food Chem.* 53 (2005) 9485–9490.
35. A. Biswas, B.K. Sharma, J.L. Willett, A. Advaryu, S.Z. Erhan, H.N. Cheng, Azide derivatives of soybean oil and fatty esters, *J. Agric. Food Chem.* 56 (2008) 5611–5616.
36. K.M. Doll, S.Z. Erhan, Synthesis of cyclic acetals (ketals) from oleochemicals using a solvent free method, *Green Chem.* 10 (2008) 712–717.
37. A. Biswas, B.K. Sharma, K. Vermillion, J.L. Willett, H.N. Cheng, Preparation of acetonides from soybean oil, methyl soyate, and fatty esters, *J. Agric. Food Chem.* 59 (2011) 3066–3070.
38. H.-S. Hwang, S.Z. Erhan, Modification of epoxidized soybean oil for lubricant formulations with improved oxidative stability and low pour point, *J. Am. Oil Chem. Soc.* 78 (2001) 1179–1184.
39. S.Z. Erhan, B.K. Sharma, Z. Liu, A. Adhvaryu, Lubricant base stock potential of chemically modified vegetable oils, *J. Agric. Food Chem.* 56 (2008) 8919–8925.
40. S.N. Khot, J.J. Lascala, E. Can, S.S. Morye, G.I. Williams, G.R. Palmese, S.H. Kusefoglu, R.P. Wool, Development and application of triglyceride-based polymers and composites, *J. Appl. Polym. Sci.* 82 (2001) 703–723.
41. L. Fu, L. Yang, C. Dai, C. Zhao, L. Ma, Thermal and mechanical properties of acrylated expoxidized-soybean oil-based thermosets, *J. Appl. Polym. Sci.* 117 (2010) 2220–2225.
42. C.K. Hong, R.P. Wool, Development of a bio-based composite material from soybean oil and keratin fibers, *J. Appl. Polym. Sci.* 95 (2005) 1524–1538.
43. L. Liu, D. Wang, Y. QI, Conductive composition and the method for producing the same, color filter and the method for producing the same, US Patent US20170146847, 2015.
44. P.A.M. Nikesh, B. Samarth, Modified vegetable oil based additives as a future polymeric material-review, *Open J. Org. Polym. Mater.* 5 (2015) 1–22.
45. N. Hernandez, R.C. Williams, E.W. Cochran, The battle for the "green" polymer. Different approaches for biopolymer synthesis: bioadvantaged vs. bioreplacement, *Org. Biomol. Chem.* 12 (2014) 2834–2849.
46. E. Goiti, M.B. Huglin, J.M. Rego, Some observations on the copolymerization of styrene with furfuryl methacrylate, *Polymer* 42 (2001) 10187–10193.
47. J.M. Sadler, A.-P.T. Nguyen, F.R. Toulan, J.P. Szabo, G.R. Palmese, C. Scheck, S. Lutgen, J.J. La Scala, Isosorbide-methacrylate as a bio-based low viscosity resin for high performance thermosetting applications, *J. Mater. Chem. A* 1 (40) (2013) 12579–12586.
48. F. Fenouillot, A. Rousseau, G. Colomines, R. Saint-Loup, J.P. Pascault, Polymers from renewable 1,4:3,6-dianhydrohexitols (isosorbide, isomannide and isoidide): a review, *Prog. Polym. Sci.* 35 (2010) 578–622.
49. B.S. Rajput, S.R. Gaikwad, S.K. Menon, S.H. Chikkali, Sustainable polyacetals from isohexides, *Green Chem.* 16 (2014) 3810–3818.
50. G.R. Palmese, J.J.L. Scala, J.M. Sadler, A.P.T. Lam, Renewable bio-based (meth) acrylated monomers as vinyl ester cross-linkers, United States Patent 9644059, 2017.

51. J. Yan, D.C. Webster, Thermosets from highly functional methacrylated epoxidized sucrose soyate, *Green Mater.* 2 (2014) 132–143.
52. E.M. Monono, J.A. Bahr, S.W. Pryor, D.C. Webster, D.P. Wiesenborn, Optimizing process parameters of epoxidized sucrose soyate synthesis for industrial scale production, *Org. Process Res. Dev.* 19 (2015) 1683–1692.
53. C. Umachitra, N.K. Palaniswamy, O.L. Shanmugasundaram, P.S. Sampath, Effect of mechanical properties on various surface treatment processes of banana/cotton woven fabric vinyl ester composite, *App. Mec. Mat.* 867 (2017) 41–47.
54. O. Nabinejad, S. Debnath, M. M. Taheri, Oil palm fiber vinyl ester composite; Effect of bleaching treatment, *Mat. Sci. For.* 882 (2017) 43–50.
55. S. Sathishkumar, A.V. Suresh, M. Nagamadhu, M. Krishna, The effect of alkaline treatment on their properties of Jute fiber mat and its vinyl ester composites, *Mat. Tod. Proc* 4 (2017) 3371–3379.
56. C. Carpenter, Mechanical characterization and corrosion effects on glass reinforced vinyl ester liners used for oil and gas production. *SPE Annual Technical Conference and Exhibition*, Dubai, September 2016.
57. J.P. Torres, L.J. Vandi, M. Veidt, M.T. Heitzmann, The mechanical properties of natural fiber composite laminates: a statistical study, *Com. Par. A Appl. Sci. Manuf.* 98 (2017) 99–104.
58. S. Pashaei, S. Hosseinzadeh, Sliding wear behaviour of walnut shell powder filled vinyl ester/WSP green composites, *Iran. Chem. Commun.* 5 (2017) 138–146.
59. K. Subramanian, R. Nagarajan, S. Saravanasankar, J. Sukumaran, P. D. Baets, Dynamic mechanical and thermogravimetric analysis of PTFE blended tailor-made textile woven basalt-vinyl ester composites, *J. Ind. Text.* 47 (2016) 1226–1240.
60. R. Nagaraja, M.K. Srinath, M. Mithilesh, Analysis of drillability of carbon nanopowder/Vinyl ester/Basalt fiber, Imp. Jou. Interd. Res. 2(2016)1577–1581.
61. A.K. HPS, M. Masri, C.K. Saurabh, M.R.N. Fazita, A, A, Azniwati, N.S. Aprilia, E. Rosamah, R. Dungani, Incorporation of coconut shell based nanoparticles in kenaf/coconut fibers reinforced vinyl ester composites, *Mat. Res. Exp.* 4 (2017) 035020.
62. C. Manickam, J. Kumar, A. Athijayamani, J. E. Samuel, Effect of various water immersions on mechanical properties of roselle fiber–vinyl ester composites, *Pol. Com.* 36 (2015) 1638–1646.
63. F.N.P. Nicolai, V.R. Botaro, V.F. Lins, Effect of saline degradation on the mechanical properties of vinyl ester matrix composites reinforced with glass and natural fibers. *J. Appl. Polym. Sci.* 108 (2008) 2494–2502.
64. P. Zhang, J. Zhang, One-step acrylation of soybean oil (SO) for the preparation of SO-based macromonomers, *Green Chem.* 15 (2013) 641–645.
65. M. Sultania, J.S.P. Rai, D. Srivastava, Studies on the synthesis and curing of epoxidized novolac vinyl ester resin from renewable resource material, *Eur. Polym. J.* 46 (2010) 2019–2032.
66. M. Sultania, J.S.P. Rai, D. Srivastava, Process modeling, optimization and analysis of esterification reaction of cashew nut shell liquid (CNSL)-derived epoxy resin using response surface methodology, *J. Hazard. Mater.* 185 (2011) 1198–1204.
67. M. Sultania, J. Rai, D. Srivastava, Kinetic modeling of esterification of cardanol-based epoxy resin in the presence of triphenylphosphine for producing vinyl ester resin: mechanistic rate equation, *J. Appl. Polym. Sci.* 118 (2010) 1979–1989.
68. M.S. Garg, D. Srivastava, Effect of glycidyl methacrylate (GMA) content on thermal and mechanical properties of ternary blend systems based on cardanol-based vinyl ester resin, styrene and glycidyl methacrylate, *Prog. Org. Coat.* 77 (2014) 1208–1220.
69. M. Sultania, J. Rai, D. Srivastava, Modeling and simulation of curing kinetics for the cardanol-based vinyl ester resin by means of non-isothermal DSC measurements, *Mater. Chem. Phys.* 132 (2012) 180–186.

70. M. Sultania, S. Yadaw, J. Rai, D. Srivastava, Laminates based on vinyl ester resin and glass fabric: a study on the thermal, mechanical and morphological characteristics, *Mater. Sci. Eng. A* 527 (2010) 4560–4570.
71. F. Jaillet, H. Nouailhas, R. Auvergne, A. Ratsimihety, B. Boutevin, S. Caillol, Synthesis and characterization of novel vinylester prepolymers from cardanol, *Eur. J. Lipid Sci. Technol.* 116 (2014) 928–939.
72. F. Jaillet, E. Darroman, A. Ratsimihety, R. Auvergne, B. Boutevin, S. Caillol, New biobased epoxy materials from cardanol, *Eur. J. Lipid Sci. Technol.* 116 (2014) 63–73.
73. T. Fouquet, L. Puchot, P. Verge, J.A.S. Bomfim, D. Ruch, Exploration of cardanol-based phenolated and epoxidized resins by size exclusion chromatography and MALDI mass spectrometry, *Anal. Chim. Acta* 843 (2014) 46–58.
74. S. Li, J. Xia, M. Li, K. Huang, New vinyl ester bio-copolymers derived from dimer fatty acids: preparation, characterization and properties, *J. Am. Oil Chem. Soc.* 90 (2013) 695–706.
75. X. Yang, S. Li, J. Xia, J. Song, K. Huang, M. Li, Renewable myrcene-based UV-curable monomer and its copolymers with acrylated epoxidized soybean oil: design Preparation, and Characterization, *Bioresources* 10 (2015) 2130–2142.
76. X. Yang, S. Li, J. Xia, J. Song, K. Huang, M. Li, Novel renewable resource-based UV-curable copolymers derived from myrcene and tung oil: preparation, characterization and properties, *Ind. Crops Prod.* 63 (2015) 17–25.
77. T. Eren, S.H. Küsefoğlu, Synthesis and polymerization of the bromoacrylated plant oil triglycerides to rigid, flame-retardant polymers, *J. Appl. Polym. Sci.* 91 (2004) 2700–2710.
78. W. Mao, S. Li, M. Li, X. Yang, J. Song, M. Wang, J. Xia, K. Huang, A novel flame retardant UV-curable vinyl ester resin monomer based on industrial dipentene: preparation, characterization, and properties, *J. Appl. Polym. Sci.* 133 (2016) 44084.
79. A.M. Atta, S.M. El-Saeed, R.K. Farag, New vinyl ester resins based on rosin for coating applications, *React. Funct. Polym.* 66 (2006) 1596–1608.

13 Kenaf-Banana-Jute Fiber-Reinforced Vinyl Ester-Based Hybrid Composites

Thermomechanical, Dynamic Mechanical and Thermogravimetric Analyses

Sangilimuthukumar Jeyaguru and
Senthil Muthu Kumar Thiagamani
Kalasalingam Academy of Research and Education

Chandrasekar Muthukumar
Hindustan Institute of Technology and Science

Senthilkumar Krishnasamy
PSG Institute of Technology and Applied Research

Suchart Siengchin
King Mongkut's University of Technology North Bangkok
Technische Universität Dresden

CONTENTS

DOI: 10.1201/9781003270997-13

13.1 INTRODUCTION

The use of natural fibers in composites made of polymers is growing every day, due to their similar functional characteristics, ease of production, low cost and wide availability [1–3]. According to the experts, the natural fiber-reinforced polymer composite (NFRPC) sector is reportedly growing significantly worldwide. The benefits mentioned above are expected to contribute significantly to a growth rate of 10% over the following 5 years [4–6]. For instance, the thermal and mechanical properties of various natural fibers, including jute, coir, sisal, bamboo, kenaf, hemp and banana, were studied for numerous advantages in composite sectors. These materials are utilized in many sectors, including aerospace, automotive, construction, etc. [7–10]. The automobile sector, in particular, has benefited greatly from the use of several NFRPCs. In order to use the NFRPCs in their automobile components, particularly in the car door panels, seat covers, parcel shelves and boot area, the majority of automobile companies throughout the world have conducted extensive study [3,4,11–13]. Natural fibers have a shorter life duration despite having greater benefits and less environmental effects [4,14–16]. Most of the time, a single reinforcing fiber cannot deliver the required properties for the application. To reduce the problem, hybrid composites, a combination of two or more fiber materials, were produced [17–20].

Researchers examined the dynamic mechanical analysis (DMA) of flax-sugar palm-epoxy hybrid composites. They reported that the storage modulus (E′), loss modulus (E″) and tan delta values increased as a result of incorporating flax into sugar palm fiber-reinforced composites [21]. In another study, the influence of TGA and DMA on hemp-sisal-bio epoxy hybrid composites was investigated. Their results revealed that the pure composites showed better DMA properties than hybrid composites. However, the Hemp, Hemp, Sisal Sisal (HHSS) hybrid composites exhibited highest E′, E″ and tan delta values. Further, TGA measurements revealed that the incorporation of sisal and hemp fibers significantly improved the thermal stability of the composites compared to pure epoxy composites [22]. Chee et al. [23] have investigated the thermomechanical (TMA) and DMA analyses of bamboo-kenaf-epoxy hybrid composites. They concluded that the bamboo fiber and the kenaf fiber in the ratio of 50:50 showed superior thermal properties on TGA and DMA analysis. The effects of thermal properties on coir-pineapple leaf fiber-Poly(lactic acid) (PLA) hybrid composites were studied by Siakeng et al. [24]. From the results, it was observed that the 1:1 ratio of coir-Pineapple Leaf Fibre (PALF) exhibited better thermal stability of the composites. The thermal behaviors of treated and untreated jute-sisal-curaua-ramie-epoxy hybrid composites were reported [25]. The results showed that treated hybrid composites had higher thermal stability of the composites. However, untreated jute-sisal-epoxy hybrid composites possess better thermal stability. In a similar study, the effect of jute-*Sansevieria cylindrica*-epoxy hybrid composites was investigated [26]. According to data derived from differential scanning calorimetry (DSC) analysis, the treated hybrid composites raised the temperatures of both the exothermic and endothermic peaks.

This research aims to develop and evaluate the performance of hybrid bio-composites based on the vinyl ester. In this study, three natural fibers (kenaf, banana and jute) were layered to produce hybrid composites. The laminates that were created were subjected to thermal testing for their TGA, TMA, DMA and DSC capabilities.

13.2 MATERIALS AND METHODS

13.2.1 Materials

The woven mats of kenaf, banana and jute fiber mats were provided by G.V. enterprises, Madurai, Tamil Nadu. The vinyl ester, accelerometer (ethyl methyl ketone peroxide), catalyst (Co naphthenate) and promoter (N-dimethylaniline) were used as matrix materials, all provided by the Sakthi Fiber Glass, Chennai, Tamil Nadu. The fibers and matrix properties are listed in Table 13.1.

13.2.2 Fabrication

The vinyl ester resin solution was made using a 1% mixture of catalyst, accelerator and promoter. The jute, banana and kenaf fiber mats were dried for 15 min at 100°C to remove the moisture before making the composites. The fiber mats were piled according to the preferred stacking pattern (shown in Table 13.2), and then a $300 \times 300 \times 3\,mm^3$ mold was used to impregnate the mats with vinyl ester matrix. After that, the mold was hot-pressed at 100°C and 200 bar pressure for 75 min. The laminates underwent post-curing at 110°C for 15 min and then taken out of the mold. The mono-fiber-reinforced composites were also fabricated for the comparison purpose.

TABLE 13.1
Fibers and Matrix Properties

Materials	Density (g/cm^3)	Tensile Strength (MPa)	Young's Modulus (GPa)	References
Jute	1.3	700	26.5	[27–29]
Kenaf	1.2	295	53	
Banana	1.35	355	33.8	
Vinyl ester	1.1	75	3.5	

TABLE 13.2
The Different Stacking Patterns of Fabricated Hybrid Composites

Stacking Patterns	Description
K-J-B-B-J-K	Kenaf-jute-banana-banana-jute-kenaf
K-B-J-J-B-K	Kenaf-banana-jute-jute-banana-kenaf
B-J-K-K-J-B	Banana-jute-kenaf-kenaf-jute-banana
J-K-B-K-B-J	Jute-kenaf-banana-kenaf-banana-jute

13.3 CHARACTERIZATION

13.3.1 Thermogravimetric Analysis (TGA)

A TGA test was performed according to the ASTM E1131 standard, utilizing a TGA/DSC Mettler Toledo thermogravimetric analyzer. The thermograms were recorded at a temperature range of approximately 25°C–600°C at a heating rate as high as 10°C/min. Each sample was put to the test in a nitrogen environment.

13.3.2 Thermomechanical Analysis

The thermal expansion or shrinkage of the composite specimen was assessed using a thermomechanical analyzer (TMA/SDTA 2+ HT/1600) according to the ASTM E831 standard. Tests were run at a 3°C/min heating rate between 25°C and 150°C and preloaded at 0.5 kN.

13.3.3 Dynamic Mechanical Analysis

All specimens with dimensions of $60 \times 12.5 \times 3\,mm^3$ and temperatures ranging from 30°C to 145°C at a rate of 5°C/min were tested to three-point bending DMA in accordance with ASTM D4065.

13.3.4 Differential Scanning Calorimetry Test

DSC (DSC 204F1) in a nitrogen atmosphere was used to examine the effect of different stacking on the glass transition of the hybrid composites. The test was conducted according to the ASTM D3418 standard. A heating rate of 10°C/min was used for testing the samples at temperatures between 0°C and 200°C.

13.4 RESULTS AND DISCUSSIONS

13.4.1 Thermogravimetric Analysis

The thermogravimetric study was carried out to evaluate the hybrid composites' thermal stability. The maximum degradation temperature ranges of several composite samples are listed in Table 13.3. Figure 13.1a and b depicts the primary and derivative

TABLE 13.3
Maximum Degradation Temperature Ranges of Several Composite Samples

Composite Samples	Maximum Degradation Temperature Ranges (°C)			Residue (%)
	Onset	Inflection	End Set	
Banana	312.03	350	409.12	16.9608
Kenaf	368.40	415.67	432.60	13.9920
Jute	333.34	363.67	406.93	15.3076
K-J-B-B-J-K	325.96	361	405	16.4095
K-B-J-J-B-K	335.32	365.33	403.64	14.9001
B-J-K-K-J-B	324.76	359.17	400.38	17.2881
J-K-B-K-B-J	321.73	358.17	402.54	16.6354

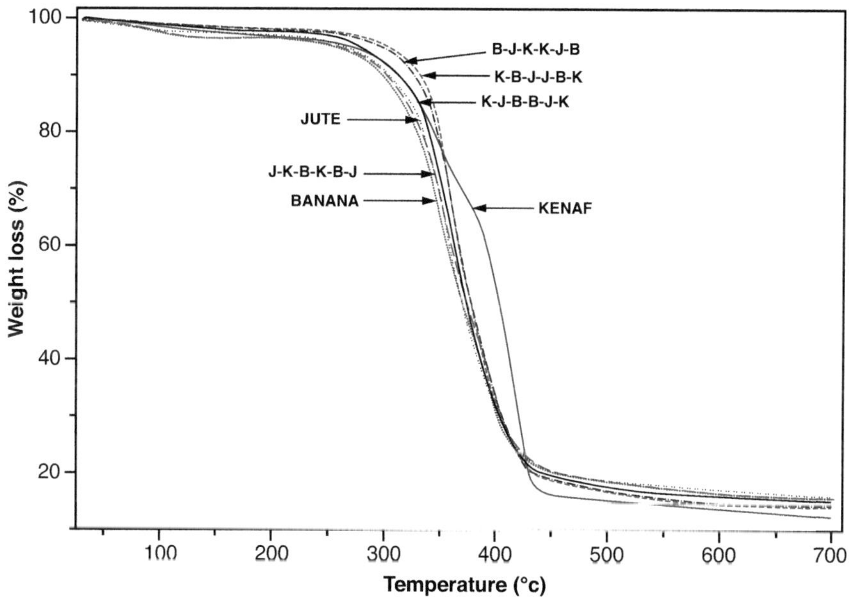

FIGURE 13.1A Primary thermograms of several composite samples.

thermograms of the pure composites and various stacking of hybrid composites. It clearly shows that there is no peak in the 30°C–80°C temperature range, demonstrating the absence of moisture content. The pure banana and jute composites and hybrid composites were decomposed in a single stage except for pure kenaf composites. The kenaf-epoxy composites possessed the highest thermal degradation range of 368°C–432°C, while the banana-epoxy composites possessed the minimum thermal degradation range of 312°C–409°C, although the pure jute composites showed the intermediate thermal degradation range of 333°C–406°C. A higher decomposition temperature indicates greater thermal stability [30]. In hybrid composites, K-B-J-J-B-K-type hybrid composites exhibited the highest thermal degradation temperature

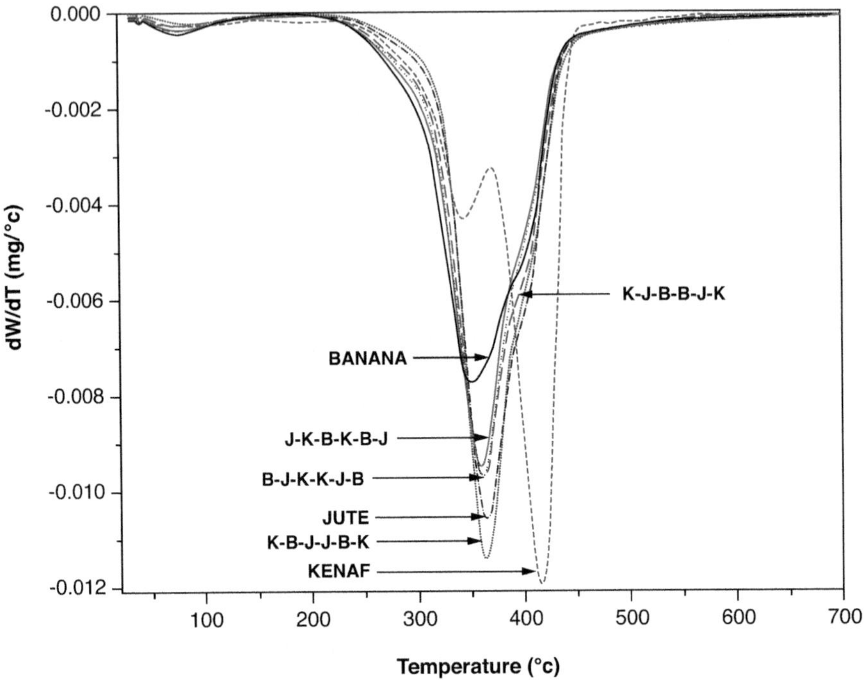

FIGURE 13.1B Derivative thermograms of several composite samples.

range of 335°C–403°C. This is because the jute fiber and kenaf fiber are stacked at the inner and external layers of the hybrid composites. Furthermore, the J-K-B-K-B-J and B-J-K-K-J-B-type hybrid composites showed the lowest thermal degradation temperature range of 321°C–402°C and 325°C–405°C, respectively. The B-J-K-K-J-B-type hybrid composites showed moderate thermal degradation temperature range of 324°C–400°C.

From the results, it was evident that pure kenaf and jute composites exhibited greater thermal stability compared to pure banana composites. Further, K-B-J-J-B-K hybrid composites possessed better thermal stability among all hybrid composites. The J-K-B-K-B-J and B-J-K-K-J-B-type hybrid composites showed lowest thermal stability than other hybrid composites, however these hybrids possessed slightly higher properties than the pure banana composites.

13.4.2 Thermomechanical Analysis

The property of dimensional stability is crucial for composite materials to be used in several applications. All materials expand or shrink in response to temperature changes. The coefficient of thermal expansion (CTE) of materials affects how they dimensionally change. Dimensional change and thermal pressures brought on by thermal variation can both be understood using the CTE. The CTE was measured by using TMA test. The thermal expansion or shrinkage percentages of fabricated composites are shown in Figure 13.2a. From the data, the kenaf-epoxy and jute-epoxy composites observed the highest shrinkage of 0.2% among all composites. The shrinkage percentages did not significantly alter when different hybrid composites were stacked. The jute-epoxy composites' thermal expansion percentage exhibited a higher value of 3.1% than the banana-epoxy composites. Further, the banana-epoxy composites had the lowest thermal expansion percentage of 1.1%. The K-B-J-J-B-K hybrid composites have a higher thermal expansion percentage of 3.7% than typical hybrid composites. Additionally, the B-J-K-K-J-B hybrid composites showed the lowest thermal expansion percentage (2%). The hybrid composites' thermal expansion behavior exhibits the following progression: K-B-J-J-B-K>J-K-B-K-B-J>K-J-B-B-J-K>B-J-K-K-J-B.

Figure 13.2b represents the CTE values of different fabricated composite samples. The maximum CTE values of jute composites and K-B-J-J-B-K hybrid composites are 327 and 391 ppm/°C, respectively. The banana composites and B-J-K-K-J-B hybrid composites exhibited minimum CTE values of 224 and 222 ppm/°C, respectively. The pure composites had superior CTE values than the hybrid composites with the exception of K-B-J-J-B-K.

The T_g values of the kenaf, jute, banana and their hybrid fiber-reinforced composites are listed in Table 13.4. The results evident that the kenaf-epoxy composites and jute-epoxy composites exhibited maximum T_g value compared to the pure banana composites. Further, J-K-B-K-B-J hybrid composites possess a maximum T_g value of 64°C, while the B-J-K-K-J-B hybrid composites showed a minimum T_g

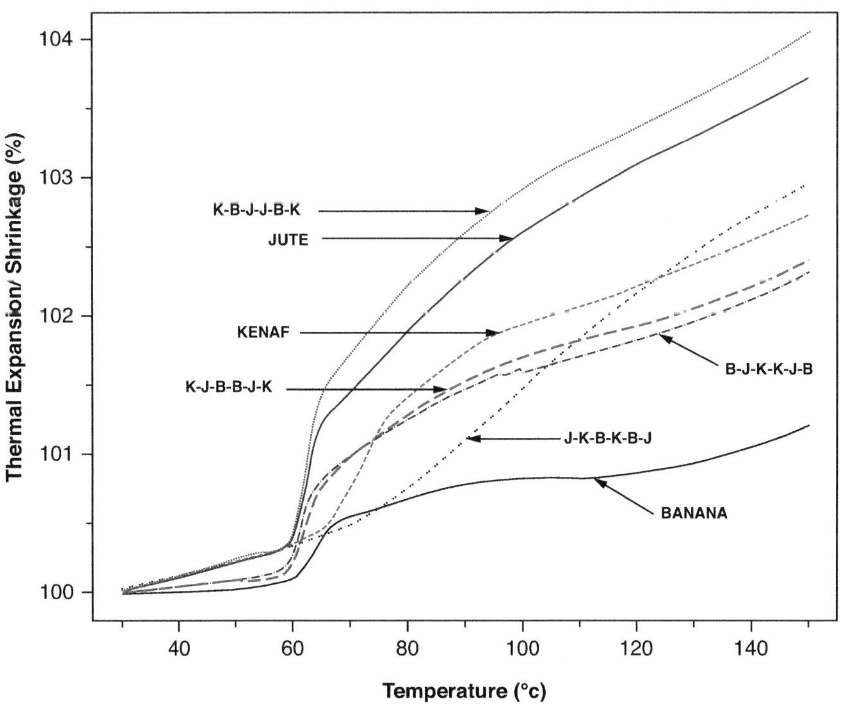

FIGURE 13.2A Thermal expansion or shrinkage on varied composite samples.

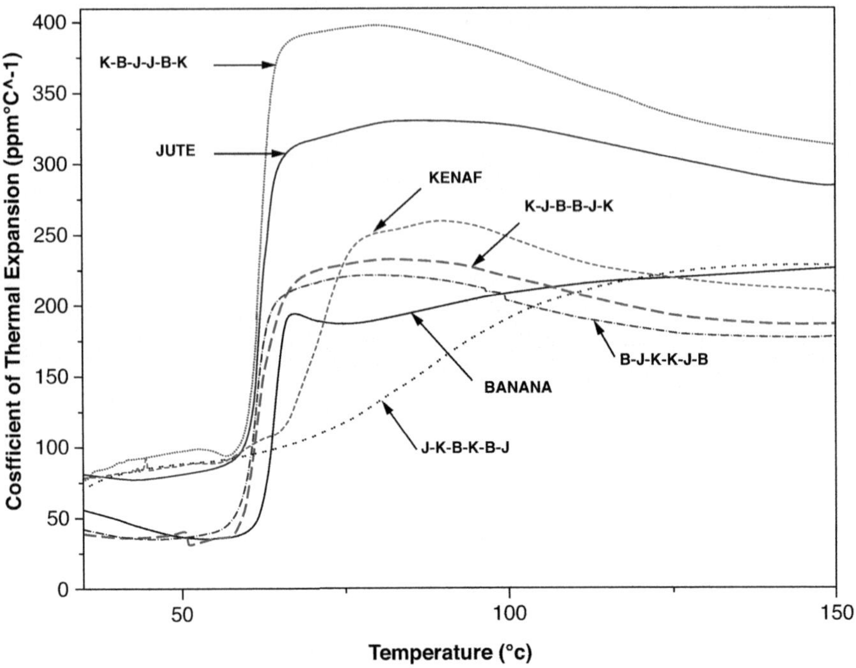

FIGURE 13.2B Coefficients of thermal expansion on varied composite samples.

TABLE 13.4
TMA Parameters of Varied Composite Samples

Composite Samples	Coefficient of Thermal Expansion (ppm/°C)	Thermal Shrinkage (%)	Thermal Expansion (%)	Glass Transition Temperature (°C)
Banana	224.81	0.1	1.1	60.60
Kenaf	259.02	0.2	2.1	65.39
Jute	327.80	0.2	3.1	60.71
K-J-B-B-J-K	232.47	0.1	2.1	59.34
K-B-J-J-B-K	391.77	0.1	3.7	60.29
B-J-K-K-J-B	222.08	0.1	2	59.30
J-K-B-K-B-J	228.13	0.1	2.2	64.20

value of 59°C. Results evidenced that the pure composites showed better T_g values compared to hybrid composites.

13.4.3 Dynamic Mechanical Analysis

DMA was used to analyze the viscoelastic characteristics of composites across a range of temperatures and frequencies. Following that, the inferences were discussed.

13.4.3.1 Storage Modulus E′

The storage modulus (E′) curves of composite materials as a consequence of temperature are shown in Figures 13.3a–c. Figure 13.3a shows that the storage modulus for the jute composites was significantly higher than for the banana composites. Among all the composites, kenaf composites exhibited the minimum storage modulus. This depends upon the mechanical properties of the composites. However, the hybrid composites with various stacking sequences showed a significant difference in storage modulus. The kenaf fiber as the skin layer and the jute fiber as the inner layer of K-B-J-J-B-K-type hybrid composites possess the highest storage modulus. Further, banana fiber as the inner layer of the K-J-B-B-J-K-type hybrid composites showed maximum storage modulus, while the kenaf fiber for the inner layer of B-J-K-K-J-B-type hybrid composites exhibited minimum storage modulus. Similar trends in the storage modulus of fabricated composites at frequencies of 5 and 10 Hz are shown in Figure 13.3b and c.

13.4.3.2 Loss Modulus E″

Loss modulus of the composites provides crucial data on its viscous behavior as a consequence of temperature. It displays how the specimens' energy dissipation properties changes as the temperature rises [21,31]. Figure 13.4a–c displays the loss modulus curves of composite materials. The pure banana and pure jute composites had the lowest loss of modulus, which indicates less amount of energy degradation, particularly during the transition zone. Additionally, the banana and jute fiber as the inner layers of K-B-J-J-B-K and K-J-B-B-J-K hybrid composites showed lower loss

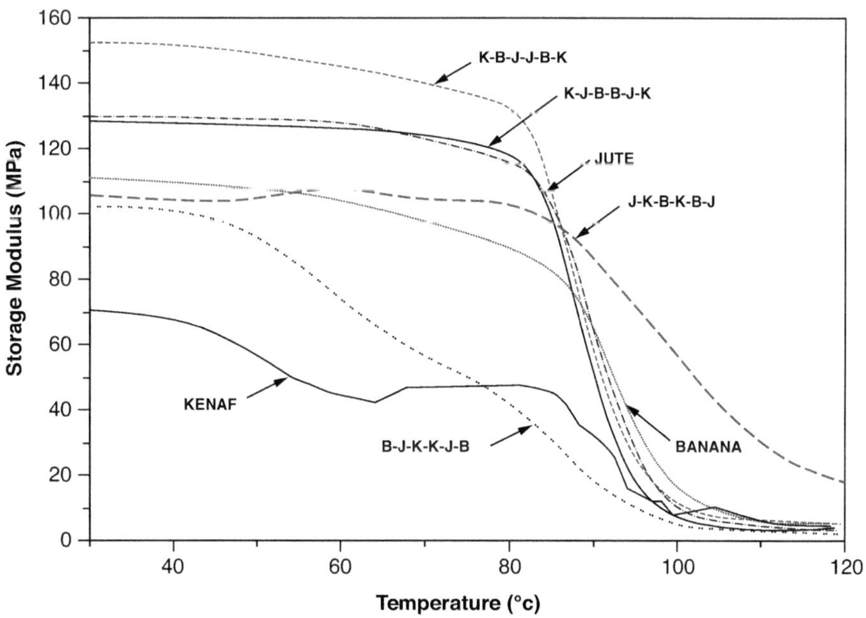

FIGURE 13.3A Storage modulus curves of several composites at 1 Hz.

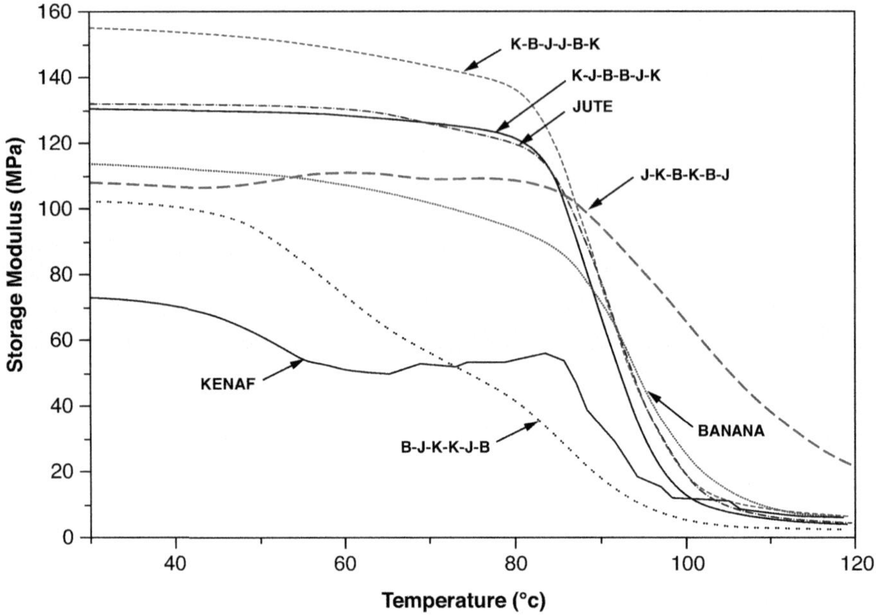

FIGURE 13.3B Storage modulus curves of several composites at 5 Hz.

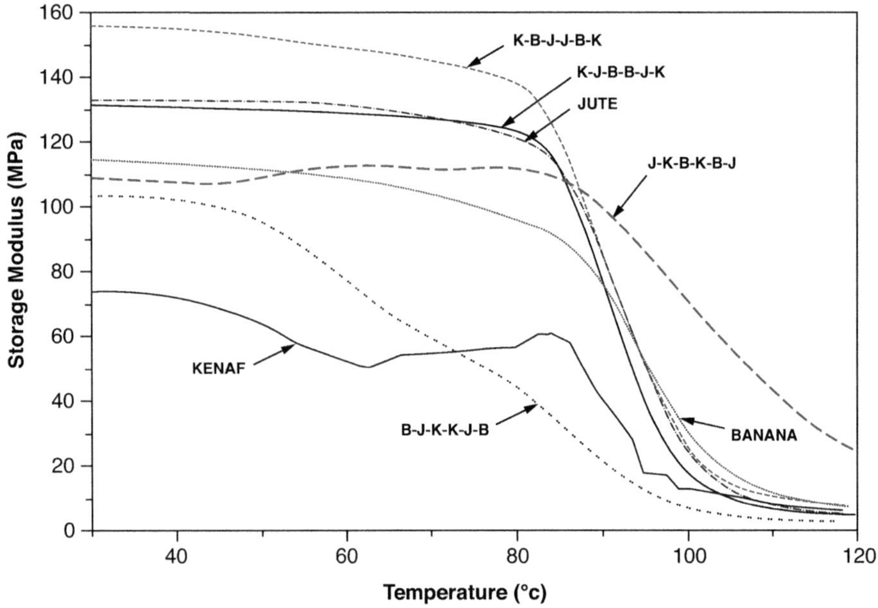

FIGURE 13.3C Storage modulus curves of several composites at 10 Hz.

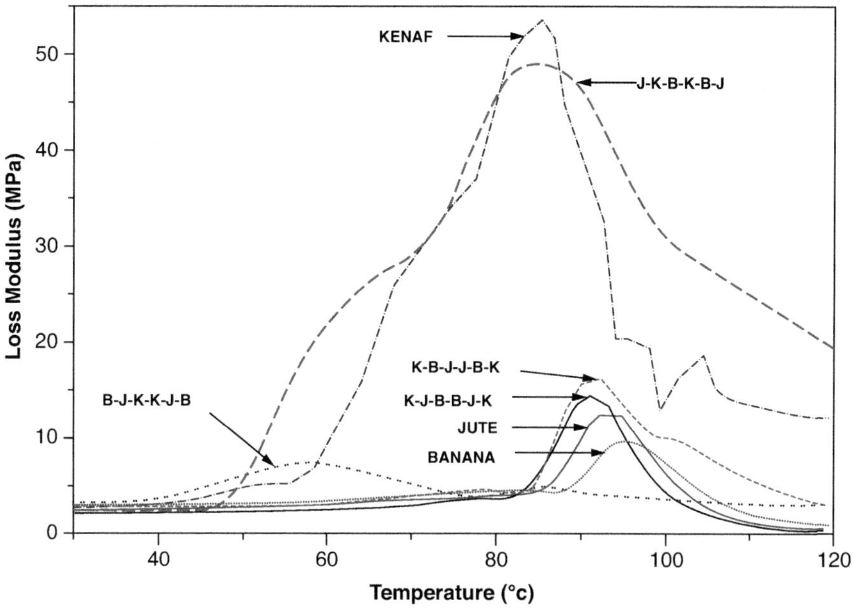

FIGURE 13.4A Loss modulus curves of several composites at 1 Hz.

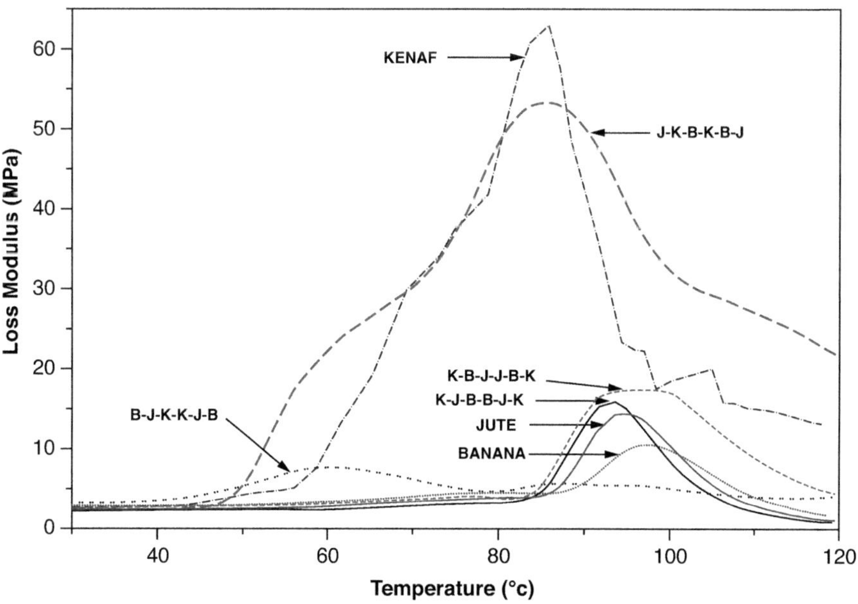

FIGURE 13.4B Loss modulus curves of several composites at 5 Hz.

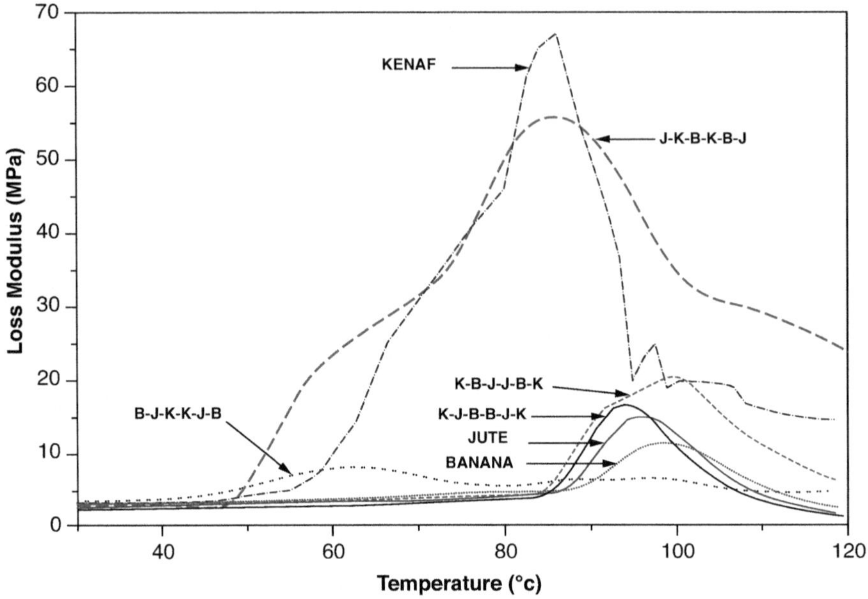

FIGURE 13.4C Loss modulus curves of several composites at 10 Hz.

modulus values. Further, when compared to all other composites, the pure kenaf and J-K-B-K-B-J composite samples both showed much greater values of the loss modulus. It could be due to the fiber properties and their energy loss behavior. Similar trends can be seen in Figure 13.4b and c for the loss modulus of composites at 5 and 10 Hz frequency.

13.4.3.3 Tan Delta

The tan delta is the ratio of the loss modulus to the storage modulus. Figure 13.5a–c shows the tan delta curves of the several composites. Generally, tan delta denotes a larger amount of energy dissipation and fewer constraints on molecule movement [22,32]. The pure jute and pure banana composites possess the lowest tan delta. According to these findings, the damping factor for hybrid composites was determined by the fiber characteristics and the order in which the fibers were stacked. The kenaf-epoxy composites and B-J-K-K-J-B hybrid composites possess the higher tan delta than the K-B-J-J-B-K, K-J-B-B-J-K and J-K-B-K-B-J hybrid composites. The tan delta of fabricated composites showed this order: Kenaf > B-J-K-K-J-B >K-B-J-J-B-K >K-J-B-B-J-K >J-K-B-K-B-J > Jute > Banana. A similar trend in the tan delta of fabricated composites at frequencies of 5 and 10 Hz is shown in Figure 13.5b and c.

13.4.4 Differential Scanning Calorimetry

The DSC analysis was conducted to investigate the effect of hybrid composites on transition temperatures. Figure 13.6 shows the DSC thermograms of the several

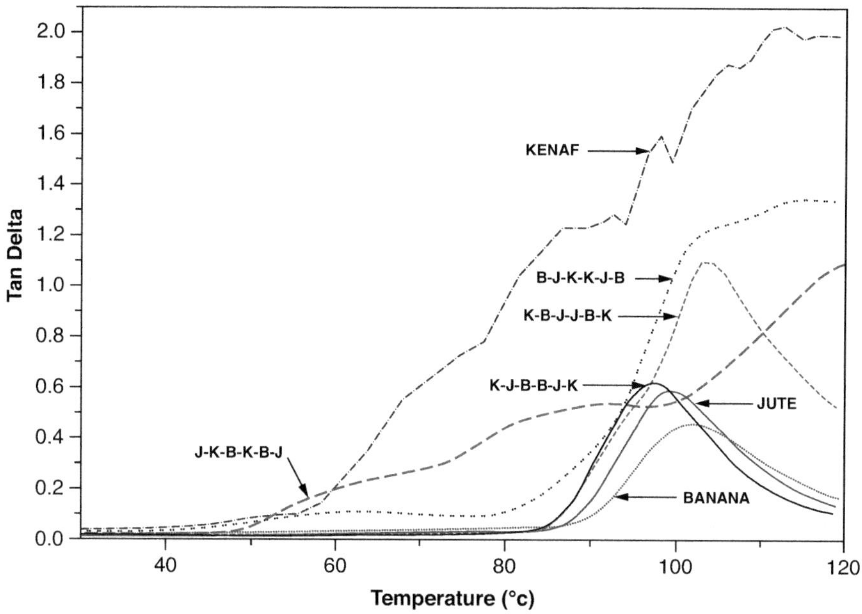

FIGURE 13.5A Tan delta curves of several composites at 1 Hz.

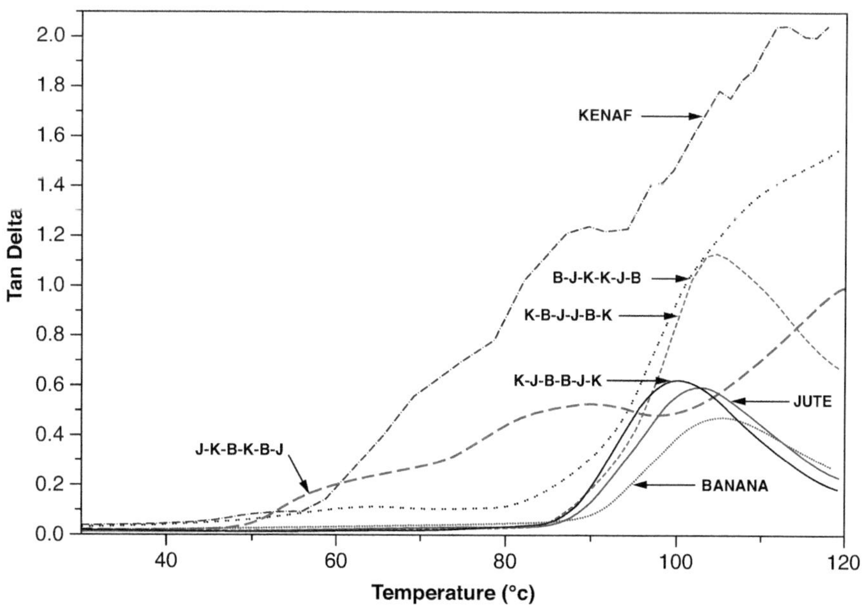

FIGURE 13.5B Tan delta curves of several composites at 5 Hz.

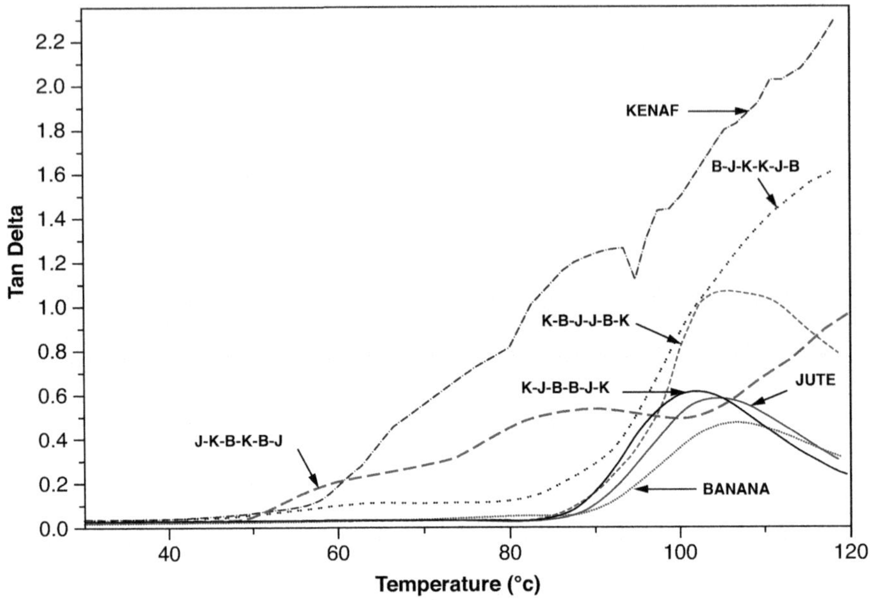

FIGURE 13.5C Tan delta curves of several composites at 10 Hz.

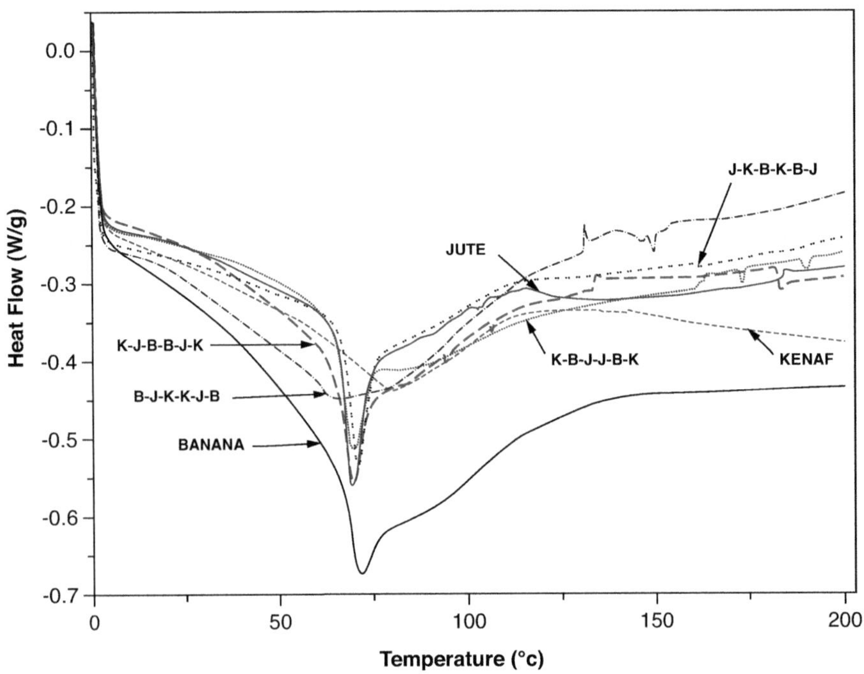

FIGURE 13.6 DSC curves of several composite samples.

TABLE 13.5
T_g Values of Several Composite Samples

Composite Samples	DSC Curve of T_g Value (°C)
Banana	72
Kenaf	80
Jute	70
K-J-B-B-J-K	71
K-B-J-J-B-K	71
B-J-K-K-J-B	66
J-K-B-K-B-J	72

investigated composites. The thermograms were used to measure the glass transition temperatures (T_g) values. Table 13.5 lists the T_g values of several investigated composites. The glass transition temperatures (T_g) for banana, kenaf and jute composites were discovered using the DSC thermograms at 72°C, 80°C and 70°C, respectively. From these results, the kenaf composites showed a higher T_g value, while jute composites showed the lower T_g value. This is because of lower moisture content in kenaf composites. Further, banana and jute composites exhibited slight changes in the Tg value. The findings show that there have been no appreciable changes in the glass transition temperature of hybrid composites with different stacking sequences. The glass transition temperatures for K-J-B-B-J-K, K-B-J-J-B-K and J-K-B-K-B-J composites were observed at 71°C, 71°C and 72°C, respectively. However, the B-J-K-K-J-B hybrid composites possess the lowest T_g value of 66°C. This is owing to the banana fibers in the external layer of these composites.

It was evident from the results that kenaf-epoxy composites showed maximum T_g value compared to other all composites. There is no significant differences in the Tg values of the hybrids when compared with the pure composites. However, the J-K-B-K-B-J composites exhibited the higher T_g value, while the B-J-K-K-J-B composites showed the lowest T_g value. This could be due to the lower and higher moisture content of J-K-B-K-B-J and B-J-K-K-J-B hybrid composites, respectively.

13.5 CONCLUSION

Experimental research has been done to determine the effect of different layering sequences of jute, banana and kenaf in a matrix of vinyl ester on the TGA, TMA, DMA and DSC properties. The following are the experimental study's concluding observations:

- From the TGA analysis, the pure kenaf and jute composites have more thermal stability than pure banana composites. Furthermore, K-B-J-J-B-K hybrid composites exhibited better thermal stability.
- The TMA results clearly showed that pure composites had higher CTE values than hybrid composites, with the exception of K-B-J-J-B-K.

- It was clear from the DMA data that the maximum storage modulus and loss modulus values were found in K-B-J-J-B-K hybrid composites at frequencies of 1, 5 and 10 Hz. However, the maximum tan delta value was found for B-J-K-K-J-B hybrid composites at all frequencies.
- DSC measurements showed that the J-K-B-K-B-J composites had a higher T_g value of 72°C. However, the pure kenaf composites displayed a higher T_g value of 80°C.

REFERENCES

1. S. M. K. Thiagamani, S. Krishnasamy, and S. Siengchin, "Challenges of biodegradable polymers: An environmental perspective," *Appl. Sci. Eng. Prog.*, vol. 12, no. 3, p. 149, 2019.
2. J. Sangilimuthukumar, T. S. M. Kumar, C. Santulli, M. Chandrasekar, K. Senthilkumar, and S. Siengchin, "The use of pineapple fiber composites for automotive applications: A short review," *J. Mater. Sci. Res. Rev.*, pp. 39–45, 2020.
3. S. Krishnasamy, S. M. K. Thiagamani, C. Muthukumar, R. Nagarajan, and S. Siengchin, *Natural Fiber-Reinforced Composites: Thermal Properties and Applications.* Weinheim: Wiley-VCH Verlag GmbH & Co. KGaA, 2022.
4. L. Mohammed, M. N. M. Ansari, G. Pua, M. Jawaid, and M. S. Islam, "A review on natural fiber reinforced polymer composite and its applications," *Int. J. Polym. Sci.*, vol. 2015, 243947, 2015. https://doi.org/10.1155/2015/243947
5. T. Senthil Muthu Kumar, M. Chandrasekar, K. Senthilkumar, J. Sangilimuthukumar, S. Suchart, and N. Rajini. "Polymers for aerospace applications," in *Encyclopedia of Materials: Plastics and Polymers, Reference Module in Materials Science and Materials Engineering,* Elsevier. 2022, vol. 4, pp. 382–391. https://doi.org/10.1016/B978-0-12-820352-1.00077-8.
6. K. Senthilkumar et al., "Performance of sisal/hemp bio-based epoxy composites under accelerated weathering," *J. Polym. Environ.*, vol. 29, no. 2, pp. 624–636, 2021.
7. A. Atiqah, M. Chandrasekar, T. Senthil Muthu Kumar, K. Senthilkumar, and M. N. M. Ansari. "Characterization and interface of natural and synthetic hybrid composites," in *Reference Module in Materials Science and Materials Engineering - Encyclopedia of Renewable and Sustainable Materials,* Amsterdam: Elsevier B.V. 2020, vol. 4, pp. 380–400. https://doi.org/10.1016/B978-0-12-803581-8.10805-7.
8. S. Jeyaguru et al., "Mechanical, acoustic and vibration performance of intra-ply Kevlar/PALF epoxy hybrid composites: Effects of different weaving patterns," *Polym. Compos.*, vol. 43, no. 6, pp. 3902–3914, 2022.
9. J. Sangilimuthukumar, S. M. K. Thiagamani, C. M. Kumar, S. Krishnasamy, G. R. Arpitha, and S. Mayakannan, "Erosion characteristics of epoxy-based Jute, Kenaf and Banana fibre reinforced hybrid composites," *Mater. Today Proc.*, vol. 64, pp. 6–10, 2022.
10. S. Jeyaguru et al., "Effects of different weaving patterns on thermomechanical and dynamic mechanical properties of Kevlar/pineapple leaf fiber hybrid composites," *Polym. Compos.*, vol. 43, no. 8, pp. 4979–4997, 2022.
11. S. Jeyaguru, S. M. K. Thiagamani, A. G. Rajkumar, S. M. Rangappa, and S. Siengchin, "Solid particle erosion, water absorption and thickness swelling behavior of intra ply Kevlar/PALF fiber epoxy hybrid composites," *Polym. Compos.*, vol. 43, no. 6, pp. 3929–3943, 2022.
12. S. M. K. Thiagamani et al., "Mechanical, absorption, and swelling properties of jute/kenaf/banana reinforced epoxy hybrid composites: Influence of various stacking sequences," *Polym. Compos.*, vol. 43, no. 11, pp. 8297–8307, 2022.

13. K. Senthilkumar et al., "Free vibration analysis of bamboo fiber-based polymer composite," in Jawaid, M., Mavinkere Rangappa, S., Siengchin, S. (eds), *Bamboo Fiber Composites*, Composites Science and Technology, Singapore: Springer, 2021, pp. 97–110. https://doi.org/10.1007/978-981-15-8489-3_6
14. M. Chandrasekar, I. Siva, T. S. M. Kumar, K. Senthilkumar, S. Siengchin, and N. Rajini, "Influence of fibre inter-ply orientation on the mechanical and free vibration properties of banana fibre reinforced polyester composite laminates," *J. Polym. Environ.*, vol. 28, no. 11, pp. 2789–2800, 2020.
15. T. Senthil Muthu Kumar, M. Chandrasekar, K. Senthilkumar, N. Ayrilmis, S. Siengchin, and N. Rajini, "Utilization of bamboo fibres and their influence on the mechanical and thermal properties of polymer composites," in Jawaid, M., Mavinkere Rangappa, S., Siengchin, S. (eds), *Bamboo Fiber Composites*, Composites Science and Technology, Singapore: Springer, 2021, pp. 81–96. https://doi.org/10.1007/978-981-15-8489-3_5
16. S. Krishnasamy, M. Chandrasekaran, T. Senthil Muthu Kumar, R.M. Shahroze, M.R. Ishak, and S. Siengchin. *Effect of Hybridization and Water Absorption Properties of Flax and Sugar Palm Fibre Reinforced Polymer Composites: A Review*. December 20, 2019. https://ssrn.com/abstract=3654047; http://dx.doi.org/10.2139/ssrn.3654047.
17. N. A. Nasimudeen et al., "Mechanical, absorption and swelling properties of vinyl ester based natural fibre hybrid composites," *Appl. Sci. Eng. Prog.*, vol. 14, no. 4, pp. 680–688, 2021.
18. S. Krishnasamy et al., "Recent advances in thermal properties of hybrid cellulosic fiber reinforced polymer composites," *Int. J. Biol. Macromol.*, vol. 141, pp. 1–13, 2019.
19. K. Senthilkumar et al., "Dual cantilever creep and recovery behavior of sisal/hemp fibre reinforced hybrid biocomposites: Effects of layering sequence, accelerated weathering and temperature," *J. Ind. Text.*, p. 1528083720961416, 2020.
20. C. Muthukumar, S. Krishnasamy, S. M. K. Thiagamani, R. Nagarajan, and S. Siengchin, "Thermal characterization of the natural fiber-based hybrid composites: An overview," *Nat. Fiber-Reinforced Compos. Therm. Prop. Appl.*, pp. 1–15, 2022.
21. M. Chandrasekar et al., "Flax and sugar palm reinforced epoxy composites: Effect of hybridization on physical, mechanical, morphological and dynamic mechanical properties.," *Mater. Res. Express*, vol. 6, 105331, 2019. https://doi.org/10.1088/2053-1591/ab382c
22. S. Krishnasamy et al., "Effects of stacking sequences on static, dynamic mechanical and thermal properties of completely biodegradable green epoxy hybrid composites," *Mater. Res. Express*, vol. 6, no. 10, p. 105351, 2019.
23. S. S. Chee, M. Jawaid, M. T. H. Sultan, O. Y. Alothman, and L. C. Abdullah, "Thermomechanical and dynamic mechanical properties of bamboo/woven kenaf mat reinforced epoxy hybrid composites," *Compos. Part B Eng.*, vol. 163, pp. 165–174, 2019.
24. R. Siakeng, M. Jawaid, H. Ariffin, and S. M. Sapuan, "Mechanical, dynamic, and thermomechanical properties of coir/pineapple leaf fiber reinforced polylactic acid hybrid biocomposites," *Polym. Compos.*, vol. 40, no. 5, pp. 2000–2011, 2019.
25. J. S. S. Neto, R. A. A. Lima, D. K. K. Cavalcanti, J. P. B. Souza, R. A. A. Aguiar, and M. D. Banea, "Effect of chemical treatment on the thermal properties of hybrid natural fiber-reinforced composites," *J. Appl. Polym. Sci.*, vol. 136, no. 10, p. 47154, 2019.
26. M. A. Kumar and G. R. Reddy, "Mechanical & thermal properties of epoxy based hybrid composites reinforced with jute/*Sansevieria cylindrica* fibres," *Int. Lett. Chem. Phys. Astron.*, vol. 19, no. 2, pp. 191–197, 2014.
27. M. Jawaid and H. P. S. A. Khalil, "Cellulosic/synthetic fibre reinforced polymer hybrid composites: A review," *Carbohydr. Polym.*, vol. 86, no. 1, pp. 1–18, 2011.
28. M. Ramesh, K. Palanikumar, and K. H. Reddy, "Plant fibre based bio-composites: Sustainable and renewable green materials," *Renew. Sustain. Energy Rev.*, vol. 79, pp. 558–584, 2017.

29. J. J. Sokołowska, "Technological properties of polymer concrete containing vinyl-ester resin waste mineral powder," *J. Build. Chem.*, vol. 1, no. 1, pp. 84–91, 2016.
30. S. Oza, H. Ning, I. Ferguson, and N. Lu, "Effect of surface treatment on thermal stability of the hemp-PLA composites: Correlation of activation energy with thermal degradation," *Compos. Part B Eng.*, vol. 67, pp. 227–232, 2014.
31. G. George, E. T. Jose, D. Åkesson, M. Skrifvars, E. R. Nagarajan, and K. Joseph, "Viscoelastic behaviour of novel commingled biocomposites based on polypropylene/jute yarns," *Compos. Part A Appl. Sci. Manuf.*, vol. 43, no. 6, pp. 893–902, 2012.
32. B. R. Mohammed, Z. Leman, M. Jawaid, M. J. Ghazali, and M. R. Ishak, "Dynamic mechanical analysis of treated and untreated sugar palm fibre-based phenolic composites," *BioResources*, vol. 12, no. 2, pp. 3448–3462, 2017.

Index

Note: Bold page numbers refer to tables; *italic* page numbers refer to figures.